Contributions in Petroleum Geology & Engineering 9

Horizontal Wells

Contributions in Petroleum Geology and Engineering
Series Editor: George V. Chilingar, University of Southern California

Volume 1: Geologic Analysis of Naturally Fractured Reservoirs
Volume 2: Applied Open-Hole Log Analysis
Volume 3: Underground Storage of Natural Gas
Volume 4: Gas Production Engineering
Volume 5: Properties of Oils and Natural Gases
Volume 6: Introduction to Petroleum Reservoir Analysis
Volume 7: Hydrocarbon Phase Behavior
Volume 8: Well Test Analysis
Volume 9: Horizontal Wells

Gulf Publishing Company
Houston, London, Paris, Zurich, Tokyo

Contributions in Petroleum Geology & Engineering 9

Horizontal Wells

R. Aguilera J. S. Artindale
G. M. Cordell M. C. Ng
G. W. Nicholl G. A. Runions

Dedicated To Our Families

Contributions in Petroleum and Engineering

Volume 9

Horizontal Wells

Copyright © 1991 Gulf Publishing Company, Houston, Texas. All rights reserved. Printed in the United States of America. This book, or parts thereof, may not be reproduced in any form without permission of the publisher.

Printed on acid-free paper (∞)

Library of Congress Cataloging-in-Publication Data

Horizontal wells / Roberto Aguilera . . . [et al.].
    p. cm.—(Contributions in petroleum geology & engineering; g)
    Includes bibliographical references and index.
    ISBN 0-87201-573-4 (acid-free paper).—
ISBN 0-87201-066-X (ser. : acid-free paper)
    1. Horizontal oil well drilling. I. Aguilera, Roberto. II. Series.
TN871.25.H67 1991
622'.3381—dc20   91-9303
                                                    CIP

10 9 8 7 6 5 4 3 2 1

Contents

Foreword vii

Preface ix

Acknowledgments xi

About the Authors xii

1. Geologic Aspects 1
   Introduction, 1; Horizontal Well Applications, 2; Core and Log Data for Hole Azimuth Determination, 11; Drill Cutting Logs and Total Gas Logs, 13; Summary, 17.

2. Drilling 18
   Introduction, 18; Selection of the Well Profile/Casing Program, 26; Drilling Depth Control, 30; Mechanical Aspects of Steerable Motors and Their Effect on KOP Section, 36; Basic Mechanics and Application of Torque and Drag to Medium Radius Horizontal Drilling, 43; Drillstring Design and Tubular Selection for Horizontal Drilling, 50; Casing Design Considerations, 61; Drilling Fluids and Cuttings Transport, 64; Summary, 75.

3. **Completions** 76
   Planning the Completion, 76; Equipping the Horizontal Well, 83; Working in the Horizontal Environment, 87; Perforating, 92; Cased Hole Logging, 99; Stimulation of Horizontal Wells, 102.

4. **Well Logging** 127
   Pipe Conveyed Logging Systems, 127; Coiled or Reeled Tubing Conveyed Wireline Systems, 133; Measurement While Drilling (MWD), 135; Log Interpretation, 136; Simulation of Resistivity Behavior, 142; Porosity and Water Saturation, 143; Case Histories, 143; Looking at the Future, 154.

5. **Well Test Analysis** 156
   Introduction, 156; Analysis Procedure, 157; Horizontal Well Model, 158; Duration of Flow Regimes, 161; Pressure Response Functions, 167; Pressure Transient Character, 175; Wellbore Storage Effects, 176; Skin Effects, 177.

6. **Well Performance and Productivity** 181
   Mathematical Model, 181; Productivity of Vertical Wells, 181; Wellbore Flowing Pressure of Horizontal Wells, P_{wf}, 182; General Review, 183; Productivity in Horizontal Wells, 185; Wellbore Storage and Skin, 201; Analytical Solution to a Gas Horizontal Well, 204.

7. **Thermal Recovery and Primary Production for Heavy Oils** 210
   Introduction, 210; Overview, 211; Basis for Studying Heavy Oil Recovery Processes, 214; Thermal Recovery Processes, 215; Steam Assisted Gravity Drainage (SAGD) Process, 236; Combination Displacement and Gravity Drainage Processes, 311; Displacement Thermal Processes, 324; Placement of Steam Along HW Injectors, 350; Primary Recovery of Heavy Oils, 352.

8. **Naturally Fractured Reservoirs** 359
   Geologic Aspects, 359; Drilling and Completion, 367; Well Test Analysis, 367; Fractured Shales, 387.

Index 393

Foreword

The first directional holes under water and other obstacles were drilled more than 50 years ago. Continually improving tools for turning sharper angles finally came together in the early 1980s to permit a ninety degree bend. Horizontal drilling became an effective technique and allowed directional drilling to advance from a method of moving the bottom-hole location of a well to an entirely new means of completion. We have learned that many reservoirs yield much higher productivity to long horizontal penetrations than to short vertical wellbores, and thus horizontal drilling is potentially the most important new completion technique since hydraulic fracturing.

The natural response of trained technical minds to a new engineering problem is to define, analyze, and solve it. In writing this book the authors have provided an important service to the whole oil industry. A decade from now, the subject will be treated with far more sophistication and experience, but this book will stand out as the pioneer effort to gather all the relevant existing knowledge under one intellectual roof. Unusual and welcome is the long chapter on the application of horizontal drilling to heavy oil recovery; much of the technology has been developed in Canada and is hence particularly familiar to the authors.

Laymen, including most petroleum management and personnel outside the engineering departments will be startled by the complexity of the mathematics and physics involved in calculation of BUR, KOP, tangent angles, flow time equations, etc. They may also be gratified to observe the integration of disciplines when they read the words of a drilling engineer discussing slower penetration and increased drilling costs in an effort to improve the ultimate completion: "The well design should be approached from the point of view that drilling the target formation is a completion operation," and "There are a number of completion subtleties that can be addressed easily in the preplanning stages that may be virtually impossible to correct once the well is drilled," are somewhat unusual statements from technical authors. The people who wrote this book are the new breed of engineers and geologists, all entirely involved in the whole process of discovery and none of them operating in isolation any longer.

As a member of the industry in which this new tool has now been made available for the exploration and development of untold new reserves, I would like to express my admiration to the six professionals who wrote this book largely at night and on weekends, not for personal gain, but to satisfy their own intellectual standards. They saw a problem and set about providing a solution.

And lastly, a gentle caution. Don't tackle this book unless you are serious. It is written for professionals who are familiar with advanced mathematics and intricate downhole equipment.

John A. Masters
President, Canadian Hunter Exploration Ltd.
Calgary, Alberta

Preface

Horizontal wells are coming of age, and are, in our opinion, here to stay. This book is the result of new technology and our hands-on experience dealing with horizontal wells.

This book is written for students, log analysts, reservoir geologists, and petroleum engineers. We have strived to make it as useful as possible by including many case histories and practical examples that the reader should find easy to reproduce.

Most horizontal wells have been drilled rather recently, and consequently production and performance are in their infancy. This indicates that care and common sense should be exercised when utilizing theoretical models.

Chapter 1 deals with geologic aspects, horizontal well applications, and the use of core data for properly drilling the horizontal well. It must be stressed that horizontal wells are not suitable for all types of reservoirs.

Chapter 2 discusses horizontal drilling methods. The chapter emphasizes that communication between all levels of personnel, from the geologist to the roughneck, is important for a sound drilling operation. This chapter highlights medium radius horizontal drilling.

Chapter 3 focuses on completion practices. Horizontal completions require more preplanning than conventional vertical wells. The completion designer should be involved right from the conceptual stage, just as the drilling engineer. A number of completion subtleties can be addressed easily in the preplanning stages that may be virtually uncorrectable once the well is drilled. Completion in homogeneous, heterogeneous, and naturally fractured reservoirs are discussed in detail.

Chapter 4 reviews the state of the art on well logging techniques including pipe-conveyed and tubing-conveyed logging systems, and measurement while drilling (MWD). The difference in log interpretation between horizontal and vertical wells is discussed, including borehole rugosity, mud cake, anisotropy, invasion, bed boundaries intersecting the wellbore and away from the wellbore, true vertical depth, and natural fractures. Case histories from various sites around the world are highlighted.

Chapter 5 concentrates on well test analysis and reviews the current theoretical understanding associated with transient pressure analysis in horizontal wells to assess the practical application of these analytical solutions.

Chapter 6 reviews well performance and productivity. Care should be exercised when using theoretical formulations as production histories from horizontal wells are still limited, and a practical corroboration of many of the equations is still lacking. Comparison of productivity in vertical and horizontal wells is presented.

Chapter 7 deals with thermal recovery and primary production for heavy oils. To employ thermal recovery methods with horizontal wells frequently requires different, more specialized approaches than are typically followed in conventional horizontal well activities. It is important that each of the disciplines involved in the drilling, completion, and operation of thermal horizontal wells be aware of the specifications required for a successful outcome.

Chapter 8 focuses on naturally fractured reservoirs. They represent excellent targets for horizontal wells, a fact that has been recognized by the oil and gas industry, as more than 70% of horizontal wells drilled to date have been experienced in naturally fractured reservoirs. The chapter discusses geologic aspects, well test analysis, and reservoir performance of horizontal wells in naturally fractured reservoirs.

Roberto Aguilera, Ph.D.

Acknowledgments

We would like to thank the Society of Petroleum Engineers of the American Institute of Mining, Metallurgical, and Petroleum Engineers, the Society of Professional Well Log Analysts, the American Association of Petroleum Geologists, Dresser Atlas, Schlumberger, Gulf Publishing Co., PennWell Publishing Co., the Petroleum Society of CIM, and the Canadian Well Logging Society for permission to draw material from their publications.

In addition, we express our gratitude to the various authors and organizations that have published material on the subject of horizontal wells.

Although we are the only persons responsible for the final form of this book, we would like to thank Mr. David Smith and Mr. Don Myers of Canadian Hunter for their help in the preparation of Chapters 1 and 2.

Last but not least, we would like to thank our wives and children for their patience and understanding during the long hours needed for the preparation of this book.

About the Authors

Dr. Roberto Aguilera is a petroleum engineering graduate from the Universidad de America, Colombia, and holds M.S. and Ph.D. degrees in Petroleum Engineering from the Colorado School of Mines. He is president of Servipetrol Ltd. in Calgary, Canada. He is extensively involved in petroleum engineering throughout the world as a consultant, and as a presenter of a course on naturally fractured reservoirs. He is also an AAPG lecturer on the subject of "Fractured Reservoir Analysis." Dr. Aguilera has developed techniques for evaluation of naturally fractured reservoirs and horizontal wells and has published his findings in leading oil industry journals. He has also authored and co-authored books on the subject.

James S. Artindale is a district level reservoir engineer with Canadian Hunter Exploration Ltd. in Calgary, Canada. His interests and responsibilities include horizontal wells, fractured reservoirs, tight gas sands, pool optimization, enhanced recovery, and regulatory issues. He obtained a B.Sc. degree in engineering in 1979 from the University of Calgary. With respect to horizontal technology, Artindale has developed expertise in selection criteria, well test interpretation, production forecasting, and development strategy. He has been invited to present talks and seminars on horizontal applications for the Society of Petroleum Engineers, the Canadian Institute of Mining, the Montana Geologic Society, the Canadain Society of Petroleum Geologists, and the American Association of Petroleum Geologists.

Gilbert M. Cordell graduated from the University of Alberta with a B.Sc. Degree in Chemical Engineering. He worked for the Alberta Oil Sands Technology and Research Authority (AOSTRA) as a reservoir engineer involved in the monitoring and evaluation of EOR projects. Field pilot projects included cyclic steam stimulation, steam pressure cycling, thermal horizontal wells, combustion drive, enriched air combustion, and CO_2 miscible flooding. He helped to design the parameters for AOSTRA's horizontal well program at the Underground Test Facility. Cordell joined Canadian Hunter Exploration Ltd. in 1985, to oversee the technical aspects of commercial scale thermal recovery projects. In conjunction with the thermal projects, he established the predicted productivities for the first 3 medium radius horizontal wells that were drilled by Canadian Hunter.

Dr. Michael C. Ng is a Senior Consultant with Servipetrol Ltd. in Calgary, Canada. Dr. Ng holds a Ph.D. degree in Applied Mathematics (Fluid Dynamics) from the University of Western Ontario, Canada. Previously he was a Post Doctoral Research Fellow at the same university, and worked for the Petroleum Recovery Institute in Calgary as a petroleum reservoir simulation scientist on computer modelling applications. He has developed well logging, well testing, horizontal well, and numerical simulation software. He has co-authored a number of papers on the area of fluid dynamics and well testing.

Gordon W. Nicholl graduated from the University of Calgary with a B.Sc. degree in Mechanical Engineering. He then spent seven years in drilling with two major Canadian oil companies, gaining extensive experience in most types of drilling in Western Canada, the Beaufort Sea, and the east coast offshore. Mr. Nicholl joined Canadian Hunter Exploration Ltd. in 1986 as a Senior Drilling Engineer. He has worked on a variety of drilling projects but most recently assumed a lead role in the planning of several horizontal wells.

Gordon A. Runions is currently the Manager of Completions at Canadian Hunter Exploration Ltd. in Calgary, Alberta. He graduated from the University of Manitoba in 1980 with a B.Sc. degree in Civil Engineering. He joined Canadian Hunter Exploration Ltd. in 1985 after spending several years in drilling and production with a major Canadian oil company. Mr. Runions' specific area of expertise is well completions, including conventional, sour, deep, horizontal, and unconventional technologies. Mr. Runions has given presentations on horizontal completions at many technical forums and conferences.

1
Geologic Aspects

INTRODUCTION

The applications where a horizontal wellbore will enhance production and economic return on investment are as broad as the choice of geologic horizons. Theoretically, all reservoirs can benefit from horizontal wells, but often these reservoirs may be just as effectively produced using conventional vertical techniques.

For example, consider a horizontal wellbore in a high permeability homogeneous gas reservoir. No natural barriers traverse through the reservoir, and the reservoir can be easily hydraulically fractured to overcome the pressure drop near the wellbore caused by drilling damage and gas turbulence. The ultimate percentage of reserves recovered will be high and the flow rate will deliver the reserves in an acceptable time frame. In this example, a horizontal well will not significantly increase return on investment, or improve on ultimate recovery from the reservoir. If the same reservoir is internally subdivided by shale drapes, as in a channel point bar system (Figure 1-1), the horizontal well will significantly increase the ultimate recovery of oil and gas from the reservoir.

To fully utilize the potential of a horizontal well, the geologist must have a clear understanding of both vertical and lateral reservoir characteristics. The drive mechanism and fluid characteristics in the reservoir system are equally important. Once a candidate has been selected, the geologist must work closely with the drilling and completion engineer to match hole azimuth to reservoir character and geometry.

Although the driller has good control of the well path, sudden changes of the well trajectory are difficult to achieve. Effectively, the hole azimuth

2 Horizontal Wells

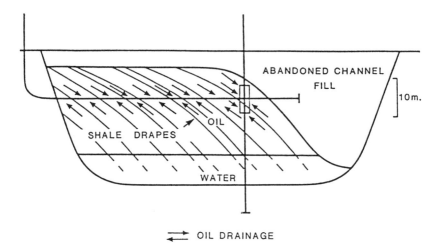

Figure 1-1. Channel point bars.

must be chosen prior to kicking the well off from vertical (Figure 1-2) and in fact, the azimuth should be chosen prior to surface lease acquisition, with the aid of regional mapping and well control. If a vertical pilot hole is drilled, data from an oriented core and electric log run can often be used to adjust the azimuth of the final hole.

As the well is drilling, an accurate geological model of the reservoir and information regarding the lithology that could be encountered are critical to operational efficiency and a successful well. Sample descriptions and mud gas logs are also valuable tools. Using these tools, the geologist can provide a detailed picture of the reservoir as the well is drilled laterally through the formation. This picture represents valuable input when making decisions regarding completion techniques, future development drilling, and enhanced recovery methods in the field.

HORIZONTAL WELL APPLICATIONS

Horizontal drilling can be applied in any phase of reservoir recovery: primary, secondary, and enhanced. A common application would be infill. Horizontal wells can be considered as an alternative to infill drilling and fracturing, with the objective in all cases being to increase the economic recovery of oil and gas. Generally, stimulation and infill drilling is driven by the need to overcome production problems. These may include low reservoir permeability, high formation fluid viscosity, high pressure drawdown in the near wellbore region, or heterogeneous reservoir characteristics resulting in low recovery efficiency.

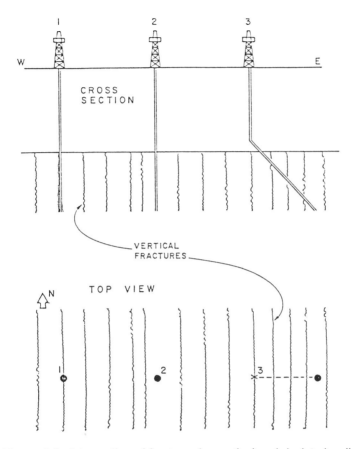

Figure 1-2. Intersection of fractures by vertical and deviated wells.

When considering horizontal well applications, two categories generally exist. These are applications that overcome reservoir fluid flow characteristics, and applications that overcome heterogeneous reservoir characteristics. Both applications can be present in the same system.

Fluid Flow Problems

Gas and Water Coning. The coning of water upwards or gas downwards, bypassing oil into the wellbore, is illustrated in Figure 1-3. With production, reservoir pressure is drawn down around the wellbore, elevating the level of the oil/water surface in the vicinity of the wellbore. The height of this rise reflects oil pressure immediately above the contact. The slope of the oil/water surface reflects the horizontal pressure gradient at the contact. As

4   Horizontal Wells

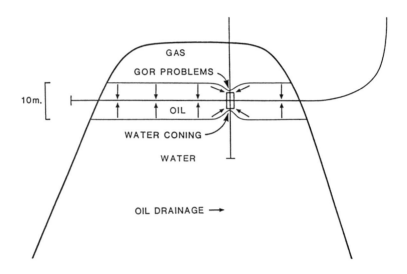

Figure 1-3. Reefs—gas/oil/water.

the oil production rate is increased, the pressure gradient is increased and the water rises even more. Eventually, the critical production rate is reached: water rises to the wellbore and water breakthrough results. The critical production rate is dependent on the density differential of the oil and water, and the oil viscosity (Butler, 1988).

Table 1-1 shows the typical effect of oil gravity on critical production rate (Butler, 1988). The critical production rate is uneconomically low for heavier oils and marginal for 30° API oil.

Horizontal drilling can place the wellbore near the top of the reservoir, well away from the oil/water contact. The volume of oil displaced prior to water breakthrough is the volume within a cone around a vertical well. In a horizontal well, the cone becomes a crest (Figure 1-4) and is able to capture a much larger volume of oil. Due to the length of the wellbore, drawdown is minimized while still maintaining production. This reduces the horizontal pressure gradient and increases the width at the base of the crest.

Problems can occur if reservoir permeability along the horizontal wellbore is variable. Water may be prematurely produced from a high permeability section and negatively effect the overall well performance. Horizontal drilling to reduce water coning in heavy oil applications has been attempted in various areas with encouraging results. Horizontal drilling in Prudhoe Bay, Alaska, has been encouraging in dealing with gas coning. Production rates have increased vertical well performance 2–4 times over, with reduced gas/oil ratios (Butler, 1988).

**Table 1-1
Typical Effect of Oil Gravity on Critical Production Rate***

°API	ΔP/μ	BBL/day
40	0.2	250
30	0.02	25
20	0.001	1.25
10	0.0001	0.13

* Butler, 1988.

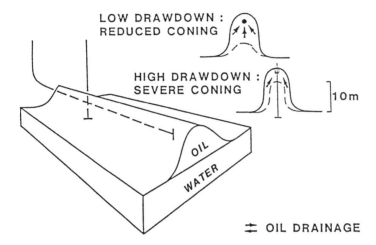

Figure 1-4. Sandstone bar ridges; heavy oil/water.

Low Permeability Reservoirs. In low permeability reservoirs, the unstimulated vertical wellbore is often not capable of economic flow rates. However, with hydraulic fracturing, low permeability reservoirs can become economically viable if a significant fracture length is placed in the formation. The fracture must overcome wellbore damage and positively stimulate the reservoir. If the reservoir is sandwiched between formations that will not contain the fracture, then successful hydraulic fracturing is difficult.

Horizontal wellbores can be used to place a flow path through the reservoir as an alternative to hydraulic fracturing. Even if the formation is easily fractured, horizontal wellbores have an inherent advantage in low permeability reservoirs that require tight well spacing. Only one wellbore, one surface facility, and one tie-in are required for a long horizontal well. The well can be fractured at intervals along its length and achieve the effect of multiple vertical completions. This concept was first tested in 1975 in the Caddo Pine Island field of northwestern Louisiana. Four fractures were placed in an inclined (52°) wellbore over a 160-foot interval. The well was reported to have produced 2–3 times the oil of an average vertical well (Bell and Babcock, 1986). More recently the technique has been attempted in horizontal wellbores with encouraging results. The wells may be fractured using longitudinal fractures (Figure 1-5a) or by placing multiple vertical fractures (Figure 1-5b). This technique was used in the Cardium Sandstone in central Alberta. Initial flow rates of 3–4 times those of a vertical fractured well in the area were achieved (Canadian Hunter, 1989).

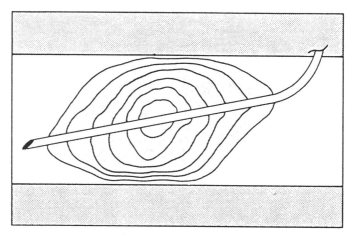

Figure 1-5a. Longitudinal fracturing out of horizontal wellbore.

Heterogeneous Reservoirs

Understanding of the heterogeneous character of a reservoir and the directional relationship of reservoir permeability (anisotropy) is the key to achieving successful horizontal completions.

Examples of heterogeneous reservoirs include:

- Carbonate systems
- Channel point bars
- Braided stream systems
- Fractured reservoirs

Geologic Aspects 7

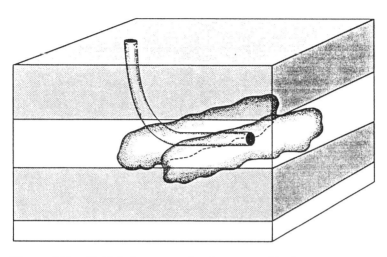

Figure 1-5b. Multiple transverse fractures out of horizontal wellbore.

Carbonate Systems. Porosity patterns in carbonate reservoirs are often complex and difficult to predict (Figure 1-6). The horizontal well can be used to develop this type of reservoir by connecting areas of high permeability rock that are separated by low permeability sections. A clear understanding of the carbonate body is important to determine the wellbore azimuth. In the

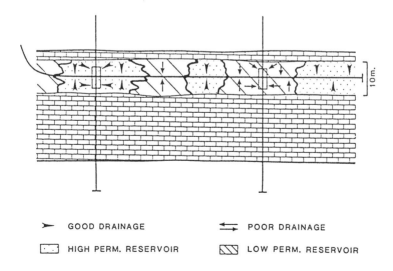

Figure 1-6. Low permeability carbonate reservoir.

8 Horizontal Wells

case of a carbonate reef (Figure 1-7), the azimuth should be selected to run parallel to the reef margin as drilling across the reef may result in a wellbore that drills into progressively poorer reservoir rock within the reef flat or lagoon.

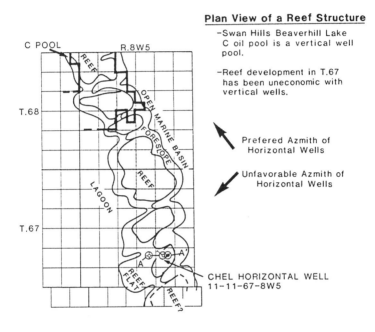

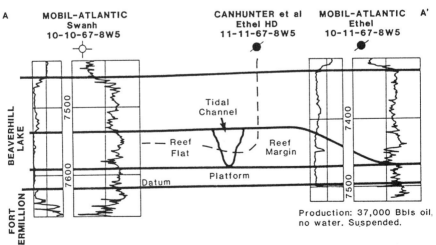

Figure 1-7. Plane view of a reef structure.

Geologic Aspects   9

Channel Point Bars. Channel point bar sands formed by lateral accretion within fluvial streams often contain non-permeable clay layers, deposited during the waning flood stage of river flow. These layers, often referred to as shale drapes (Figure 1-1), present problems in achieving high recovery factors because they inhibit lateral flow to the wellbore. Development by water flooding often shows complicated production performance, poor development results, and low ultimate recovery. If the orientation of the point bar can be determined, a horizontal well can be directed to penetrate multiple sand wedges which would normally be isolated from the wellbore by impermeable shale drapes. This can lead to increased ultimate recovery from the reservoir (Peihua, 1986).

Braided Stream System. Braided streams often contain low permeability lenses created by pods of finer sediment within the channel fill (Figure 1-8). Such pods or lenses are usually oriented with their long axis parallel to the stream direction. Consequently, the direction of fluid flow will generally be along the primary axis of the stream system. In such a case, a vertical well will produce reserves from a long narrow section of the reservoir with minimal contribution from channel sands located adjacent to the wellbore, but in a direction normal to the stream's direction of flow. A horizontal wellbore drilled normal to stream flow will increase recovery by accessing more channel segments.

Fracture Systems. Natural fracture systems within the reservoir will greatly enhance the production of oil and gas by providing natural flow paths for reservoir fluids (Figure 1-9). Some reservoirs depend exclusively on this system to produce oil and gas (i.e., Austin Chalk). In such cases, if a fracture system is not encountered in the wellbore, the well is not capable of economic production. Horizontal wells greatly increase the probability of encountering a fracture system and significantly reduce the dry hole risk. Extensive fracture systems which extend into underlying water zones may be detrimental to production and should be avoided.

The importance of understanding the heterogeneous nature of the reservoir cannot be overstressed. Drilling parallel to shale drapes or parallel to a braided stream axis will not optimize horizontal well performance. In fracture systems the wellbore must be drilled perpendicular to the fracture direction to encounter new fracture systems and new reserves. Regional core and log data combined with vertical pilot hole data should be used to determine directional reservoir properties before selecting the optimum hole azimuth of the horizontal well.

10   Horizontal Wells

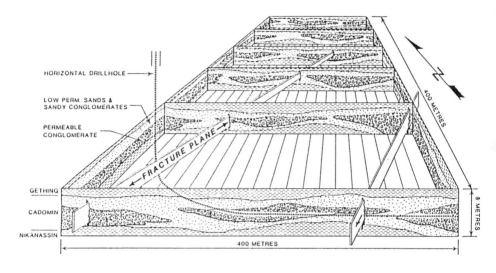

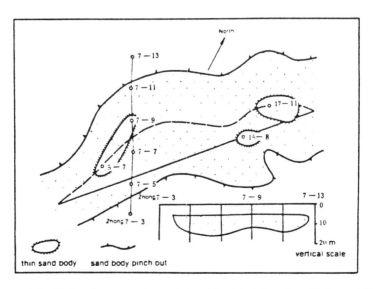

Figure 1-8. Cadomin reservoir model and braided river bar (Peihua, 1986).

Geologic Aspects 11

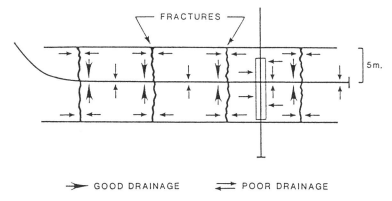

Figure 1-9. Fractured low permeability sands.

CORE AND LOG DATA FOR HOLE AZIMUTH DETERMINATION

The use of cores has long been used to solve exploration and production problems. If the core has been oriented, directional reservoir properties such as permeability and in situ stress measurements can be made. If an oriented core is fortunate enough to capture a natural fracture, the fracture direction can be determined.

Oriented core tools combine an oriented barrel with a multishot survey instrument. The knife scribes orientation lines on the core. The orientation of the knife is recorded by the survey tool so the core can be realigned to a reference direction for analysis (Nelson, Lenox, and Ward, 1987).

Fracture direction determination using oriented cores is dependent on identifying natural fractures in the core. Induced fractures may occur during coring as rock stress is relieved. Fractures that show evidence of secondary cementation or crystal growth on the fracture surface are most desirable. One technique of increasing the number of fractures encountered in the core is to cut the core at an inclined angle (10-20°). This increases the lateral exposure of the core. If the bedding plane is known, the method also provides backup to the multishot tools in orienting the core, provided the hole direction and azimuth are known (Figure 1-10).

Anisotropic reservoir parameters can be determined using oriented cores. The fact that maximum permeability generally occurs in a particular direction is widely recognized and documented (Bell and Babcock, 1986). If a horizontal well is drilled perpendicular to the direction of maximum permeability, the production will likely be enhanced. In some basins, the direction of maximum stress has remained generally constant throughout geologic

12 *Horizontal Wells*

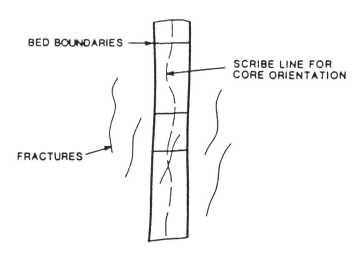

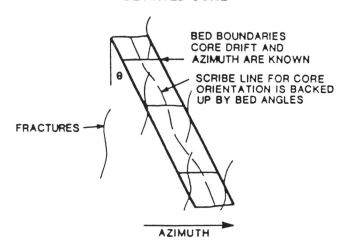

Figure 1-10. Inclined oriented core increases cored fractures.

time. In such cases, many sediments would have been exposed to an anisotropic horizontal compression throughout their history. The Western Canadian Basin, for example, has had a near constant stress regime since Middle Jurassic time. This compression clearly has had some effect on the diagenetic fabric of the reservoirs in this basin (Bell and Babcock, 1986). If the direction of maximum permeability can be determined from oriented core or regional stress information, the hole azimuth can be optimized if placed perpendicular to this direction.

Understanding the stress regime is essential for planning an artificially fractured inclined wellbore. Wellbore breakout is one method of determining the stress regime within a basin. Bell and Babcock (1986) showed that the orientation of wellbore breakout could be used to determine the direction of maximum and minimum stress in the wellbore. This direction is important because natural and hydraulically induced fractures will often propagate in directions parallel to the maximum principle stress. Breakout refers to the process of borehole walls spalling to relieve in situ stress. The spalling produces an elliptical borehole and wells in a given area often share a common breakout orientation. In areas where reliable in situ stress measurements are available, the main breakout axis is parallel to the minimum stress (S direction minimum) and perpendicular to the maximum stress (S direction maximum). Breakout has reliably been used to indicate the orientation of principle horizontal stresses affecting the wellbore in which they were measured. Figure 1-11 shows the normal breakout pattern and several possible causes of anomalous breakout patterns. The possible stress trajectory inferred from breakout data for western Canada is shown in Figure 1-12.

Analysis of existing core and log data is important in planning a horizontal well. Normally wellbore placement is constrained by regulated spacing units. If we consider the case of a one section drainage unit and a 3,000-foot horizontal wellbore, some choice of hole azimuth will have to be made prior to choosing a surface location. This choice will depend on regional stress patterns, reservoir size and shape, and anticipated completion practices.

DRILL CUTTING LOGS AND TOTAL GAS LOGS

Inspection of drill cuttings by the wellsite geologist provides an important source of information in horizontal wells. All reservoirs have variation in lithology and grain size, and estimates of porosity and production potential are first made using these observations. While the horizontal well is drilling, observations regarding lithology and grain size are important in estimating the vertical position of the wellbore within the reservoir. If the wellbore is

14   Horizontal Wells

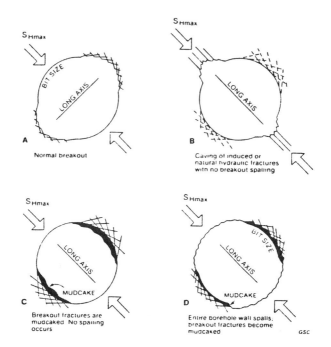

Figure 1-11. Possible causes of anomalous breakout orientations.

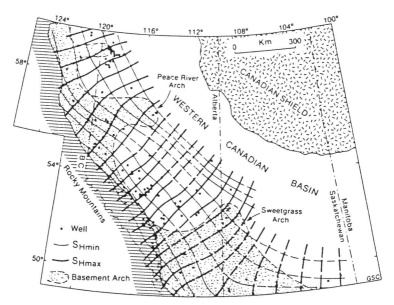

Figure 1-12. Postulated stress trajectories inferred from breakout data, Western Canadian Basin.

accidentally drilled out of the reservoir, samples will be important in determining whether the bit is above or below the reservoir or whether the reservoir has ended. Thus, samples will be important in the decision to continue or cease drilling operations.

Experience has shown that the quality of drill cuttings using a slow-speed motor without drillstring rotation is excellent up to approximately 60°. After this inclination, fluid erosion, drillpipe milling, and particle collisions begin to reduce cutting size at surface, but sample quality is excellent. At 90° the sample size rapidly erodes to individual grain size. In the horizontal hole, porosity estimates must often be made from careful scrutiny of the individual grain surfaces to determine the size of the contact points and evidence of secondary crystal growth (indicating the presence of pore spaces). Occasional cores cut along the horizontal hole are very useful in calibrating the porosity and permeability estimates made from samples.

The lag time to surface of cuttings leaving the bit is dependent on the efficiency of cutting transport along the horizontal wellbore. Cutting transport is analogous to two-phase flow in horizontal pipes, with the drilling fluid as the liquid phase and the cuttings as the solids phase.

Normally, the fluid phase and the solids phase do not travel at the same velocity in the wellbore (solids are slower). Thus, using carbide gas (acetylene) to check lag time can be misleading, as the acetylene travels in the fluid phase. A more accurate method of determining lag time is to use a synthetic marker sand sized to match the cuttings. The sand will give a true indication of lag time.

Field experiments using high-strength sorted sand as a lag time indicator showed that lag time was related to grain size. In fact, sand sizes significantly greater than the formation grain size were not lifted out of the hole. The difference in transport velocity of the gas show traveling in the fluid phase and cuttings traveling in the solids phase should be considered when calibrating hydrocarbon shows to depth and lithology.

Mud gas logs are very useful in estimating what sections of a horizontal well will be most productive. This estimate must often be made prior to the setting of production casing, because production equipment may include tools that eliminate certain sections of the wellbore from production.

Total mud gas logs in vertical wells are commonly used to estimate the potential of a thin reservoir. The gas response is of short duration and is used in conjunction with samples and logs to pinpoint the reservoir.

In like manner, total mud gas logs in horizontal wells can be used to identify fractures or sections of high porosity and permeability. But since the well is drilled for long periods in the reservoir, a different approach to mud gas logging must be used. The liberated gas response will be a function of porosity, bit size, rate of penetration, and the efficiency of the equipment recording the gas.

16 Horizontal Wells

The recorded total gas (V_{gas}) can be defined as:

$$V_{gas} = \text{constant} \times \text{porosity} \times A \times ROP$$

where: V_{gas} = total gas recorded on gas log (units)
porosity = estimated porosity (%)
A = area of wellbore constant (ft^2)
ROP = rate of bit penetration (ft/hr)
constant = constant depending on gas log equipment

If fractures or streaks of good porosity are to be identified from the log, then the total gas reading must be corrected for the rate of bit penetration. For example, if the recorded gas measures 1,000 units at 5 ft/hr penetration, it does not necessarily indicate a better show if the log were to record 2,000 units at 10 ft/hr.

A more reasonable estimate of total gas between two points in the horizontal well can be made if the total gas is corrected by the ratio of ROP and porosity at the two points. For example, if at 7,700 ft-md the total gas measures 3,500 units at 10 ft/hr penetration with an estimated porosity of 9%, how would that compare to a show at 7,900 ft-md of 2,000 units while drilling at 5 ft/hr with an estimated porosity of 10%?

$$TG\ 7,700\ (corr) = TG\ 7,700\ \frac{(ROP\ 7,900)}{(ROP)\ 7,700)} \times \frac{(\text{porosity}\ 7,900)}{(\text{porosity}\ 7,900)}$$

TG 7,700 (corr) = 1,944 units
TG 7,900 = 2000 units

In this case, the corrected gas log would indicate a similar show quality at both depths. In horizontal drilling the ROP is significantly affected by friction between the drillstring and the hole. If pipe rotation is used, the rate of bit penetration (ROP) can easily double from the rate achieved while drilling oriented (stationary drillstring). The ROP increase is the result of drag reduction due to rotation (reduced friction) which has the effect of increasing weight on the bit. This has no relationship to rock drillability but the ROP increase will result in a higher total gas response. If not corrected for ROP, the gas log will be overoptimistic.

Gas shows should always be reported alongside ROP. The ROP should include reference to whether the drilling mode was rotation or orientation.

SUMMARY

- The geology of a reservoir is a critical component in the selection of reservoirs where the benefits of horizontal drilling can be fully realized.
- Reservoir heterogeneities and in situ stress patterns are critical in determining hole azimuth and completion strategies.
- As the well is drilled, the geologist must be aware of the mechanics of drilling horizontal wells so the correct interpretation of wellsite data is made.
- The geologist is an important team member from project inception through production.

REFERENCES

1. Butler, R. M., "The Potential for Horizontal Wells for Petroleum Production," Petroleum Society of CIM, June 12, 1988.
2. Bell, J. S. and B. A. Babcock, "The Stress Regime of the Western Canadian Basin and Implications for Hydrocarbon Production," *Bulletin of Canadian Petroleum Geology,* Vol. 34, No. 3, September 1986.
3. Economides, M. J., et al., "Horizontal Wells," Dowell Schlumberger Technical Presentation, 1989.
4. Canadian Hunter Exploration Well Tests, 1989.
5. Peihua, Xwe, SPE 14837, "A Point Bar Facies Reservoir Model—Semi-Communicated Sand Body," 1986.
6. Nelson, R. A., L. C. Lenox, and F. J. Ward, "Oriented Core: Its Use, Error, and Uncertainty," *AAPG,* Vol. 71, No. 4, April 1987, pp.357-367.
7. Aguilera, Roberto, "Exploring for Naturally Fractured Reservoirs," Presented at the SPWLA twenty-fourth annual logging symposium, June 27-30, 1983.

2
Drilling

INTRODUCTION

Medium radius horizontal drilling is the term used to describe the techniques discussed in this chapter. The medium radius well turns from vertical to horizontal at 8 to 20°/100 ft, which defines a radius of 285 to 700 ft (Figure 2-1). Horizontal wellbores have been extended more than 3,200 ft (Karlsson et al., 1989).

The long radius, extended reach well achieves a large horizontal displacement prior to reaching the target zone. The well builds inclination at 2 to 6°/100 ft which defines a radius of 1,000 to 3,000 ft. Rotary equipment for long radius drilling is often larger than the equipment required for a vertical well at a similar true vertical depth.

The short radius well builds inclination very rapidly at 1.5 to 3°/ft and can reach 90° in 20 to 60 ft. Horizontal lengths are limited to a few hundred feet and special downhole tools are required to drill the curve. Conventional downhole tubulars, evaluation tools, and completion tools often cannot pass the tight radius section.

A medium radius horizontal well often calls for the same rig size, hole size, and drillstring as a vertical well drilled to the same true vertical depth (TVD). Most conventional vertical evaluation and completion tools can be used with some modifications in medium radius applications. By using similar tubulars and evaluation methods the medium radius approach is generally the most cost effective for horizontal holes.

Methods used to drill a medium radius horizontal well include a combination of tools and techniques that first appeared in the early 1900s and have since evolved into today's methods. Nearly all survey methods used up

Drilling 19

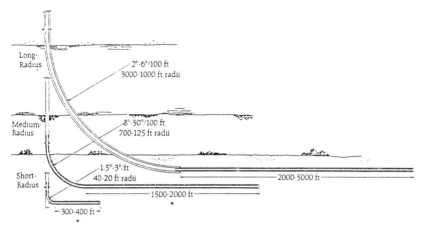

*Depends on formation type

Figure 2-1. Horizontal well types. Source: Eastman Christensen, 1988-89.

to the advent of the electronic measurement while drilling (MWD) tools had been anticipated prior to 1910. These include photographic single and multishot tools and gyro compasses. By 1931, the basic principles of well surveying were established and wells had been surveyed with up to 60° of hole inclination (Figure 2-2). These wells often crossed lease boundaries and lawmakers soon forced operators to control unplanned hole deviation. As some drillers were being forced to drill straight down, controlled directional drilling was being developed in Huntington Beach, California, by 1933 (Brantley, 1971). Here wells were spudded above the high water mark and drilled to targets hundreds of feet offshore. Techniques employed included single shot surveys, whipstocks, and stabilizers. In the 1960s, offshore platform drilling began to employ downhole motors with bent subs which were oriented using a Universal Bottomhole Orientation sub (UBHO). This permitted drillers to orient survey tools downhole using surface data displays which were linked downhole with multiconductor wirelines. By the 1970s, steerable drilling techniques in long radius wells had been well established. Survey calculation methods were refined, and the minimum radius of curvature method was in common use. Drillstring fatigue limits in doglegs had been studied carefully and long reach high angle wells were being drilled routinely.

As with all disciplines within the petroleum industry, the computer has enhanced the design of downhole tools and permitted the efficient handling of data related to drilling operations. Demand for lower speed motors coupled with computer aided design and machining resulted in the development of dependable low speed motors. Previously, motors turned the bit at high speeds which limited bit life (8-10 hr). It is now common to get 50 or

20 *Horizontal Wells*

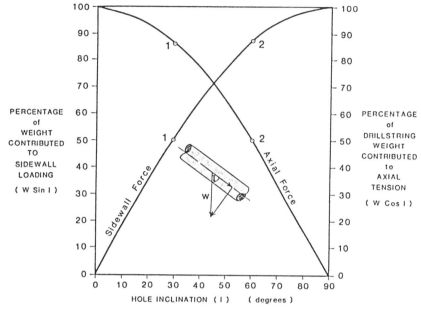

Figure 2-2. Relationship of hole inclination to drillstring axial and sidewall forces.

more hours of bit life at lower rpms. This allows the economic drilling of the entire horizontal section using motors. These motors can be oriented to correct hole trajectory, or rotated to drill straight ahead.

Field portable computers made possible the development of survey tools that transmit survey data via pressure pulses in the mud system. This eliminated the need for a wireline connection downhole. Handling survey data with a computer permits constant updating and projection of survey information. At high angle buildup rates (BUR), accurate hole trajectory prediction is critical. The computer does in minutes what would take days by hand calculation.

By the 1980s, steerable bottomhole assemblies incorporating slow speed motors and MWD survey tools were routinely used on directional and long reach wells. In 1984, the first medium radius wells were drilled (Eadland, 1987), and the number of medium radius horizontal wells drilled annually has increased rapidly since.

Successful medium radius wells require careful teamwork in the planning phase and on the wellsite. The drilling supervisor, engineer, geologist, and directional driller all contribute to the decision-making process. Mechanical problems can be avoided by applying the principles and concepts of drillstring design and hole cleaning developed in long radius drilling. Although the loading may differ, the mechanical principles are still valid. The high BUR of medium radius wells eliminates many problems inherent in long radius wells. Understanding why the high BUR is advantageous is critical to understanding medium radius horizontal drilling.

A typical medium radius well path is shown in Figure 2-3. The vertical well is drilled to kickoff point (KOP). Hole inclination is increased at a constant BUR which defines an arc with constant radius (R). A tangent section is normally drilled at a constant inclination (I_{tan}) to correct for variations in BUR or make corrections in target TVD (T_{tgt}). The second angle build section increases the hole inclination to the desired lateral hole angle (I_f). This point is often referred to as the end of curve (EOC). The lateral hole is drilled to the final total depth (FTD). The following formulas are based on a true circular axis with no changes in hole azimuth (Karlsson et al., 1989).

A minimum build rate is chosen and the KOP is defined. The KOP is calculated with the assumption that no tangent is drilled.

$$T_{kop} = T_b - \frac{k_1}{BUR_{Min}} (Sin(I_f) - Sin(I_i)) \qquad (2\text{-}1)$$

In the build interval, the following equations are used to determine TVD, MD, and horizontal displacement (D).

$$T_2 = T_1 + \frac{k_1}{BUR_{Exp}} (Sin(I_2) - Sin(I_1)) \qquad (2\text{-}2)$$

$$MD_2 = MD_1 + \left(\frac{\pi k_1}{180}\right) \cdot \left(\frac{I_2 - I_1}{BUR_{Exp}}\right) \qquad (2\text{-}3)$$

$$D_2 = D_1 + \left(\frac{k_1}{BUR_{Exp}}\right) \cdot (Cos\, I_1 - Cos\, I_2) \qquad (2\text{-}4)$$

In a tangent section of length ΔMD drilled at an inclination (I), the following equations are used to determine the changes in depth and displacement.

$$T_3 = T_2 + \Delta MD\, Cos\,(I) \qquad (2\text{-}5)$$

$$MD_3 = MD_2 + \Delta MD \qquad (2\text{-}6)$$

$$D_3 = D_2 + \Delta MD\, Sin\,(I) \qquad (2\text{-}7)$$

If the BUR of a particular set of tools in a particular area has been established and the target is defined, the well can be drilled without a tangent

22 Horizontal Wells

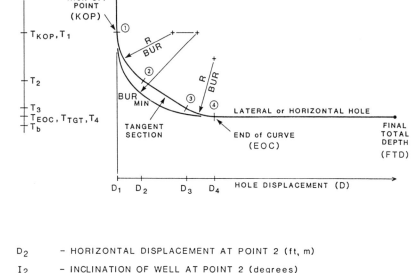

D_2	–	HORIZONTAL DISPLACEMENT AT POINT 2 (ft, m)
I_2	–	INCLINATION OF WELL AT POINT 2 (degrees)
I_i	–	INITIAL WELL INCLINATION (degrees)
I_f	–	FINAL WELL INCLINATION (degrees)
K_1	–	CONSTANT : 5730 if BUR is degrees/100ft
		1719 if BUR is degrees/30m
BUR_{EXP}	–	EXPECTED or PLANNED BUILDUP RATE (degree/100ft, degree/30m)
BUR_{MIN}	–	MINIMUM EXPECTED BUILDUP RATE (degree/100ft, degree/30m)
BUR_{MAX}	–	MAXIMUM EXPECTED BUILDUP RATE (degree/100ft, degree/30m)
MD_1	–	MEASURED DEPTH AT POINT 1 (ft, m)
MD_2	–	MEASURED DEPTH AT POINT 2 (ft, m)
T_1	–	TRUE VERTICAL DEPTH AT POINT 1 (ft, m)
T_2	–	TRUE VERTICAL DEPTH AT POINT 2 (ft, m)
T_{KOP}	–	TRUE VERTICAL DEPTH AT KOP (ft, m)
T_t	–	TRUE VERTICAL DEPTH AT TARGET TOP (ft, m)
T_b	–	TRUE VERTICAL DEPTH AT TARGET BASE (ft, m)
T_{TGT}	–	TRUE VERTICAL DEPTH AT EOC (ft, m)

Figure 2-3. Medium radius well path definitions and formulas.

Drilling 23

section. For the first well in an area where the actual BUR is not known, a well design with one or more tangents is recommended.

Table 2-1 shows the importance of selecting as high a BUR as circumstances permit. Maximizing BUR reduces the length of costly directional hole required to reach 90° (1,125 ft at 8°/100 ft vs 450 ft at 20°/100 ft).

Table 2-1
Summary of Buildup Rates (BUR), True Vertical Depth (Radius), and Measured Depth (Arc Length)

Buildup Rate (BUR) (degrees/100 ft)	True Vertical Depth or Radius of Curve (ft)	Measured Depth or Arc Length (ft)
8	716	1125
10	576	900
12	477	750
13	440	692
14	409	642
15	382	600
16	358	562
17	337	529
18	318	500
19	301	473
20	286	450
BUR	$TVD = \dfrac{MD}{1.57}$	$MD = \dfrac{90}{BUR} \times 100$

Assumption: Build section starts at 0° and ends at 90°

A second advantage of selecting a high BUR is the relative change in TVD variation with a given variation in BUR (Figure 2-4). The BUR will always vary from theoretical design due to formation interaction and drilling conditions. A 1° increase in BUR at 19°/100 ft decreases the TVD by 15 ft; a 1° increase in BUR at 12°/100 ft decreases the TVD by 37 ft.

It is the ability of modern angle-build motors to drill at a high BUR with predictable performance that has made horizontal drilling possible in thin zones. The high BUR reduces costs by reducing the hole length through the curve, and variations in BUR do not result in large TVD variations that require costly correction runs.

Variations in BUR can be corrected using tangent sections. Figure 2-5 shows how a tangent section can place a wellbore in the same TVD position using both 18°/100 ft and 15°/100 ft buildup rates. The difference in TVD between the two BURs is 64 ft. A 90 ft tangent drilled at 45° will deepen the well 64 ft TVD:

24   Horizontal Wells

$$(MD_{Tangent} = \Delta\, TVD/Cos\, I_{Tangent})$$

The length and inclination of a tangent section is site specific. Factors such as geological markers, bit trips, and desired vertical depth accuracy at the beginning of the lateral hole will all influence the inclination and length of a tangent section. The vertical incremental depth of an inclined hole becomes smaller as the inclination increases, due to the sinusoidal relationship $\Delta\, TVD = \Delta\, MD\, Cos\, (I_t)$.

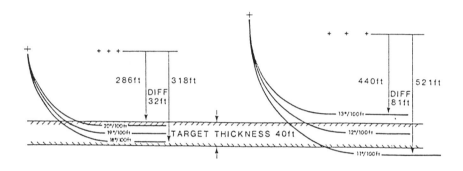

Figure 2-4. The effect of buildup rate (BUR) on true vertical depth variation.

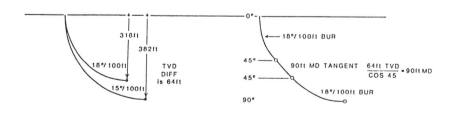

Figure 2-5. Use of a tangent to adjust buildup rate.

Drilling 25

Placing a tangent section early in the well profile, at perhaps 30°, will result in large variations in TVD at 90° as variations in BUR occur. By drilling the tangent later in the well profile, say 60°, variations in TVD at 90° will be reduced for a similar variation in BUR. Figure 2-6 shows how increasing tangent inclination (I_{tan}) significantly increases the accuracy of TVD at 90°.

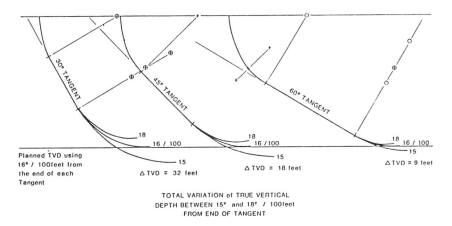

Figure 2-6. Effect of tangent inclination on true vertical depth control at 90°.

Normally the tangent is placed between 45° and 60°. Less than 45° does not give the desired accuracy and the likelihood is high that a second tangent will be required. A tangent placed at greater than 60° requires excessive length to achieve significant depth correction. The high angle tangent will increase the accuracy of the TVD at 90°, but such accuracy is achieved at great expense and is often not required.

Between 45° and 60°, 70–86% of the vertical section from KOP to 90° has been drilled. This vertical section is normally of sufficient depth to correlate geological control and estimate a TVD depth from the bit to the target zone. The drilled interval from KOP to 45° provides a good average BUR to project to the end of the curve at 90°. Between 45° and 60° the change in TVD is still significant for the length of tangent drilled, and good confidence can be placed in the expected BUR.

Once maximum and minimum BURs have been established for a given field area, the optimum tangent angle can be found for a particular target thickness.

26 *Horizontal Wells*

$$I_{tan} = Sin^{-1} \left[Sin\ (I_f) - \frac{Bur_{Max}\ Bur_{Min}\ (T_b - T_t)}{k_1\ (BUR_{Max} - BUR_{Min})} \right] \quad (2\text{-}8)$$

where I_{tan} = inclination of tangent (degrees)
 I_f = final well inclination (degrees)
 BUR_{Max} = maximum expected BUR
 BUR_{Min} = minimum expected BUR
 k_1 = Constant 5,730 if BUR °/100 ft
  1,719 if BUR °/30 m
 T_b = base of target TVD
 T_t = top of target TVD

The corresponding target TVD (T_{tgt}) within the zone is calculated using Equation 2-9.

$$T_{tgt} = T_t + k_1 \left[\frac{1}{BUR_{Exp}} - \frac{1}{BUR_{Max}} \right] \left[(Sin(I_f) - Sin(I_{tan}) \right] \quad (2\text{-}9)$$

where T_{tgt} = target TVD at EOC
 BUR_{Exp} = expected or planned BUR

Equations 2-8 and 2-9 are very useful in preliminary well planning and continue to be important as actual maximum and minimum BURs are observed in the wellbore.

These basic geometric relationships are used to design a well profile that will minimize the hole length and number of bottomhole assembly changes, while maximizing the certainty of hitting the target. The penalty is severe if the zone has been missed by more than a few feet. At 92°, 30 ft must be drilled to gain 1 ft of vertical depth.

SELECTION OF THE WELL PROFILE/CASING PROGRAM

Drilling laterally through a producing formation should be considered a completion operation as well as a drilling operation. To maximize hydrocarbon production rates and recoveries, the interaction of drilling fluids, various damage mechanisms, and variations in the reservoir characteristics must be considered. Reservoir variations will occur both vertically and laterally as up to 3,000 ft of lateral hole is drilled.

Three basic well types are shown in Figure 2-7. The site specific circumstances will determine which casing profile is used. The overall project economics will determine what completion method is optimum for the well-

Drilling 27

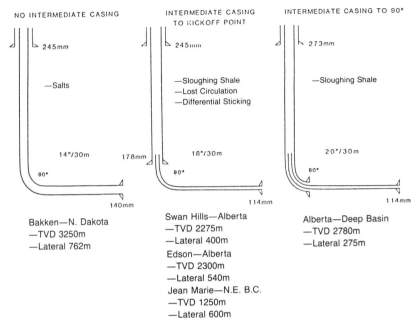

Figure 2-7. Basic horizontal well casing programs.

bore. A casing program selected to minimize drilling cost without considering horizontal completion cost may ultimately reduce the project economics by increasing total well costs.

Often a reservoir system will be damaged by drilling a well with circulating fluid that is incompatible with the reservoir or reservoir fluid. Examples include:

- Water sensitive formations
- Formations sensitive to relative permeability changes and fluid loading
- Formations that require stimulation fluids incompatible with drilling fluids
- Fluids that contain products insoluble in reservoir fluids and stimulation fluids

In a vertical well, damage is removed by stimulating the wellbore. It is not always economically possible to stimulate a horizontal wellbore using the same techniques used in a vertical wellbore. What works economically on a 10-ft vertical zone may not achieve the same result economically when 2,000 ft of the same zone is exposed laterally. In horizontal wellbores, formation damage control and reduced stimulation are often preferable to drilling a wellbore at the lowest cost. Isolating sections of the lateral wellbore and the increased volumes required in stimulations often result in total costs greater than the cost of controlling the damage mechanism while drilling.

The completion program may require a minimum diameter or pressure rating for the production casing. High volume fracture treatments may require high burst ratings and larger casing diameters to reduce circulating pressures; wells with sand control problems may require very large production casing to allow for sand control tools. These considerations influence the lateral hole size and production casing size.

After determining the preferred mud system and production casing to optimize the horizontal well completion, uphole conditions can be considered.

Factors that influence the casing design can occur above and below the planned kickoff point. Examples include:

- Lost circulation zones
- Sloughing hole conditions
- Plastic salt zones
- Secondary producing zones
- Over pressure and under pressure zones
- Hole washouts and ledges that prevent passage of bent-housing directional drilling tools.

Any of these conditions may warrant setting intermediate casing. Casing can be set above, at, or below kickoff point. The cost of intermediate casing is offset by reduced problem time after casing is run, when daily operating costs are high due to the directional services.

It is frequently possible to drill the entire wellbore with one fluid type, although the fluid is often not compatible with the producing formation. If intermediate casing is set, the fluid can often be changed for a nondamaging fluid to drill the horizontal section of the hole. Intermediate casing can be justified if the overall project cost is considered:

- The volume and downhole loss of expensive completion fluid is minimized.
- Annular clearance is reduced, which optimizes hole cleaning at lower mud rheology.
- Solids control is optimized, further reducing damage.

The casing designs for two horizontal wells in two different areas are compared in Figure 2-8. It is useful to compare these wells to show how different wells must overcome different problems.

Drilling 29

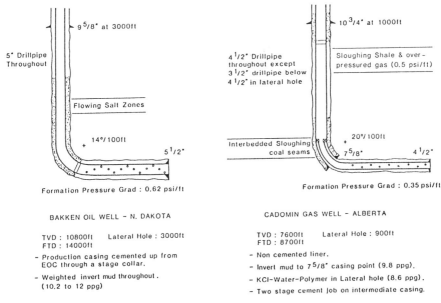

Figure 2-8. Comparison of two horizontal wells—casing and drilling fluid programs.

Bakken Oil Well, Williston Basin, North Dakota

The objective formation in this case is an oil wet, overpressured, fractured shale at a true vertical depth of about 10,500 ft.

Thick salt zones above the kickoff point can cause problems during drilling and plastic flow has resulted in collapsed production casing in numerous wells. To prevent this casing collapse, operators have concentrated on limiting hole enlargement, achieving good annular cementing, and running heavy wall casing designed for 1.2 psi/ft or more.

A weighted invert oil mud has proven successful for drilling the lateral hole without significant damage. This mud system also controls uphole salts and helps achieve good cement jobs. Although mud costs are rather high due to significant seepage losses, this system allows these wells to be drilled to TD without intermediate casing.

The wells are complete with 5½-inch production casing run to TD, with the lateral section left uncemented below an annular packer and stage collar.

Running intermediate casing is an expensive alternative in this case because of the high collapse design gradients required and the drillstring and hydraulics limitation with drilling small hole at this depth.

30 *Horizontal Wells*

Cadomin Gas Well, Deep Basin, Alberta

Mud System. Invert oil mud is required for shale control in the vertical well and sloughing coal beds in the angle build section. Overpressured gas is often encountered high in the vertical well and requires increased mud density.

The producing zone is an underpressured gas reservoir, which is damaged using invert mud and requires fluid loss control. The mud system selected was a Water-KCL-Polymer fluid using $CaCo_3$ as an acid-soluble bridging agent for fluid loss control.

Casing Design. The last coal beds are located immediately above the producing zone. Casing is set at 85° to permit the mud system change. The 7⁵/₈-inch intermediate casing is set through the 20°/100 ft radius, and cemented at 85°.

The overall well design (MD 8,700 ft, TVD 7,600 ft) requires a rotary rig equipped with 4¹/₂-inch drillpipe. The minimal rheology, density, and circulating pressures of the polymer fluid system allow the operator to run 4¹/₂-inch drillpipe inside the 7⁵/₈-inch intermediate casing. This minimizes the volume of the system and reduces circulating rates: this further optimizes solids control. The high annular velocities reduce rheology requirements for hole cleaning. The overall results are good hole cleaning and solids control, and the cost of changing to a 3¹/₂-inch drillstring is saved.

The well was completed by placing an uncemented 4¹/₂-inch production liner in the lateral below the intermediate casing. A light acid wash/squeeze was successful in cleaning up the well to meet economic production rates. Stimulation costs to fracture past mechanical and fluid-induced damage of weighted invert mud would have cost more than intermediate casing.

The overpressured Bakken oil well and the underpressured Cadomin gas well are both medium radius horizontal wells drilled to maximize the economic production of oil and gas. Each well is different, yet their designs meet specific well requirements in cost-effective ways.

After a well design is selected, based on overall well objectives, the drilling can be optimized through careful depth control, tubular design, and mud system design. These factors reduce the well cost by decreasing the time required to drill the well.

DRILLING DEPTH CONTROL

The actual well path through the curve and the lateral section will be influenced by vertical and lateral geologic considerations, including:

Drilling 31

- Target TVD and thickness
- Desired hole azimuth
- Marker zones
- Formation dips
- Vertical and lateral variations affected by deposition and subsequent geological history

The following techniques are used to correct the well path for geological variations encountered while drilling the curve. Using these techniques, the lateral well can be accurately placed in thin zones.

Marker Beds (Continuous vs Discontinuous)

Some marker beds will be continuous to a control well, and some will be discontinuous. The discontinuous marker may change depth relative to the target. Markers with isopachs constant to the zone are the most desirable (Figure 2-9).

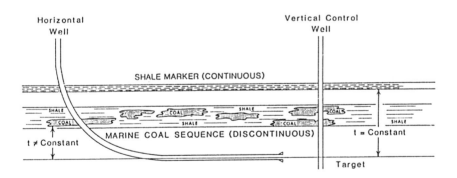

Figure 2-9. Continuous vs discontinuous marker beds.

Marker Bed Dip Angle Does Not Equal Target Dip Angle

When the dip angle of a marker bed does not equal the dip angle of the target, as at an unconformity, the relative dip relationship must be determined from other controls (Figure 2-10).

32 Horizontal Wells

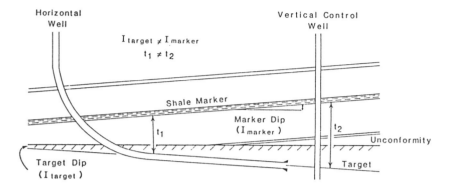

Figure 2-10. Marker bed dip angle is unequal to target dip angle.

Faults

The presence of faults can change the relationship of marker beds to the target zone. If faulting is anticipated in an area, the depth and displacement should be anticipated (Figure 2-11).

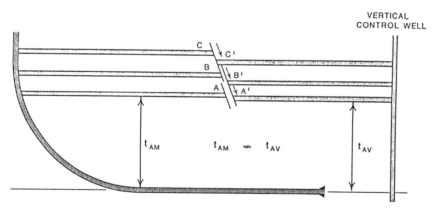

Figure 2-11. Faults in uphole marker beds.

Discontinuous Reservoir

A discontinuous reservoir may be one of three types. Figure 2-12a shows a laterally heterogenous reservoir with no variation vertically. The lateral hole would simply be drilled at the established depth until the reservoir is encountered.

If the reservoir varies laterally as well as vertically (Figure 2-12b), and poor control exists to determine whether the reservoir is above or below the bit, then an average depth should be followed that has the highest probability of encountering the reservoir. This can be accomplished by taking the average isopach thickness from some continuous marker above the objective to the center of the good pay in offset control wells.

If a fault is encountered (Figure 2-12c), knowing the magnitude and direction of the displacement is useful. Making depth corrections in horizontal holes is time-consuming. The well should be placed in the top of the reservoir if an upward displacement is anticipated. This will reduce the vertical displacement required to get back into the reservoir.

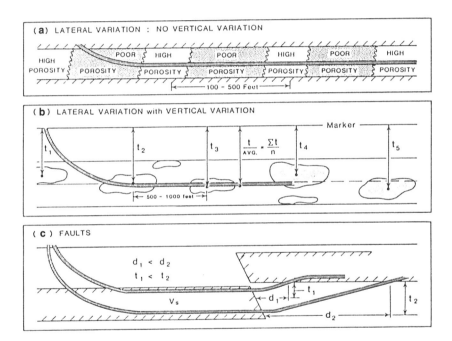

Figure 2-12. Discontinuous reservoirs.

Incorrect Formation Dip

Knowing the correct formation dip is critical to controlling the lateral wellbore over a long lateral hole (see Figure 2-13). A method of recognizing the depth in the reservoir is important. Incorrect dip determination of $3/4°$ could place the wellbore 13 ft deeper in the reservoir over 1,000 ft. If accu-

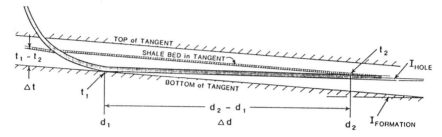

Figure 2-13. Calculating formation dips.

rate dip data is not available, reservoir characteristics such as vertical grain size distribution, shale or chert beds, and reservoir boundaries can all be used to check the local formation dip. As the hole is drilled laterally, the calculated bit depth change and distance between encounters with markers is noted. The formation dip should be recalculated whenever possible, using the simple equation:

$$I_{form} = I_{Hole} - Tan^{-1}\left(\frac{\Delta t}{\Delta d}\right) \qquad (2\text{-}10)$$

Tangent Angles

If the only depth correlation marker is close to the target with little geological control above the marker, a high angle tangent section can be useful. In the example shown in Figure 2-14, the lithology change at the bit will occur as the MWD gamma ray (GR) sensor reaches a change in lithology. If these events are to occur at the same time, the tangent inclination is determined by the TVD difference between the events (i.e., marker thickness) and the spacing from the bit to the MWD-GR sensor. If the tangent section ends near the lithology change and the last build section is started, the GR will confirm the sample pick. At high angles, sample changes can be gradual as the bit drills both lithologies. The exact point of the lithology change is also masked if the drill cuttings are logged incorrectly due to poor hole cleaning at the high angle. The MWD-gamma ray tool is very useful in confirming sample depths for marker beds and also provides a valuable gamma ray correlation log.

Drilling 35

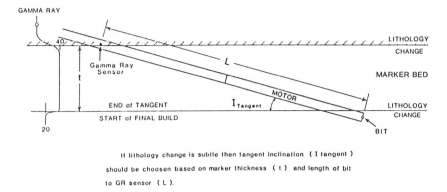

Figure 2-14. Simultaneous gamma ray/sample change for marker beds at high tangent angles.

Hole Inclination at Target Zone

It is often preferable to penetrate the target zone at a lower angle (80°–85°) and reduce the BUR in the target zone to finish the angle build section (Figure 2-15a). The advantages of this approach are twofold. The last section of the arc is drilled in the producing zone, which increases the effective length of lateral hole, and, more important, if the zone is not encountered as planned, a tangent can still be drilled to find the zone (Figure 2-15 b).

If the well is designed to encounter the target zone at 88° and a final tangent is required to find the zone, the tangent will be excessively long. If the target zone dip is incorrectly estimated and in fact dips away from the hole at 1 or 2°, the hole will run along the top boundary of the target. If the well is designed to encounter the target zone at 80° and the zone is not encountered by 85°, a tangent at 85° will be much shorter than the 88° tangent. The tangent will be less likely to run along the top of the zone if the target dip is underestimated. The inclination chosen to enter the target zone is strongly influenced by the target thickness. Thin targets will be approached at higher angles than thicker targets. If the top and bottom of the zone can be contacted in the early part of the lateral hole by approaching the target at lower angles, say 80°, the formation dip can be confirmed. This may be important in lateral wells.

In summary, the well design should be approached from the point of view that drilling the target formation is a completion operation. After choosing a casing program, the well profile must be determined by working closely with the geologist to determine formation dips and characteristics through the angle build section and in the target zone. Maximizing BUR will reduce

36 Horizontal Wells

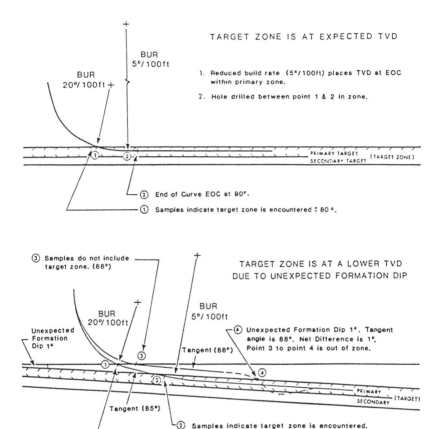

Figure 2-15. Hole inclination at the target zone.

costly tangent sections and reduce the hole length from kickoff point to the producing zone.

MECHANICAL ASPECTS OF STEERABLE MOTORS AND THEIR EFFECT ON KOP SECTION

Modern steerable drilling systems use positive displacement mud motors (PDM). These motors operate on the principle of a Moineau pump (Tiraspolsky, 1985). As fluid is pumped into the motor, the rotor is forced to turn, allowing fluid to pass. This rotation is transferred to the bit. The operating parameters which include flow rate, bit rpm, and torque are governed by the rotor/stator configuration. The stator is a molded-rubber spiral helix. The rotor is a helicoidal steel shaft. The shaft is made up of z_1 lobes and is run inside the stator with z_2 indentations. In cross section, the shaft and sheath have a wavelike outline. Each wave corresponds to a lobe. Motors are referred to by their lobe count ($z_1 : z_2$) where $z_2 = z_1 + 1$. The ratio of pitch length between the shaft (t) and the sheath (T) is similar to the ratio of lobes, ($z_1/z_2 = t/T$). Power collected on the rotor shaft is hydraulic power expended by fluid working its way between the rotor and stator, causing the rotor to turn (Tiraspolsky, 1985).

The rotational speed and available torque is related to the rotor/stator geometry. The pitch angle and number of lobes on the rotor/stator combination can be selected to match torque and rpm requirements. Rotational speed is proportional to flow rate and torque is proportional to the pressure drop in the fluid as it flows through the motor. Motors with more lobes generally turn more slowly and with higher torque ratings (Figure 2-16).

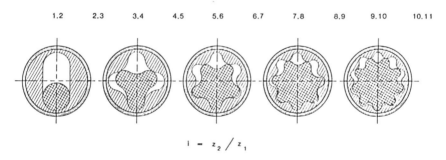

Figure 2-16. Cross section through helimotors with different lobe patterns (according to M. T. Gusman and D. F. Baldenko, 1985).

Downhole motors generally include 3 sections: the rotor/stator, universal joint, and bearing housing. (Figure 2-17) The rotor is connected to the bit through a universal joint which removes any offset or eccentric motion in the system. The bit is screwed into the bearing housing which carries the axial and lateral loads.

The advantages of modern multilobe motors are high torque output, slow bit speed, and short length. The short motor length is important: it permits the addition of deflection points above and below the motor. As the distance between deflection points increases, the bit offset and associated side forces increase. Figure 2-18 shows the relation of motor length and bit offset. If the total deflection of 4° had been placed in the bent sub above the motor the bit offset would be approximately 22.5 inches (Tan(4) × 26.8 ft × 12). By

38 Horizontal Wells

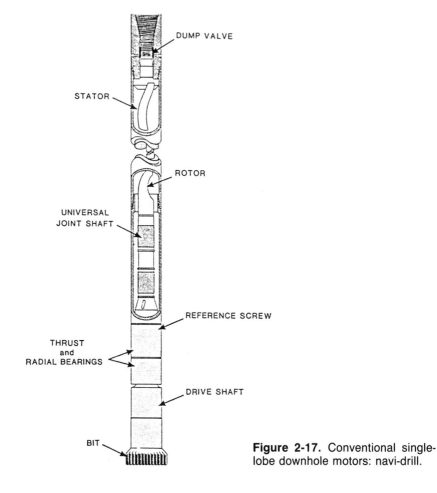

Figure 2-17. Conventional single-lobe downhole motors: navi-drill.

placing 2.5° of deflection below the motor, the bit offset is reduced to 13.38 inches (Kerr and Lesley, 1988).

If the first deflection point is near the bit, a small deflection will cause a build tendency similar to a larger deflection further away from the bit. The near bit deflection creates less bit offset and lower side forces on the bit. A simple three-point geometry is shown in Figure 2-19. Any three points not on a line describe an arc which is related to BUR.

When a bent housing assembly is run in a straight hole, the assembly will be exposed to high bending stresses and high sidewall contact forces. In the initial phase of kicking off a straight hole, the wall contact forces will differ from those acting on the assembly when the motor is seated in a curved hole and has established a BUR. The maximum bending stresses through a motor

Drilling 39

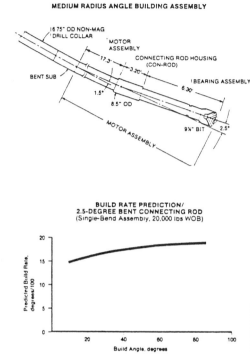
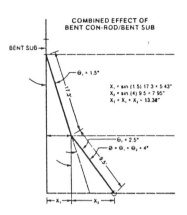

Figure 2-18. Medium radius angle build assembly and combined effect of bent con-rod bent sub. Source: Kerr and Lesley, 1988.

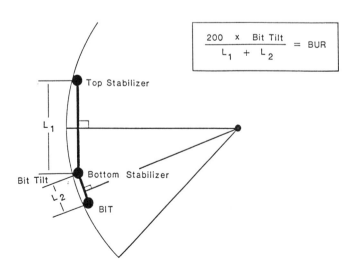

Figure 2-19. Three point geometry.

40   *Horizontal Wells*

have been calculated as high as 13 times greater than those present when the motor is relaxed in its designed curve (see Figure 2-20).

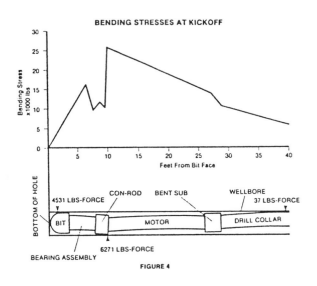

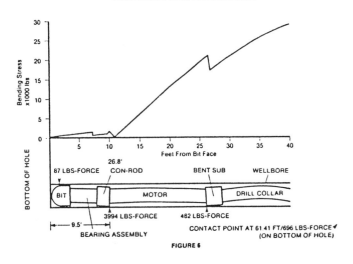

Figure 2-20. Motor bending stresses at kickoff and equilibrium. Source: Kerr and Lesley, 1988.

Drilling   41

The high side forces on the bit and first fulcrum point are important considerations when selecting a kickoff point out of the vertical hole or at the end of tangent sections. The fulcrum point will always maintain high side forces but these will reduce by roughly half when the motor is in a relaxed state. The bit side forces drop exponentially (Figure 2-21) from very high values (4,500 lb) to very low values (100 lb) when the motor is in equilibrium. The high initial side forces can create severe doglegs and accelerate gauge wear on the bit in abrasive formations.

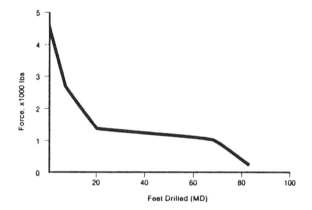

Figure 2-21. Lateral forces on bit from kickoff to equilibrium. Source: Kerr and Lesley, 1988.

In a soft formation the bit will quickly cut sideways, creating a high initial dogleg. The bottomhole assembly can easily wipe out the dogleg in soft homogeneous formations (Figure 2-22b). What should be avoided is kicking off in a very soft bed that lies immediately below a very hard bed (Figure 2-22a). In a heterogenous formation, the dogleg may result in a ledge or keyseat which will hang up the bit and stabilizers. Interbedded lithology with very soft beds underlying harder beds should be avoided as a kickoff point or tangent end points. If a motor is seated in a curved hole defined by its BUR prior to entering interbedded lithology, the dogleg variation will be minimal as the bit crosses bed boundaries. This is largely due to minimal bit side forces.

In hard abrasive formations, the high side forces at kickoff can cause severe bit shank wear. If the bit is turned at a stationary point while the motor is oriented or survey data is obtained, the cutting structure will cut into the hole wall. The abrasive formation will wear the bit shank above the bit cone. Wear protection on the shank area is worthwhile, as is extra gauge

42   Horizontal Wells

(a)   KICKOFF in a SOFT FORMATION
      UNDERLYING a HARD FORMATION

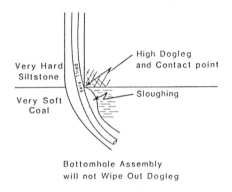

(b)   KICKOFF in a SOFT
      HOMOGENEOUS FORMATION

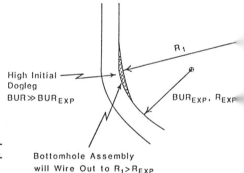

Figure 2-22. Kickoff in a soft homogeneous formation; and in a soft formation underlying a hard formation.

protection on the cutting structure. The shank wear can be reduced by keeping the bit moving axially along the hole wall while circulating. The bit will make wall contact on the cutting structure rather than cutting into the formation and making wall contact on the shank.

The kickoff point should be selected in a nonabrasive homogeneous formation. The motor should be seated prior to major formation changes, particularly if rock drillability changes. Bit sideloads are an important consideration in abrasive formations and are dependent on the particulars of the motor design (see Figure 2-23).

Drilling 43

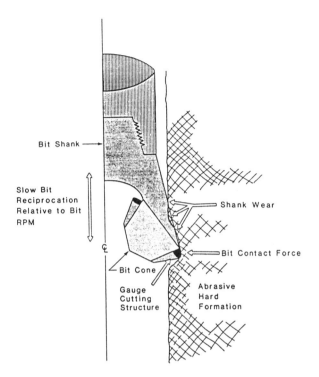

Figure 2-23. Bit shank wear in abrasive formations using bent housing motors.

BASIC MECHANICS AND APPLICATION OF TORQUE AND DRAG TO MEDIUM RADIUS HORIZONTAL DRILLING

Reduction and control of torque and drag is a very important objective in horizontal well design and drilling. The presence of torque and drag cannot be eliminated, but if ignored, drilling efficiency will be reduced and the final well objectives will not be met economically. The importance of monitoring torque and drag cannot be overstressed. Monitoring can be used to foretell the onset of hole-cleaning problems and stuck pipe.

Hole drag is the result of contact between the drillstring and the hole wall or casing. In its simplest form, sliding friction force (F) is the product of the friction factor (ff) and the force acting normally to the contact surface (N).

$$F = ff \cdot N \tag{2-11}$$

44 Horizontal Wells

In a straight vertical hole with minimal deviation, the sliding friction drag is negligible due to a lack of normally acting forces. In an inclined hole or holes with sudden doglegs, forces acting normally to the pipe wall result in sliding friction drag between the hole and the drillstring.

Sliding friction force will be observed as drillstring torque, drillstring drag, or as a combination of both. If the drillstring is pulled without rotation, the friction force will be seen as drag. If the drillstring is only rotated, the friction force will be observed as torque. Sliding friction forces oppose motion, and therefore the friction force acts in a direction opposite to the velocity vector (Dellinger et al., 1980).

Figure 2-24 shows the velocity components due to drillpipe rotation (Vc) and downward motion (Va). The total velocity vector (Vr) is easily calculated. Frictional drag occurs along the vector (Vr) but in the opposite direction. The torsional and axial friction drag components are directly proportional to their respective velocity counterparts.

When the drillstring is rotated and reciprocated the torque and drag will vary depending on the rotational and axial velocity of the pipe. The benefit of pipe rotation is easily seen in the lateral hole section where the penetration rate may double when the pipe is rotated after drilling without pipe rotation. The drag which reduces effective WOB is overcome using increased torque through pipe rotation. The sliding friction force is observed

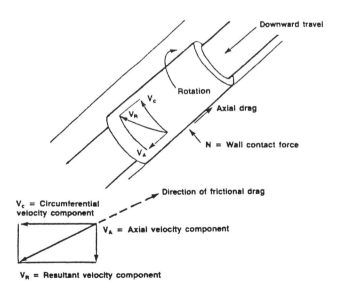

Figure 2-24. Friction resultant with pipe movement. Source: Dellinger et al., 1980.

Drilling 45

as torque or drag depending on the mode of pipe movement, although the source of the friction force is the same for both torque and drag.

A drillstring in tension, acting in a dogleg, creates side forces on the inside of the dogleg as the pipe contacts the wall (Figure 2-25). This side force acting at the pipe/wall contact surface sets up a friction force acting parallel to the direction of movement of the drillstring. If pipe is pulled out of the hole, the friction force would be parallel to the axis of the pipe and is observed as drag or an increase in hook load. If the pipe is rotated, the friction force would be observed as rotary torque. The friction force is related to the drillpipe tension and dogleg severity.

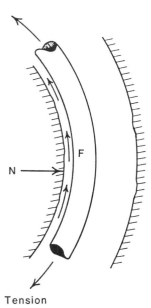

Tension

Figure 2-25. Drag through doglegs.

Drag due to hole inclination depends on the weight (W) of the tubular and the hole inclination (I) (Figure 2-26). Only the component of weight acting normal to the hole wall contributes to friction force. The component of weight acting along the axis of the drillstring contributes to tension in the drillstring.

The combination of these two sliding friction forces accounts for much of the torque and drag experienced in horizontal wells. The torque and drag resulting from bottomhole assembly drag is normally negligible and constant if hole cleaning is not a problem. The bottomhole assembly drag in a clean hole will be maximum when bent housing motors are run in the

46 Horizontal Wells

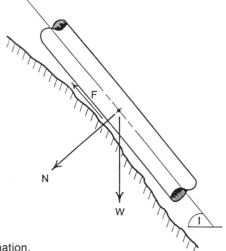

Figure 2-26. Drag due to hole inclination.

straight hole. Based on the example in Figure 2-20, the motor exerts a total of 10,000 lb sidewall contact force, and the sliding friction drag expected would be 4,000 lb (ff = 0.4). If the motor was run at the end of a 2,000-ft lateral hole for a correction run, using a 41-lb/ft drillstring, the total sliding friction drag would be approximately 33,000 lb from the drillstring. Total drag in the lateral hole would be approximately 37,000 lb.

The bit, stabilizer, and tubular OD profiles acting on ledges and cutting beds can contribute significant drag which is not easily calculated. These forces are often responsible for stuck pipe. Recognition of impending problems is difficult unless the drag due to dogleg and drillstring weight is monitored.

Computer models of torque and drag are very useful in the design and drilling phases of highly deviated wellbores. These programs are often based on a simple model developed by Exxon Production Research (Johancsik et al., 1984). This model assumes that the loads on the drillstring result only from gravity and sliding frictional forces that come about through contact between the hole wall and pipe body. The models also consider friction through inclined wall contact and side forces in doglegs. Pipe bending is not considered. Bottomhole assembly drag must be estimated and inputted directly. These models are very useful for quickly comparing hole configurations in the design stages, and optimizing string design. Using a computer model to recognize the onset of hole problems while drilling is important. Observed torque and drag values throughout the well are inputted to determine the friction factor. A rising friction factor may indicate the onset of developing hole problems (Figure 2-27).

Drilling 47

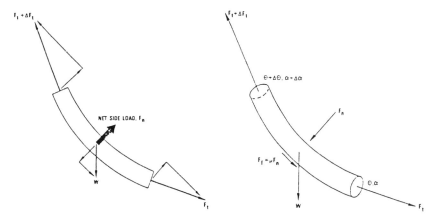

Figure 2-27. Forces acting on a drillstring element. Source: Johancsik et al., 1984.

Torque drag models require a complete set of drillstring and wellbore survey data. The program can calculate torque and drag forces directly if the friction factor is inputted directly. If the friction factor is desired as output, the torque and drag data is inputted directly and the friction factor is solved by iterating to match the data.

The program calculates forces on the drillstring in increments based on the following formulas (Johancsik et al., 1984).

$$F_n = [(F_t \alpha \sin \bar{O})_2 + (F_t O + W \sin \bar{O})_2]^{1/2} \qquad (2\text{-}12)$$

$$\Delta F_t = W\cos \bar{O} \pm \mu F_n \qquad (2\text{-}13)$$

$$\Delta M = \mu F_n r \qquad (2\text{-}14)$$

Torque/drag design in medium radius horizontal wells is done in three sections: the vertical hole, angle build hole, and the horizontal hole.

The vertical wellbore down to kickoff point is treated as a conventional wellbore. If the hole is not deviated, hole drag is the result of minor doglegs and tension in the drillstring. For planning purposes, hole drag from a similar vertical well in the area is used. This hole section may be openhole or cased, in either instance a lower friction factor is used in this section of the hole due to the absence of cutting beds in a vertical hole.

The angle build hole section from kickoff point to the end of curve is influenced by dogleg drag and inclined hole drag. The hole drag in this section is lower in a medium radius well than a long radius well. By maximizing the BUR, the dogleg is increased but the hole length is reduced. This reduces the

48 Horizontal Wells

weight of tubulars in the angle build section (inclined hole drag) and minimizes tension below the kickoff point (dogleg drag). Designing excessive tangent sections and conservative BURs will increase the length of hole from KOP to EOC. This increases drag from pipe contact in the inclined hole, and dogleg drag due to higher tension in doglegs. It has been shown that the overall torque and drag of a long reach deviated well can be reduced by drilling an underreach type well (Sheppard et al., 1988). Horizontal reach is not gained by drilling long tangents at shallow angles. The underreach well uses a catenary curve at higher angles to achieve the reach. Wall support of the drillstring at higher angles reduces the tension in doglegs higher in the hole.

Drag in the horizontal section is dominated by the weight of the drillstring and bottomhole assembly drag. Dogleg in the lateral hole does not greatly contribute to drag because the pipe is on the low side of the hole and pipe tension is low.

For design purposes, the sliding friction drag in the horizontal hole can be approximated using Equation 2-15.

$$F = ff \times BHA \text{ weight} \tag{2-15}$$

where F = sliding friction force, lb
ff = friction factor, (0.2 − 0.4)
BHA = weight of tubulars in the horizontal hole, lb

For example, in a 2,000-ft lateral section of hole, 41 lb/ft HWDP would create a sliding friction drag of approximately 24,600 lb using a 0.3 friction factor. Minimizing the BHA weight is one approach available to permanently reduce drag. Reductions in the friction factor will reduce drag, but the reduction may be lost if hole-cleaning problems develop. Drag design must be done at the highest anticipated friction factor.

Although sliding friction drag caused by tool deflection in a straight hole is negligible, the drag of bottomhole assemblies in cutting beds is not. The worst case is when an angle build motor is run in a long lateral hole for a correction run. The high side forces can disturb filter cake and cutting beds, which can pack around the motor assembly. Cutting bed development and poor hole cleaning can lead to a reduction in the effective diameter of the hole. A cutting bed develops when cuttings fall to the low side of the hole. If the bed is not circulated out of the hole prior to tripping the bit and bottomhole stabilization, the bit will ride up on the cutting bed. A buildup of cuttings will not fall past the bit, as in a vertical hole. Figure 2-28 shows the effect of a cutting bed on the bottomhole assembly. The worst case would be if the assembly became stuck. If the assembly sticks when pulling out of the hole, it can be jarred downward to free the string. The buildup of cuttings

Drilling 49

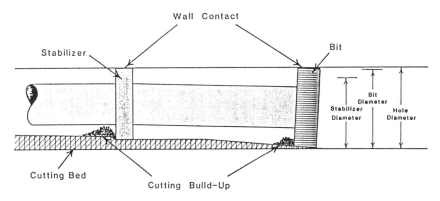

Figure 2-28. Bottomhole assembly drag in lateral cutting beds.

will not follow the assembly due to the lack of gravity along the axis of the wellbore.

The monitoring of torque and drag can be a valuable tool in recognizing cutting bed development. The development of a cutting bed normally increases the friction factor; this increases the torque and drag in the hole. The increase in drag may be noticeable, but it is not enough to cause concern. For example, consider the effect of a friction factor increase in only the lateral portion of the hole:

2,000 ft of 41 lb/ft HWDP using an increase in friction factor from 0.3 to 0.4 would increase drag from 25,000 lb to 33,000 lb.

This drag increase would be noticed first on connections. The drag increase would be due to sliding friction increasing along the lateral hole. On a trip, the drag would increase and possibly level out as the bit and stabilizer are pulled through the cutting bed. As pipe is pulled out of the lateral hole, the sliding friction drag is reduced as the pipe weight in the lateral hole is reduced. If the observed drag does not decrease as the bit approaches the end of the curve, the cutting bed drag on the bottomhole assembly is increasing and may eventually lead to stuck pipe.

An example of how this may occur follows:

1. 2,000 ft of lateral hole is drilled with an observed hole drag of 40,000 lb at TD; 15,000 lb at EOC; the friction factor is 0.3; and the pipe weight 41 lb/ft. Neglect buoyancy.

50   Horizontal Wells

2. While drilling 200 ft of lateral hole the drag increases to 51,000 lb, when theoretically the drag should be 42,000 lb. The effective friction factor is now 0.4.
3. On the trip out of the hole, the drag increases to 60,000 lb at the start of the trip. The drag due to the cutting bed is 9,000 lb and the effective friction factor is 0.5, assuming all drag is sliding friction drag.
4. After tripping 1,000 ft of pipe, drag is still 60,000 lb when actually the drag should be reduced by 12,000–24,000 lb, i.e., 41 lb/ft × 1,000 ft × 0.3 or 0.6. The drag through the cutting bed has increased and the effective overall ff would be 0.90.
5. A decision is made to circulate and rotate the hole to attempt to clean up the cutting load at the bit. The trip is resumed and the drag is 50,000 lb. The trip is finished.
6. The mud check indicates a significant drop in mud properties and the decision to run a hole-opener clean out trip is made. Cuttings are seen over the shaker, and the hole drag is reduced to an effective friction factor of 0.32.

On longer lateral wells that are drilled with weighted muds, cutting beds can be a significant source of drag. The drag is difficult to monitor on trips. The indicated weight of the string will change with drag and as heavy tubulars enter the vertical hole, if lighter tubulars are removed from the top of the string. Computer torque/drag models are useful in monitoring the drag present in the hole because they consider the string configuration.

The friction factor is a global value for the entire wellbore. The ability of some programs to use multiple friction factors can be useful. A single friction factor is used down to a point where cutting bed accumulation begins to occur (approximately 60°). A second friction factor is monitored below this point to foretell cleaning problems in the hole.

Friction factors have been determined by matching models to actual measured torque and drag values. This global approach matches research and results in friction factors from 0.2–0.4 for most mud systems. Factors influencing the friction factor include mud type, formation lithology, tool joint configuration and the casing program.

DRILLSTRING DESIGN AND TUBULAR SELECTION FOR HORIZONTAL DRILLING

Conventional oil field tubulars have been evolved to withstand the rigors of both straight hole and deviated hole drilling. The affect of fatigue on drillpipe has been well documented (Lubinski, 1960; Hansford and Lubinski, 1965; Lubinski, 1977). The understanding of fatigue damage and its control led directional drillers to long reach, and eventually long radius, horizontal drilling. The concept of inverting the drillstring evolved from long reach drilling. Inverting the drillstring is the practice of structuring a drillstring to keep the heavy tubulars required for bit weight in the vertical portion of the well. The tubulars below this weight section are run in compression. The medium radius horizontal drilling method uses an inverted drillstring design to keep the drillstring in compression while rotating the string in regions of high dogleg. This is one reason why medium radius drilling has been able to use average BURs of 20°/100 ft without severe drillstring damage.

A typical medium radius inverted drillstring configuration is shown in Figure 2-29.

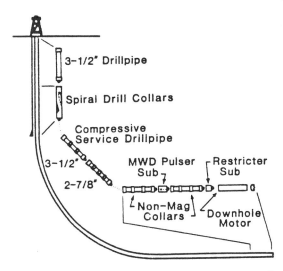

Figure 2-29. Generalized inverted drillstring configurations. Source: Edland, 1987.

The material published by Lubinski (1960) and Lubinski and Hansford (1965) must be considered if the operator is to avoid costly drillstring failures. If the operator wishes to optimize hole drag, hydraulics and drillstring cost then work published by Paslay and Bogy (1964) and Dawson and Paslay (1980) is invaluable in understanding compressive drillstring design in confined high angle holes.

Drillstring design for horizontal wellbores is very dependent on hole geometry. After the well path has been defined, the loads on the drillstring can be defined. The loading will vary with different operations; and the loading must be considered at different points in the well.

Horizontal Wells

Operations
• Tripping without rotation • Tripping with rotation • Rotating off bottom • Drilling without rotation • Drilling with rotation • Stuck pipe

Loading

Vertical Hole

- Tension above the weight section
- Compression below the weight section while drilling
- Tension below the weight section while off bottom

Angle Build Section

- Compression while drilling
- Tension while off bottom
- Fatigue

Horizontal Hole

- Compression on bottom
- Tension while pulling pipe

Torque/drag models are very useful in designing drillstrings for deep medium radius horizontal wells. A model can calculate tensile loads at discrete increments along the string under different loading conditions. Without a model, drillstring design optimization is difficult, and costly overdesign often results.

Tension. When considering the top joint tension for a vertical well, the total weight of the drillstring is the design load. In high angle and horizontal drilling, the wellbore supports much of the pipe weight and the top joint tension is not equal to the weight of the drillstring below the top joint.

Hole drag is a definite factor in drillpipe tension design. In the previous section, it was shown that sliding friction drag in the horizontal hole was a function of drillstring tubular weight. Bottomhole assembly drag is normally negligible, but can become significant if hole cleaning is poor. The top joint tension is a combination of drillstring weight, axial sliding friction drag, and bottomhole drag.

Often a drillstring will be configured in such a way that the maximum tensile load occurs when only a portion of the drillstring is in the hole and the bit is off the bottom. Consider the case in Figure 2-30:

- The well has a long horizontal section (3,000 ft).
- The vertical hole extends 4,000 ft to the KOP.
- The angle build section is 500 ft long and has an overall build up rate of 18°/100 ft.
- The vertical drillstring is 4½-inch, 16.6 lb/ft drillpipe.
- A vertical drillcollar section (100 lb/ft) is placed from 3,500–4,000 ft and 4½-inch, heavy-walled drillpipe 41 lb/ft is run from KOP to TD.
- Discounting buoyancy, the top joint tension will be 193,000 lb when the bit is at KOP, but only 122,000 lb when the bit is at TD. If this string were run in a vertical well, the top joint tension would be 251,000 lb.

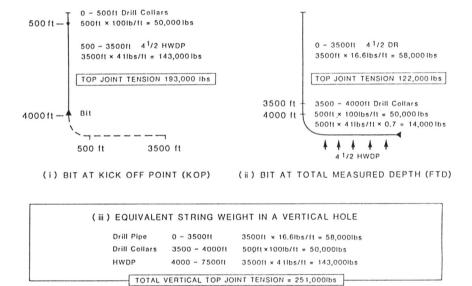

Figure 2-30. The effect of hole support on string tension.

While tripping, the top joint tension will be reduced if the tubular entering the horizontal section is heavier than the tubular added to the drillstring at the surface. This change in string tension is important when considering hole drag and downhole tool settings. The effect also makes the monitoring of hole drag difficult on trips because drillstring tension may increase as pipe is pulled out of the hole. This increase of indicated weight is due to drillstring configuration and not to increased drag.

54   *Horizontal Wells*

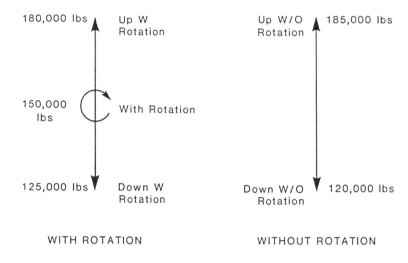

Figure 2-31. Geolograph record of string weight on connections.

To record the top joint tension without axial sliding friction, the string is rotated and lowered slowly. The frictional drag is overcome as torque and a true drillstring weight is recorded. Oftentimes it is not desirable to rotate the pipe after picking up off bottom. An example would be when drilling oriented, tool orientation can often be maintained throughout a connection. In this case, only pickup and slack-off loads are available. Hook loads should be recorded and trends examined to foretell hole-cleaning problems. Inputting these values into a torque/drag model can generate friction factors which are useful in casing design and future drillpipe design. Pickup and slack-off values are normally recorded on the geolograph, along with other pertinent information. This information should be recorded on all connections. Figure 2-31 shows a suggested format.

Figure 2-32 is a geolograph chart for a trip into the hole using $3^{1}/_{2}$-inch, heavy-walled pipe for a bottomhole assembly. The assembly is run on $3^{1}/_{2}$-inch, 13.3-lb/ft drillpipe in the vertical hole. The geolograph shows a weight reduction greater than the theoretical reduction expected, caused by the sliding friction drag. As the string is rotated prior to drilling, the axial drag is seen as torque and the string weight increases to the theoretical weight expected.

Fatigue. Whenever a drillstring is run around a dogleg, the situation exists where cyclical fatigue wear on the drillstring may occur. If we again consider a variety of drilling operations and the corresponding drillstring loads throughout the angle build section where doglegs are high, it becomes ap-

Drilling 55

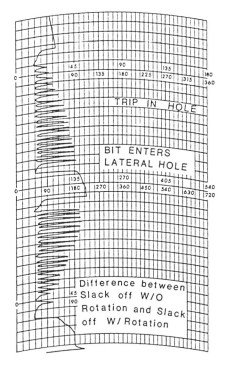

Figure 2-32. Geolograph record of a trip in the hole with 3½-in. drillpipe, heavy weight.

parent that cyclical fatigue loading can occur in medium radius horizontal wells.

Drilling

- Rotation with the entire drillstring in compression from above KOP to the bit
- Low risk of cyclical fatigue damage

Slack-off with rotation

- At slack-off speeds of 30–90 ft/min (typical while cleaning to bottom), and slow rotating speeds (20–30 rpm), a net axial drag will exist, putting the string in compression below KOP
- Low risk of fatigue damage.

Rotation off bottom

- The drillpipe in the curve will be subject to tension from tubulars between KOP and 90°. The advantage of medium radius horizontal drilling is that the arc length from 0–90° is short. The limited length minimizes the tension in the drillstring below the kickoff point. If the well is designed with excessive tangents high in the build section, tension at the kick off dogleg may be increased to levels where cyclical fatigue could become a concern.

Tripping out of the hole with rotation (top drive)

- While pulling pipe, rotation can be used to reduce axial hole drag. The axial sliding friction reduction is dependent on rotation speed and pulling speed. Axial friction drag and bottomhole assembly drag are not eliminated. These axial loads will increase the tension in the drillstring below KOP and can lead to fatigue damage. High-strength compressive service drillpipe is definitely useful in this application because of the short distance between wall contact points (Lubinski, 1977; Lubinski and Williamson, 1983).

Tubular design in the curve section, using concepts presented by Lubinski and Williamson (1983) for fatigue in doglegs, can reduce the potential for fatigue failure. Even at high buildup rates the bending stress present in the drillpipe will be less than the endurance limit of high-strength drillpipe. As increasing axial tension is coupled with cyclical bending stresses, fatigue damage to drillpipe can occur. Modified Goodman (Lubinski, 1960) diagrams for different drillpipe grades are useful to define the endurance limit of the material. If a corrosive environment or H_2S is not present, the operator then generally chooses a higher strength tubular with a higher endurance limit, below which fatigue damage will be minimal.

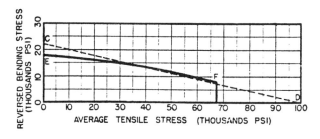

Figure 2-33. Modified goodman diagram for grade "E" drillpipe. Source: Lubinski, 1960.

The usefulness of drillpipe protectors to reduce wear and fatigue damage in doglegs has been documented by Lubinski and Williamson (1983). Placing contact points between tool joints reduces bending stresses concentrated at tool joints. Figure 2-34 shows that reducing the distance between contact points (2L) increases the permissible tension in the drillstring below which fatigue damage will be minimal (Figure 2-33).

Drilling 57

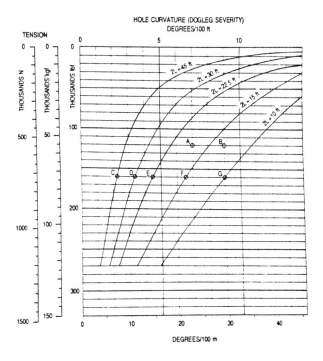

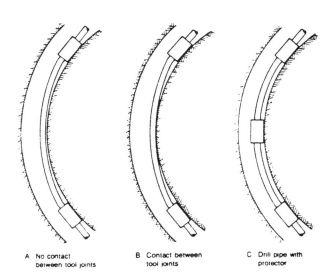

Figure 2-34. The effect of contact points of fatigue curves for gradual doglegs. Source: Lubinski and Williamson, 1983.

58 Horizontal Wells

The tubular selection for the angle build section should consider the drillstring weight that increases tension, the grade that increases endurance strength, and the wall contact points that increase the permissible tension allowed in a dogleg. Standard drillpipe lacks the lateral resistance to buckling in vertical wells and is not used in the angle build section. Heavy-walled drillpipe is often used in the angle build section and offers the advantage of increased outside diameter or contact points between tool joints. The increased weight reduces buckling and reduces stress in the tube. Compressive service drillpipe is made by placing contact points along the tube at 10-ft intervals between tool joints. Lightweight, high-strength tube, when stabilized, is ideally suited for the angle build section. Its use should definitely be considered in wells with long tangents high in the curve or when significant off bottom rotating time or top drive use is planned. Compressive service drillpipe does not restrict the passage of tools run inside drillpipe, nor increase circulating pressure by reducing the inside diameter. These are two disadvantages of heavy-walled drillpipe.

Buckling. When the need arises to run drillpipe in compression, the normal practice is to use heavy-walled drillpipe. This practice has merit in vertical wells, but it is often not necessary in horizontal wells, and indeed may hinder operations. Using lighter tubulars in the horizontal hole offers the following advantages:

- Torque and drag in the horizontal hole are reduced.
- In deep wells with long horizontal sections, overpull margins are increased on trips when the BHA is in the vertical hole.
- The overall drilling hydraulics can be optimized by reducing the pressure drop in the bottomhole assembly.
- The cost of drillpipe rental and the value of lost inhole tubulars is reduced.

Determining the operating limits of a drillpipe run in compression in the lateral hole is easily done using the following drillpipe analysis.

Paslay-Bogy Drillpipe Analysis

Drillstring design for the lateral section of the hole can be done by applying the theory of stability in circular rods lying in an inclined hole. This work was developed by Paslay and Bogy (1964) and applied to drillpipe by Dawson and Paslay (1980). The limits of stability defined by this analysis are then compared to drillstring loads calculated using a torque drag model.

The analysis shows that running drillpipe in compression is allowable at high angles. This practice has proven valid in high angle long-reach drilling.

Drilling 59

Paslay-Bogy Formulas

The stability of a circular rod in an inclined hole is defined by:

$$F_{crit} = \frac{4EIpAgSinx^{1/2}}{r} \quad (2\text{-}16)$$

where F_{crit} = axial load beyond which buckling occurs (lb)
  E = Young's Modulus (psi)
  g = gravitational force
  I = moment of inertia (in^4)
  r = radial clearance between pipe and hole (in)
  A = cross sectional area of pipe (in^2)
  p = density of pipe (lb/in^3)
  x = hole angle (degree)

The term (pAg) is simply the nominal drillpipe weight in lb/inch. This formula does not include the increased outside diameter at tool joint connections and is a conservative approximation.

The radial clearance (r) is the difference between the radius of the hole and the radius of the pipe. If the equation is solved for the radial clearance between the pipe and hole (r), the maximum hole size can be calculated for particular drillpipe size. The maximum axial load (WOB) is assumed to equal the critical force to initiate buckling (F_{crit}), and the hole diameter is solved for hole inclination between 0° and 90°. The equation was solved in this manner to generate Figure 2-35. These curves can be generated for any load or combination of pipe using a simple computer program. The results show that the 5-inch, 19.5-lb/ft drillpipe would be adequate for axial loads of 30,000 lb. The increased stability of 4½-inch, 41 lb/ft heavy-walled, drillpipe would have to be offset against increased sliding friction drag and cost.

Example

The maximum compressive loading in the horizontal section of the hole occurs at the end of the curve while drilling on bottom.

$$F_{EOC} = WOB + \text{Hole Drag} \quad (2\text{-}17)$$

The maximum WOB is often limited by the mud motor. An 8½-inch rock bit can safely be loaded with 50,000 lb but the maximum allowable load on

60   Horizontal Wells

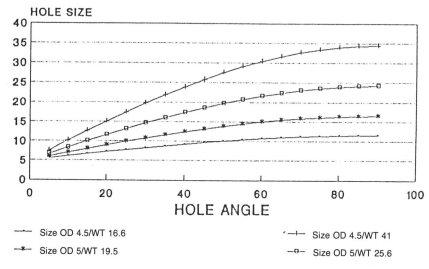

Figure 2-35. Critical angle for buckling (WOB 30,000 lb).

the motor may only be 30,000 lb. The hole drag between the drillpipe and horizontal hole varies from 600 lb, while drilling with rotation, to 18,000 lb, while drilling without rotation (5-inch, 19.5 lb/ft DP, assuming 3,000 ft of lateral hole with a 0.3 sliding friction factor). Several cases must be considered to determine the maximum load on the drillpipe at the end of the curve (Equation 2-17).

Case 1. Running Pipe in Hole Without Rotation
The axial load while tripping is equal to the hole drag. Without rotation this ranges from 36,000 lb using $4^{1}/_{2}$-inch, 41 lb/ft HWDP to 18,000 lb using 5-inch, 19.5 lb/ft DP.
Figure 2-35 shows that 5-inch drillpipe is adequate to prevent buckling of tubulars in the lateral hole. Heavy-walled or compressive service drillpipe would be considered between KOP and EOC.

Case 2. Drilling with Drillpipe Rotation
While drilling with rotation (WOB 30,000 lb, string RPM 40, Bit RPM 220) the axial hole drag is only 600 lb. Therefore, the maximum axial force is only 30,600 lb. Again, Figure 2-35 shows that 5-inch drillpipe is adequate for this load.

Case 3. Oriented Drilling Without Drillpipe Rotation
This is the worst case of axial loading because the axial drag increases dramatically without rotation. The total force is 30,000 lb WOB and 18,000 lb drag for a total of 48,000 lb. Figure 2-35

Drilling 61

shows that this approaches a region where buckling may be initiated. This buckling can be tolerated because the pipe is not rotating, and therefore, there is no fatigue damage. As soon as the pipe is rotated, the axial drag load decreases and the pipe operates in a region of stability (Case 2).

Safety Factors

The tool joint is not considered in this analysis. The equation is independent of length and is considered valid for lengths over 200 ft. The actual effective radial clearance is considerably less if one considers a $6^{3}/_{8}$-inch tool joint every 30 ft, or 7 nodes every 200 ft. The calculations are based on nominal pipe weight. If the weight of tool joints is included (approximately 6–8% increase) the stability will increase.

The use of drillpipe in the horizontal section of the hole offers advantages by reducing the torque and drag due to sliding friction. This reduction can be as high as 50% and will result in higher rates of penetration while drilling oriented, and greater control of tool orientation.

Drillstring design and optimization is greatly assisted by a torque/drag model that slows the axial and torque load on the string at increments from the bit to the surface. By using this program as a simulator, the effect of various operations and tubular selections can be anticipated. This allows the engineer to optimize the drillstring while avoiding costly failures.

CASING DESIGN CONSIDERATIONS

After drilling a well, the next costliest operation is often running production casing. The casing is designed to withstand tensile loads, collapse loads, and internal yield loads, which are defined by considering the conditions that the well will be exposed to throughout its life.

For vertical wells, casing design is thoroughly defined in API bulletins and industry literature. The vertical portion of a horizontal casing string is designed using these guidelines. The point where the maximum tensile load occurs may not be when the last joint is run and care must be taken to correctly define the maximum running load (see Figure 2-30).

Design considerations for casing that is run below the kickoff point should include loads induced by bending the casing around the angle build section. If it is assumed that the casing is free to contact the wall of the hole, either top or bottom, and that the standoff at casing connections is minimal with casing collars and integral joints, then the maximum bending stress in the

62 Horizontal Wells

casing is defined by the casing diameter and the angle buildup rate. If standoff at the casing connections is significant, the bending stress would not be constant and stresses would be concentrated at the connections (see Figure 2-34) (Lubinski and Williamson, 1983).

The maximum bending stress (S) in the casing is easily calculated using Equations 2-18 and 2-19 to yield Equation 2-20 (Popov et al., 1952).

$$\frac{1}{R} = \frac{M}{EI} \qquad (2\text{-}18)$$

$$S = \frac{My}{I} \qquad (2\text{-}19)$$

combine to yield

$$S = \frac{Ey}{R} \qquad (2\text{-}20)$$

where S = maximum bending stress (psi)
R = bending radius (inches)
M = bending moment
E = Modulus of elasticity (30×10^6 psi for steel)
y = distance from natural axis or radius of casing (inches)

From this simple formula the maximum bending stress in tubulars running through high doglegs can be estimated. The bending stress in drilling tubulars and casing must be considered as BUR or tubular size is increased. For example, if a well was drilled with 4½-inch heavy-walled drillpipe through a 12°/100 ft dogleg, and a subsequent well is being designed with a 20°/100 ft average dogleg, the operator might consider 3½-inch compressive strength drillpipe. The bending stress of 4½-inch tube run in a 12°/100 ft dogleg would be 11,775 psi: in a 20°/100 ft dogleg, the bending stress would increase to 19,626 psi. The 3½-inch tube would have a maximum bending stress of 15,264 psi in the 20°/100 ft dogleg. The tube yield strength may also be increased which will increase the endurance strength of the material and reduce fatigue damage.

According to the maximum distortional energy theory, yielding occurs when a combination of axial, tangential, and radial stresses reach the yield strength. It is beyond the scope of this book to investigate the effects of bending stress on collapse for casing. By using the stresses listed in Table 2-2, an estimate of the reduced collapse rating can be made using an ellipse of biaxial yield stress (Figure 2-36). A rigorous analysis titled "Influence of Neutral

Drilling 63

Table 2-2
Bending Stress in Casing at Various Buildup Rates (BUR) (psi)

BUR	Casing Diameter (inches)			
(°/100 ft)	4.5	5.5	7	7⅝
4	3,925	4,797	6,100	6,651
6	5,888	7,196	9,159	9,976
8	7,850	9,595	12,212	13,302
10	9,813	11,994	15,265	16,628
12	11,775	14,392	18,318	19,953
14	13,738	16,791	21,371	23,279
16	15,701	19,190	24,421	26,605
18	17,663	21,589	27,477	29,930
20	19,626	23,988	30,530	33,256

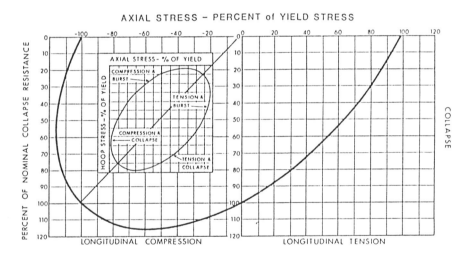

Figure 2-36. Ellipse of biaxial yield stress. Source: Popov et al., 1976.

Axis Stress on Yield and Collapse of Pipe," by Arthur Lubinski (1974), is a good reference for further information.

By using Table 2-2 as a guide, several rules of thumb can be used to design casing through the curve.

1. The bending stresses imposed on casing in a smooth curve are not dependent on hole size for a given dogleg; rather, bending stress is dependent on casing size.
2. As the pipe size increases through the angle build section, so does the bending stress. Larger pipe size requires higher yield strength casing grades.

3. The use of buttress-type thread forms and metal-to-metal seals should be considered in larger casing sizes due to a reduction in leak resistance with tension in 8 Rnd connections (Schwind, 1987).

It should be noted that although the theoretical tensile stress in Table 2-2 exceeds the endurance limit of these tubulars in some cases, this is permissible because rotation of casing is not common for extended periods of time. This same method can be used to compare different drillpipe sizes. Smaller tube diameter is preferred at higher doglegs.

DRILLING FLUIDS AND CUTTING TRANSPORT

The primary functions of a drilling fluid are:
1. To carry drill cuttings and cavings out of the hole
2. To control subsurface pressures
3. To provide drillstring lubrication
4. To provide fluid loss control to reduce formation damage

In any drilling operation, controlling the drilling fluid rheology to achieve these primary functions is the key to a successful operation.

Hole-cleaning capability is an essential function of the mud system. Drill solids have a specific gravity of approximately 2.5. The average drilling fluid used in horizontal drilling has a specific gravity of 0.9 to 1.5. Since drilled cuttings are heavier than the mud, they will slip down through it. This slip velocity is governed by fluid rheology, particle size, and particle shape. In a vertical well, the slip direction is along the axis of the wellbore and directly opposite the mud flow; in a horizontal well, the slip direction is across the axis of the wellbore and perpendicular to the mud flow (Figure 2-37).

Particle slip direction in a vertical well is opposed by the mud flow and the particle will be lifted out of the well if the mud velocity is greater than the slip velocity. In a horizontal well, the particle will slip to the low side of the hole. The particle is cleaned only if the mud flow can continually pick up and move the particle along the axis of the hole prior to the particle settling to the low side of the hole. This transport mechanism is known as saltation. In an inclined hole, a cutting can actually slide down the hole as a cutting bed if rheology and flow rate are not adequate to keep the cutting bed stationary or moving forward. This downward slide of cutting beds is usually observed between 50–70° (Seeberger et al., 1989).

The velocity profile of a fluid in the annulus is an important characteristic of any drilling fluid. In laminar flow, the velocity in the center of the annulus will be greater than the velocity along its boundaries. Turbulent flow

Drilling 65

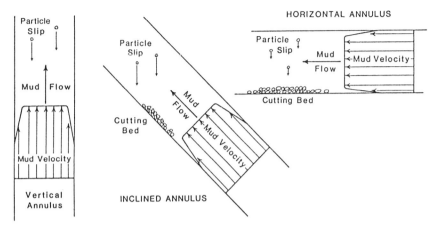

Figure 2-37. Direction of cutting slip relative to mud flow direction.

regimes exhibit crossflow and eddies which result in an average fluid velocity over the entire annulus. In horizontal and inclined wellbores, solids will settle out of the main fluid stream and to the low side of the hole, where their weight holds them stationary. Cuttings in vertical wellbores will slip downward; they may accumulate along the side of the hole, but eventually fall back into the main fluid flow due to crossflow and lack of forces holding them against the hole wall. This difference in cutting behavior requires that the velocity profile be carefully controlled to maintain flat, even velocity across the entire annulus in horizontal and inclined wellbores.

Most drilling fluids are shear thinning, which means the proportionality between shear stress and shear rate is reduced as shear rate increases. This relationship between shear rate and shear stress is what defines the shape of the velocity profile for a particular drilling fluid (Figure 2-38).

Equations have been developed to determine the velocity profile of fluids using assumed Power Law or Bingham Plastic flow behaviors. The equations do not consider pipe rotation or hole rugosity and may only be used as a guide to actual field conditions. The most common method of evaluating the drilling fluid rheology in vertical wells is to study the yield point or "n" factor using the Fann rotating viscometer. Generally, an increasing yield point or decreasing "n" factor at a constant plastic viscosity will result in a flatter velocity profile. This data is compared over a period of time, and the trends are used to determine changes in mud flow behavior.

The shear stress of a fluid used in horizontal drilling must be calculated across the full shear rate range to which the fluid will be exposed. Normally, a drilling fluid will be evaluated by using Fann 300–600 rpm deflections to evaluate the yield point. The initial and 10-minute gel strengths evaluate the

66 Horizontal Wells

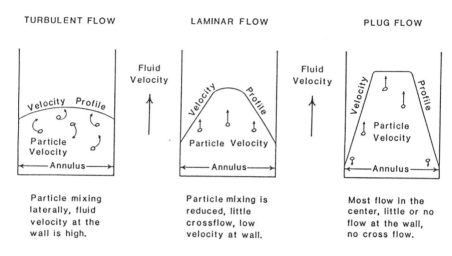

Figure 2-38. Flow regimes and velocity profiles.

consistency of the fluid after the mud movement has stopped. By using a multispeed viscometer, the shear stress over the entire shear rate range can be evaluated.

Shear rate in the Fann viscometer changes with the machine rpm. The shear rate of fluids in the well varies, and Fann viscometer speed (rpm) sets up shear rates that correspond to the following locations in the wellbore:

>600	Bit nozzles
300–600 rpm	Solids control equipment
10–100 rpm	Drillpipe annulus
1–10 rpm	Mud tanks

Mud product additions that increase the shear stress in the 300–600 rpm range but not in the 10–100 rpm range will do little to improve hole cleaning in the low side of the annulus of a horizontal hole. If barite settling out of a thin turbulent mud system causes downhole tool plugging, a product that raises the 1–3 rpm shear stress without raising the 100–200 rpm shear stress would be desirable.

Products that increase shear stress only in the 300–600 rpm range will reduce solids control efficiency without increasing hole-cleaning capability; the resulting increase in plastic viscosity is detrimental to the velocity profile. The shear stress vs shear rate relationship for several common mud additives is shown in Figure 2-39.

Drilling 67

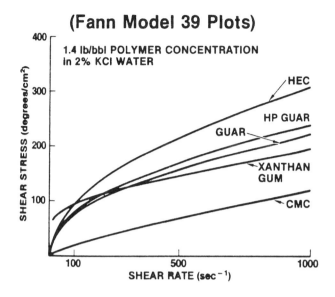

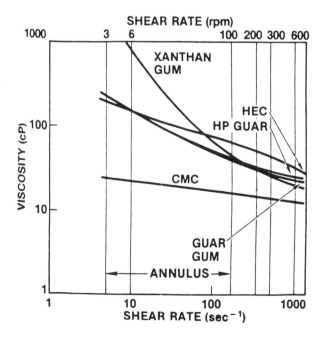

Figure 2-39. Shear rate vs shear stress and viscosity of water soluble polymers.

Two-phase Flow in Horizontal Pipes

Several analogies between hydraulic transport of solids in pipelines and solids transport in a horizontal wellbore can be made.

One characteristic of two-phase flow is that the solid and liquid phases remain identifiable. The physical character of the liquid phase is not affected by its association with the solid phase. In cases where ultrafine particles are held in homogeneous suspension by the liquid phase, the liquid phase must be reevaluated in terms of viscosity and density. The fine particles are considered part of the liquid phase, and not as a separate solid phase. If all particles were carried as ultrafine particles, hole cleaning would be easily controlled but mud rheology would not. In a drilling operation, attempts are made to remove drilled cuttings as a second phase before they breakdown in size and become permanently entrained in the liquid phase.

In general, solids smaller than 40μ will form a colloidal suspension; solids from $40-120\mu$ will settle out of a fluid (however, removal from fluid is often done using centrifugal forces); and solids larger than 120μ will settle out of a fluid due to gravitational forces. From observation of solids transported in horizontal pipelines, it is known that particle size is a definite factor in cutting transport mechanics (Bain and Bonnington, 1970).

The conveyance of particles in horizontal wells can be via heterogenous suspension maintained by turbulence, or by saltation, which is the action of particles being picked up by fluid and deposited further along the hole (Figure 2-40).

Bain and Bonnington (1971) classified particles with respect to their size and mode of transport as follows:

1. Particles less than 40μ

    - Homogeneous suspension
    - Example: unremovable drilled solids and mud products

2. Particles from 40μ to 0.15 mm

    - Heterogeneous suspension maintained by turbulence

3. Particles between 0.15 mm and 1.5 mm

    - Heterogeneous suspension maintained by turbulence and saltation
    - Increased rheology will increase the particle size transported by heterogeneous suspension

4. Particles greater than 1.5 mm

    - Transported by saltation

Drilling 69

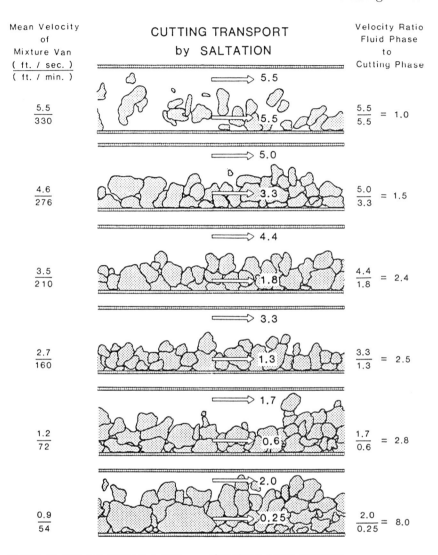

Regime of coal and water in a pipe. – pipe size 1.5 inch diameter.
– coal partical size 3/8 to 5/8 inch.

Clearly discernable are the thickening of the sliding bed and increasing velocity differentials between the two phases as the mean velocity decreases. The increase in the fluid velocity in the bottom frame is due to a reduction in area avalible for flow above the coal bed. (Bain & Bonnington, 1970)

Figure 2-40. Cutting transport by saltation. Source: Bain and Bonnington, 1970.

The behavior of solids transport in the cuttings burden is affected by the degree of turbulence, the rheology of the fluid, and the physical character of the particles.

Decreasing the cutting size increases the modes of transport available to move the cutting along the wellbore. In a homogeneous suspension, the particle surface is enveloped in a film of fluid that effectively reduces the relative density of the particle and increases the particle size. This reduces the slip velocity as the mass of fluid enveloping the particle begins to approach the mass of the particle. Figure 2-41 shows the flow mechanisms of graded sand in a 6-inch pipe using water.

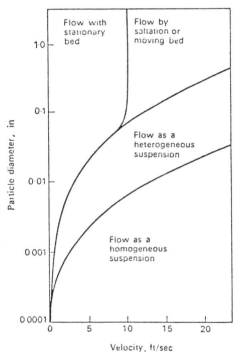

Figure 2-41. Transport mechanism of graded sand in a 6-in. pipe using water as a fluid. Source: Bain and Bonnington, 1970.

The size of drilled cuttings generated at the bit is dependent on the bit type, rock type, applied weight/rpm, and the efficiency of cutting removal at the bit. Generally, the more drillable the formation, the larger the cutting that leaves the bit. Once the cutting leaves the bit, it is subjected to mechanical breakdown and fluid erosion. The action of drillpipe rotation and parti-

Drilling 71

cle collision in a horizontal hole with turbulent flow will reduce particle size. These smaller particles may then be removed more readily, but an increase in fines that cannot be removed from the mud may also occur.

Saltation and heterogeneous suspension through turbulence can be optimized by increasing flow rate and maximizing fluid viscosity while still maintaining turbulent flow. Increased flow rate increases energy available for saltation and maintains turbulent crossflow at higher viscosities. The higher viscosity reduces the slip velocity and increases the distance traversed during one cycle of saltation.

Pipe rotation serves a dual purpose in horizontal hole cleaning. The pipe can mechanically crush cuttings to reduce cutting size and tools such as stabilizers act to break up cutting beds. Pipe rotation tends to induce crossflow in laminar flow profiles. Rig experience has shown that cutting size on the shale shaker is often larger with pipe rotation. Hole drag often increases if the drillpipe has been oriented for any length of time.

Turbulent vs Laminar Flow

The basic rheology and mud type selected for a horizontal well are often dictated by formation pressure and the compatibility of the drilling fluid to reservoir fluids. In wells where an unweighted drilling fluid is used, rheology and flow rate should result in turbulent flow to optimize cutting transports in the horizontal hole. Hole cleaning in the curved and vertical section of the hole must be carefully monitored. Low shear rate rheology can be increased by selective addition of mud products. Running the maximum drillpipe size for the hole drilled can greatly increase hole cleaning if flow rate is limited. An example is running $4^{1}/_{2}$-inch drillpipe in $7^{5}/_{8}$-inch intermediate casing and tapering down to $3^{1}/_{2}$-inch drillpipe in the $6^{1}/_{4}$-inch horizontal hole below the casing.

Wells requiring that weighted fluids run in laminar flow need careful control of the rheology over the entire shear rate range. The rheology profile must be controlled independently of the calculated yield point and plastic viscosity (Seeberger et al., 1989). Cutting suspension in laminar fluids is important: cuttings which are not immediately suspended will settle to the low side of the hole in the absence of crossflow. Drillpipe rotation is important because it agitates flowstream and breaks up cutting beds. Increasing shear stress at the shear rate in the annulus is desirable; after a particle has been picked up into the liquid flow, the higher shear stress will reduce the slip velocity and increase the distance the cutting is transported prior to settling. Products that increase the rheology only at higher shear rates will be detrimental. Product additions that raise the yield point (300–600 rpm shear

72 Horizontal Wells

rate) will often raise the entire rheology profile of the fluid. This can lead to higher circulating pressures and reduced flow rates, which in turn reduce cutting transport. A preferable approach is to raise the rheology only in the shear rate range where the increase is desired. Figure 2-39 shows the shear rate vs shear stress relation for various mud addditives.

Drilling rate is an important factor in effective hole cleaning. The formation drilled horizontally is often porous reservoir rock and hence drillable. High bit weights in drillable formations may result in high cutting volumes and large cutting sizes. If the drilling fluid cannot transport these cuttings effectively, an immovable cutting bed may develop.

Care must be taken when a formation change is observed or the ROP is increased. A hole that cleaned effectively while drilling at 30 ft/hr in siltstone may not clean at 45 ft/hr when the formation changes to a drillable shale. Often, a well-cemented sandstone drills slowly with small cuttings. Unconsolidated sands are transported as individual sand grains.

Cutting transport is a complicated mechanism dependent on many variables. Conditions change while drilling in the horizontal hole and the engineer must be careful to anticipate what effect the change will have on cutting transport in the well.

Formation Damage

Drilling fluids can induce formation damage which inhibits primary production and can lead to costly stimulation.

Three damage mechanisms to consider are:

1. Relative permeability damage and fluid compatibility
2. Mechanical damage caused by plugging with drill solids and mud products
3. Fluid loading of the reservoir with drilling fluids

To minimize the cost of stimulation, horizontal well drilling fluid programs should include input from both reservoir and completion engineers.

Many mud products intentionally or unintentionally change the surface tension and wettability of the rock surface. By introducing these products, the fluid saturation history can be adversely altered and imbibition of irremovable fluids can cause damage in the region of the wellbore (Anderson, 1989). In some cases, the drilling fluid may be incompatible with fluids used in the completion operation, i.e., invert oil emulsions and acid.

Another form of damage may be fluid loading in an underpressured formation. If fluid loss is not controlled and the reservoir energy is low, reservoir fluid flow into the well may be restricted.

Drilling 73

Fluid loss is commonly controlled by building a cake of solids and polymers to reduce fluid leak-off. Often these products are insoluble, and cause mechanical plugging. Acid soluble solids and polymers such as calcium carbonate and HEC should be considered for fluid loss control and minor increases in density. Fluid systems using calcium carbonate and polymer additions in 3% KCl water have been used to drill low pressure gas reservoirs. The mud solids were 75% acid soluble using calcium carbonate additions. Had bentonite been used for fluid loss control and barite on trip pills, the solids left in the formation would have been 100% insoluble.

SUMMARY

Medium radius horizontal drilling is an extension of high angle extended reach drilling. High angle drilling requires that the operator pay close attention to details regarding geology, hydraulics, and strength of materials. Throughout this chapter an attempt has been made to highlight drilling practices that would not be considered if a vertical infill well were drilled instead of a horizontal well in the same area. It is the attention to detail that will lead to a successful operation.

Once the well plan has been developed and the operation begins drilling, all the detailed planning is worthless if the importance of detail and changing conditions in the wellbore are not relayed to rig personnel. Horizontal drilling is only successful if everyone understands the objective of different operations. Communication between levels of personnel from the geologist to the roughneck is important. Major decisions should be made in the field whenever possible, because the staff in the field has access to all available operations information.

As more wells are drilled, more experience will become available which will further enhance the concepts outlined in this chapter. This will lead to more efficient, economical, and safer horizontal drilling operations.

REFERENCES

1. Karlsson, H., Cobbley, R., and Jaques, G. E., "New Developments in Short, Medium and Long Radius Lateral Drilling," SPE/IADC Paper 18706 presented by Eastman Christensen at the SPE Conference held in New Orleans, LA, February 28–March 3, 1989.
2. Brantley, J. E., "History of Oil Well Drilling," A Project of the Energy Research and Education Foundation, Gulf Publishing Company, 1971, pp. 221, 1167–1209.

3. Edland, P. A., "Application of Recently Developed Medium—Curvature Horizontal Drilling Technology in the Spraberry Trend Area," SPE/IADC Paper 16170 presented by ARCO Oil and Gas Co. at the SPE Conference held in New Orleans, LA, March 15-18, 1987.
4. Tiraspolsky, W., "Hydraulic Downhole Drilling Motors," Gulf Publishing Company, 1985, pp. 219-228.
5. Kerr, D. and Lesley, K., "Mechanical Aspects of Medium Radius Well Design," SPE Paper 17618, presented by Smith International at the International Meeting on Petroleum Engineering Conference held in Tianjin, China, November 1-4, 1988.
6. Eastman Christensen, 1988-89, unpublished material.
7. Dellinger, T. B., Gravely W., and Tolle, G. C., "Directional Technology Will Extend Drilling Reach," *Oil and Gas Journal*, September 15, 1980.
8. Johancsik, C. A., Dawson, R., and Friesen, D. B., "Torque and Drag in Directional Wells—Prediction and Measurement," *Journal of Petroleum Technology*, 1984.
9. Sheppard, M. C., Wick, C., and Burgess, T., "Designing Well Paths to Reduce Drag and Torque," SPE Paper 15463 presented at the SPE Conference held in New Orleans, LA, October 5-8, 1988.
10. Lubinski, A., "Maximum Permissible Doglegs in Rotary Boreholes," presented at the SPE Conference held in Denver, Colorado, October 2-5, 1960.
11. Hansford, J. E. and Lubinski, A., "Cumulative Fatigue Damage of Drillpipe in Doglegs," presented at the SPE Conference held in Denver, Colorado, October 3-5, 1965.
12. Lubinski, A., "Fatigue of Range 3 Drillpipe," Revue de L'Institut Français Du Pétrole, March and April, 1977.
13. Paslay, P. R. and Bogy, D. B., "The Stability of a Circular Rod Laterally Constrained to be in Contact with an Inclined Circular Cylinder," *Journal Applied Mechanics*, 1965, pp. 606-610.
14. Dawson, R. and Paslay, P. R., "Drillpipe Buckling in Inclined Holes," SPE Paper 11381, *Journal of Petroleum Technology*, October 1980.
15. Lubinski, A. and Williamson, J. S., "Usefulness of Steel or Rubber Drillpipe Protectors," SPE/IADC Conference held in New Orleans, LA, February 20-23, 1983.
16. Popov, E. P., Nagarajar, S., and Lu, Z. A., "Mechanics of Materials—Second Edition," Prentice Hall, Englewood Cliff, New Jersey, 1976 (1952), pp. 356.
17. Lubinski, A., "Influence of Neutral Axis Stress on Yield and Collapse of Pipe," presented at the American Society of Mechanical Engineers Conference held in Dallas, TX, September 15-18, 1974.

Drilling   75

18. Schwind, B. E., "Equations of Leak Resistance of API 8-Round Connectors in Tension," OTC Paper 5509 presented in Houston, TX, April 27–30, 1987.
19. Seeberger, M. H., Matlock, R. W., and Hanson, P. M., "Oil Muds in Large-Diameter, Highly Deviated Wells: Solving the Cuttings Removal Problem," SPE/IADC Paper 18635 presented at the SPE Conference held in New Orleans, LA, Febraury 28–March 3, 1989.
20. Bain, A. G. and Bonnington, S. T., "The Hydraulic Transport of Solids by Pipelines—International Series of Monographs in Mechanical Engineering," Pergamon Press, London, England, 1970.
21. Anderson, W. G., "Wettability Literature Survey—Part 4: Effects of Wettability on Capillary Pressure," *Journal of Petroelum Technology*, October 1989.

3
Completions

PLANNING THE COMPLETION

Horizontal completions require much more preplanning than do conventional vertical wells. The completion designer should be involved right from the conceptual or prognosis stage, as is the drilling engineer. There are a number of completion subtleties that can be addressed easily in the preplanning stages that may be virtually incorrectable once the well is drilled.

There are three main factors that impact the type of completion designed. The factors are identified in the following discussion with insights given into how each impacts the completion.

Formation Considerations

The formations penetrated fall into three broad categories:

1. Homogeneous formations
2. Heterogeneous formations
3. Naturally fractured formations

Homogeneous formations are most often found in heavy oil sands. They are easiest to accommodate in the completion design as they don't normally require extensive segregation of the wellbore. Most times they can be treated with the entire wellbore exposed, using coil tubing or perforating technology to overcome near wellbore damage.

Heterogeneous formations are more challenging, as both reservoir quality and pressure can vary between sections of the wellbore. These formations require segments of the wellbore to be isolated to accommodate selective stimulation. If the permeability is very heterogenous, some segments of the wellbore may require little stimulation while others require significantly more. Depending on the overbalance during drilling, it may be the high permeability sections that require larger stimulations to overcome deep mud or filtrate invasion. Cemented, perforated, and uncemented liners segmented with ECPs are applicable in horizontal wells in heterogeneous formations (Figure 3-1).

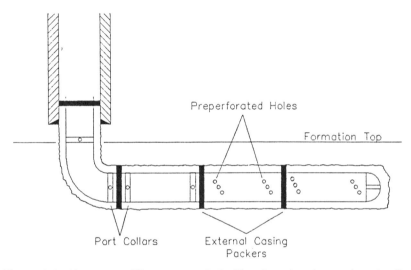

Figure 3-1. Uncemented liner segregated with external casing packers to allow selective testing and stimulation of the lateral.

Naturally fractured formations pose the greatest challenges in completion design. If the proper wellbore orientation has been picked, and if the natural fractures are present in sufficient density, then an uncased wellbore or a preperforated liner is all that is required. If the fracture intensity is low, or the wellbore has been misoriented, the completion becomes much more involved.

The main focus of a horizontal completion in a naturally fractured reservoir is to leave the fractures in an undamaged state. This precludes cementing the liner, as the cement not only invades, but seals off the fractures. On the other hand, if the fractures are not present in sufficient numbers to make the well economically productive, then it is unlikely a near wellbore stimulation such as an acid stimulation will correct the situation. A hydraulic fracture will probably be required, which requires a cemented liner to allow segregation of sequentially induced fractures.

78 Horizontal Wells

The completion should be set up to allow the horizontal interval to be evaluated in an openhole state, if there is a concern about the extensiveness of the natural fractures. The lateral can always be cemented and stimulated at a later date, if necessary. This can best be done by running an uncemented liner with perforations at the toe of the well only. The liner can be set on bottom without a packoff at the liner top. Figure 3-2 shows the liner as landed during the evaluation stage prior to cementing.

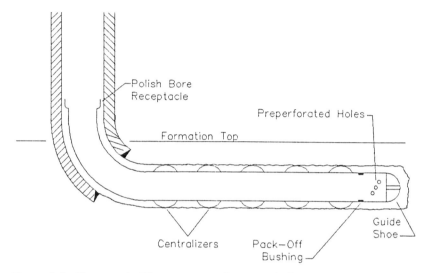

Figure 3-2. Uncemented liner set up to allow open-hole evaluation of the lateral. In this configuration it can be subsequently cemented and selectively stimulated.

In this configuration the well is brought in and will flow along the open annulus to the perforations at the toe of the well or to the liner top. If the productivity is acceptable, a packoff assembly can be installed at the liner top, and the liner perforated as required. If the zone requires a hydraulic fracture the liner can be latched onto and cemented as any liner in a horizontal hole would be. To facilitate a good cement bond, the liner needs to be run with centralizers in place.

Well Type

The type of hydrocarbon produced has a major impact on how the horizontal wellbore is configured. A well can be drilled into the following types of reservoir:

1. Gas
2. Oil
3. Heavy oil
4. Gas over oil
5. Gas over oil over water

There are unique problems associated with horizontal wellbores in each reservoir type.

Gas, oil, and heavy oil reservoirs all face the concern as to what extent the well will have to be stimulated to make it economical, and whether the extent of stimulation will be known before the well is cased.

Based on Canadian Hunter's experience, you can count on the stimulation going one step beyond what is planned for. If a small acid matrix treatment is planned it will probably require a large acid treatment to bring the well in, and if a large acid treatment is planned it will probably require an acid or sand fracture. Maximum flexibility needs to be preplanned into the casing program to allow an openhole evaluation of the formation before proceeding with whatever stimulation is required. The ideal setup allows initial evaluation of the horizontal lateral uncased, or with an uncemented liner that can be later pulled and rerun setup for cementing and selective stimulation.

If the possibility exists that the lateral will be fracture stimulated, then the completion engineer needs to be involved in the selection of the wellbore orientation. Either transverse or longitudinal fractures can be initiated from a horizontal wellbore, depending on the orientation of the wellbore relative to the formation's least principal stress directions. Normally, a low permeability formation will require transverse fractures, while the higher permeability formations can best be stimulated with longitudinal fractures.

Oil/water and gas/oil/water reservoirs must have all the previously discussed concerns addressed, and be designed to allow remedial action to be taken should water or gas breakthrough occur. Fracturing treatments are not applicable in this situation, as an induced fracture will accelerate water or gas breakthrough.

The liner or casing needs to be set up to allow selective isolation of segments of the wellbore. This allows abandonment of intervals experiencing early breakthrough.

Casing Alternatives

There are two major casing alternatives for a horizontal wellbore: openhole or cased and cemented. Included in openhole completions are pseudo openhole completions equipped with uncemented casing or liners.

Openhole Completions. The three types of openhole completions are as follows:

1. True openhole completions
2. Slotted or preperforated liner/casing
3. Segmented uncemented liner/casing

Normally, an uncemented completion is utilized when a completion with little or no stimulation is anticipated. With this completion approach it can be difficult to control placement of stimulation fluids to any significant degree.

The true openhole completion is used in medium to high permeability competent formations, with no water or gas coning concerns. It also can be used in competent formations as an intermediate step in evaluating reservoir quality. In this scenario the well is brought in and evaluated with no casing in the lateral. Based on the well's performance it will either be left as it is, or cased to accommodate whatever stimulation is required.

The potential for wellbore failure should be checked whenever a wellbore is left uncased. Borehole failure becomes a major concern when the formation is hydraulically fractured. Also, after fracturing and during production, low borehole pressure may tend to cause borehole compressive failure (Hsiao, 1987).

Slotted or preperforated liners can be run to protect the borehole from collapse. Slots are generally used in unconsolidated formations, while predrilled holes are applicable to consolidated formations. Predrilled holes are significantly cheaper than slots. If the liner needs to be circulated into place, a workstring can be run through the liner and attached to the liner running string as shown in Figure 3-3.

The workstring is landed in a packoff bushing just above the liner shoe and retrieved with the drillstring once the liner is landed. The downside of this approach is the awkwardness of running concentric strings. A much simpler solution to this problem is one currently used in the Williston Basin and illustrated in Figure 3-4. In this area the Bakken shale is being drilled horizontally at 10,000 ft TVD. No intermediate casing is run, and as much as 15,000 ft of 5^1/$_2$-inch casing is run at once. Only the vertical section of the hole is cemented and the 2,000–3,000 ft lateral is left uncemented to maximize productivity from the fractured shale. The 5^1/$_2$-inch casing in this section is predrilled with 1-inch holes, and the holes are threaded and plugged with hollow aluminum plugs capable of holding 1,500 psi differential pressure. This allows the pipe to be circulated in place without running a concentric string through the lateral.

The plugs project 1-inch into the casing and are milled off after the casing in the vertical section has been cemented. Because they are hollow, they allow flow through their centers once milled. This procedure eliminates a $50,000–$100,000 perforating job.

Completions 81

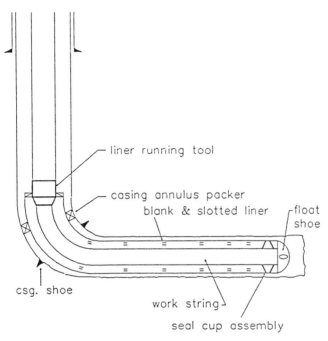

Figure 3-3. Slotted liner being run with inner workstring strung into a packoff at the float shoe, allowing it to be circulated in place.

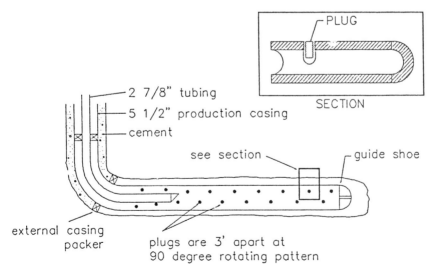

Figure 3-4. Single string of predrilled production casing run to case a horizontal well. The casing is cemented from the end of the radius to surface, after which the uncemented predrilled casing in the lateral is opened to flow by milling hollow aluminum plugs used to block each predrilled hole while running in.

82 Horizontal Wells

The uncemented and slotted or preperforated approach should only be considered when little or no stimulation is anticipated. Although a paper has been written addressing matrix stimulations in uncemented slotted liners (Economides et al., 1989), there really is no way to control stimulation fluids in an uncemented liner.

Segmented uncemented liners are the most sophisticated of the pseudo openhole completion approaches. They allow openhole evaluation of the wellbore and subsequent selective stimulation, if necessary. They also provide the luxury of isolating segments of the wellbore if port collar tools are run in conjunction with the ECPs. Figure 3-1 is an example of this type of completion. Preperforated or slotted liner/casing segmented with ECPs can have intervals isolated through use of a patch, as shown in Figure 3-5.

Segmentation is normally achieved with cement or fluid-filled ECPs. In one case, the ECPs were filled with air and effectively maintained isolation during a fracture stimulation (Yost and Overbey, 1989). However, to ensure long-term reliability it is recommended they be cement inflated.

Another option in segmenting the lateral is to use alternating sections of cemented and uncemented casing. These are difficult to put in place and not often used.

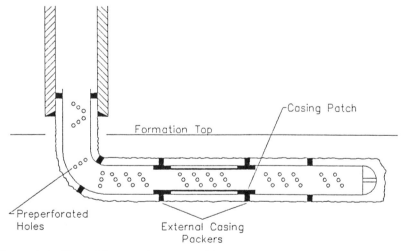

Figure 3-5. Slotted liner equipped with ECPs in which a casing patch has been run to permanently isolate a segment of the wellbore.

Cased and Cemented Completions. This is the premium completion approach. Implicit in the decision to case and cement is a commitment to perforate and stimulate the well. Costs associated with these operations are so significant that there must be problems associated with the wellbore/forma-

Completions 83

tion combination that cannot be addressed in any other way. An example of this would be a low permeability reservoir that requires transverse fractures to provide adequate drainage and deliverability from the reservoir. Only with a cased and cemented wellbore can sufficient isolation be achieved to allow sequential transverse fracturing of the lateral. Apart from this type of completion, only in extenuating circumstances can a cased and cemented completion be recommended.

EQUIPPING THE HORIZONTAL WELL

The majority of tools required to complete a horizontal well are available off the shelf with few or no modifications. Most tools are set with rotation, movement, or hydraulic pressure, which must be applied around the bend to the horizontal segment. The movement required to set or unset a tool should be minimized.

The tools must be able to travel around the curve, and this needs to be planned for in the initial well design. Few tools can make it around a short radius bend. If cementing the lateral is planned to provide interzonal isolation, the radius needs to be designed to accommodate packers and perforating guns.

External Casing Packers (ECP)

One tool integral to many horizontal completions is the external casing packer. Not all completions require the total zonal isolation supplied by cementing, and ECPs are one way to segregate sections of the wellbore without cementing off permeability or open fractures.

There have been several cases of satisfactory fluid inflation of ECPs, which subsequently leaked upon pressure testing. This has led to the conclusion that cement-filled ECPs supply the best zone isolation technique (Cooper and Troncoso, 1988). Elf Aquitane and the French Petroleum Institute carried out a joint research project in which they demonstrated that even with a short, 3 ft ECP, an effective pressure seal can be brought about quickly (Millich et al., 1988).

A cost comparison of an uncemented preperforated liner and a cemented liner shows that the uncemented liner can be completed for about 10% of the cost of a cemented and perforated completion (Isvan et al., 1986). This makes the option of selective isolation with ECPs very attractive.

Tool Size

The length and diameter of a rigid tool that can be transported around a curve is dependent on the curve radius and the tool clearance (see Figure 3-6).

The maximum length of any specific diameter of tool to be transported around a curve can be calculated using the equation:

$$L = 1/6 \, [R^2 - (R - \Delta d)^2]^{1/2} \tag{3-1}$$

where L = tool length (ft)
 R = curve radius (in)
 Δd = $ID_{casing} - OD_{tool}$ (in)

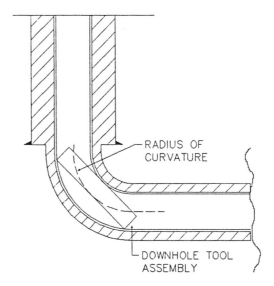

Figure 3-6. How the radius of curvature impacts on the length of tool that can be run in the well.

Tool Operation

All tools are set with mechanical, hydraulic, or passive action. Mechanical actions include rotation and reciprocation, both of which are easily transmitted to bottom. Hydraulic tools require only pressure to be transmitted around the bend.

Completions 85

Cup-type tools are labelled passive tools as they require no manipulation to be set or unset. They are also the most sensitive to mechanical wear as their packing elements are always engaged. Wire-reinforced cup elements are normally used to increase durability. A casing string of uniform internal diameter can cause significant wear on cup-type tools by the time they reach bottom, due to the cups being engaged with the casing walls for the entire trip in. The ideal set-up is to have a liner in the horizontal section. The larger intermediate casing allows the cups to be transported to the liner top with no cup wear.

To increase cup life it is important to firmly centralize the tool with star guides. If the cups are close together ($\pm$ 20 ft) a star guide at each end should suffice. If they are farther apart (in one completion this author ran cups spaced 330 ft apart), a star guide should be run on both sides of each cup. As well as minimizing wear this will ensure the weight of the tubing will not pull the cups away from the upper side of the hole.

Tools

Bridge Plugs. Both permanent and retrievable bridge plugs have been run in horizontal wells. They are run primarily to isolate previously stimulated sections during multiple stimulations. While both types set reliably, the retrievable bridge plugs (RBP) can lead to significant milling problems if unsetting problems are experienced. On the other hand, permanent bridge plugs are readily drilled but must be removed prior to running the next one. Using a flat-bottomed mill, 5½-inch bridge plugs have been milled and pushed to the toe of the well in 5½ hours. In one case, four permanent bridge plugs were set, milled, and pushed to bottom without mechanical difficulty (Zimerman et al., 1989).

Packers. The packer will normally be the largest-diameter tool run. The shortest packer available should be selected to minimize problems in running it around the bend. Minimizing the packer length precludes use of millout extensions, seal bore extensions, and other large OD additions to the tool.

Packers can be set and unset in a variety of ways. In selecting the packer, place emphasis on minimizing rotation or movement necessary to activate the tool. Mechanical set packers that require rotation present the problem of transmitting rotation to bottom without backing off the tubing connections. Compression set packers overcome this by requiring rotation to the right to set, and straight pickup to retrieve.

Hydraulic set and pickup to release hydraulic packers are very easy to-work. The packers are set by pumping plastic, brass or steel balls down the tubing to a seat in the packer. A preset pressure is held on the tubing to activate a hydraulic piston which compresses the packing element and engages the slips. The balls are then flowed to surface or pumped out the bottom of the packer through an expendable seat.

If a liner is to be run and there is no need for a packer in the horizontal section, consider using a seal nipple in a polish bore receptacle (PBR) attached to the top of the liner. Run a minimum 6 ft polish bore receptacle to allow some tubing movement without disengaging the seal nipple. The seal nipple engages by setting into the PBR and can hold differential pressure up to the liner pressure rating, depending on the amount of compression it is landed in. This is a very simple and effective way to isolate the annulus in the vertical section from the horizontal section.

We have used the PBR in situations where we needed to isolate a section of the horizontal hole. Rather than trying to set two packers independently, we ran a packer/PBR combination that allowed a right-hand set packer to be run, and used the PBR at the liner top in place of a less than ideal left-hand set packer.

Pumping Equipment. Horizontal oil wells may require pumping equipment. The three main differences between pumping vertical and horizontal wells are:

1. Depending on radius and dog leg severity, pumps may have to be landed above the zone, adding to the back pressure the well will flow against
2. Inability to land pumps below the production interval interferes with using the annular space for gas separation
3. Pumps are required to perform at an angle off vertical which may decrease pump efficiency.

Four pump types currently in use are: gas lift, hydraulic jet, electric submersible pumps, and sucker rod pumps. All are applicable to horizontal wells with some modifications.

Normally, the pump will be run in the vertical or near vertical section of the wellbore unless the well has been drilled with long radius technology. Pump design can be approached as with any other vertical well, with the exception of not being able to sump the pump.

If there is a need to run the pump around the radius to keep the hydrostatic pressure on the formation to a minimum, then the following concerns need to be addressed with each artificial lift system.

Gas Lift. It has been shown that deviated wells require more gas flow to maintain uniform oil flow rates (Brown and Beggs, 1977). For long radius wells, gas lift may become uneconomical due to high gas flow rate requirements in the radius. However, there is no reason to run the gas lift through the horizontal section, and the low lift efficiency is only a concern through the radius. This allows gas lift to remain a viable option, especially in medium or short radius wells.

Hydraulic Jet Pumps. These have been designed with knuckle joints for use in high angle applications. Jet Pumps are only slightly affected by hole inclination and can be used in horizontal wells. Their main disadvantage is they cannot be used at low reservoir pressures, as the pumps will cavitate.

Electric Submersibles. These pumps do not run well at high angles of inclination. Their labyrinth seals are not designed to work at significant angles. As the pump motors are cooled by fluids passing over the motor section, the pumps need to be well centralized. This can be difficult at high angles of deviation.

These pumps are very sensitive to high gas/oil ratios. In many cases, sections of the horizontal wellbore act like gas/oil separators, allowing the gas to break out of the oil and move ahead. This results in slugs of oil and gas arriving at the pump in alternating stages, with the gas slugs causing the pump to rev too high and trip itself out.

Despite these problems, electric submersibles with modifications are being used on some horizontal wells because of their ability to lift large amounts of fluid.

Sucker Rod Pumps. Sucker rod pumps can be used through the bend in long radius wells only. The rods will not withstand the repetitive bending loads in shorter radius wells. When rods are run through deviated sections of the wellbore, rod rollers need to be used to prevent excessive rod and tubing wear, as shown in Figure 3-7.

WORKING IN THE HORIZONTAL ENVIRONMENT

A major difference in operating in horizontal wellbores is the right angle shift in gravity direction. Sand, metal cuttings, port plugs, etc., which normally drop to the bottom of the hole, now sit on the low side of the wellbore, waiting to pile up in front of and around tools run into the well. This change in gravity direction requires a change in mindset for the completion designer and his operations staff.

Gravity-powered tools such as wireline tools will function normally in holes deviated to a maximum of 65°. Conventional tools can be used through to 70° by installing wheels on the tools. Beyond 70°, tools have to be pushed or pumped.

88 Horizontal Wells

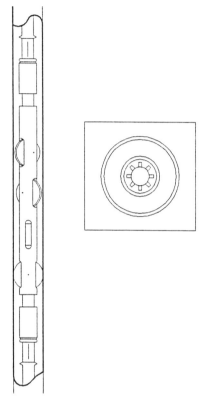

Figure 3-7. Wheeled rod guide couplings are used to minimize wear on sucker rods run in the deviated sections of the well.

Downhole junk that normally settles to the bottom of a wellbore will be encountered in the lower third of the radius while running in. In one completion equipped with a liner we found pieces of stage tool, logging tool, centralizers, and other miscellaneous bits of metal inside the liner at the bottom of the radius. The majority of this material had been lost in the horizontal section while drilling and logging. While circulating the liner in place, the liner annular velocity was sufficient to move the junk to the liner top where the fluid velocities decreased in the larger annular area. The top of the liner acted as junk catcher in this lower velocity area. When the drill pipe was disconnected from the liner top, the material dropped inside the liner to await the first tubing run. The first run was with a bit, which pushed the junk unnoticed to the end of the hole. We became aware of the junk when we ran in with a magnet to ensure that the hole was clean prior to stimulating, and recovered several metal pieces, some as big as matchbook covers.

Gulf Canada Resources Ltd. routinely runs magnets in their horizontal wells and invariably recovers significant amounts of metal.

Hole Cleaning

Hole cleaning is much more difficult in horizontal holes. There is a tendency for the solids-laden portion of the fluid to travel along the low side of the hole and be overridden by the cleaner, less dense portion of the fluid. This initial segregation leads to settling and incomplete cleaning. The best way to ensure efficient cleaning is to pump hole-cleaning fluids in turbulence.

Completions 89

The use of high shear thinning polymers, such as Xanthum gum, offers the ability to suspend the solids in turbulent flow and reduce settling (Zaleski and Spatz, 1988). At low shear rates an equal blend (50/50) of Xanthum and guar gums has almost twice the viscosity of Xanthum gum alone, but only one-eighth the modulus of elasticity for more efficient transport of solids (Motley and Hollanby, 1987). In larger diameter casing or liners it may not be possible to achieve turbulent flow. In such cases, a debris pickup gel (Figure 3-8) which utilizes complex flow rheology and gel adhesiveness may be used to pick up and carry large volumes of solids with minimal friction (Cooper and Troncoso, 1988).

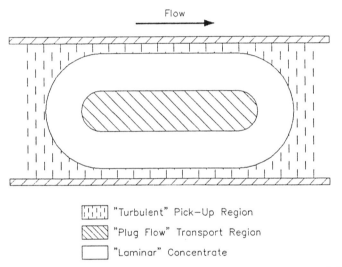

Figure 3-8. Pick-up gel coats debris and assists in its transport out of the horizontal section of the well. (Cooper and Tranosco, 1988).

During sand fracturing the stimulation rates and fluid viscosity should be sufficient to keep the proppant in suspension. Figure 3-9a and 3-9b are plots of the critical deposition and resuspension velocities for various fluids containing 6 lb/gal; 20/40 mesh in various size tubulars. In the figures it is noted that critical deposition and resuspension velocities are dependant on pipe velocities and carrier fluid type (Shah and Lord, 1989).

Tubing Bottom Location

Horizontal wells are never perfectly horizontal, and will always have vertical deviation along their length. The vertical deviation can vary from as little as 5 ft to dozens of feet along the length of the wellbore, depending on lithology, structure, and the ability of the drillers to control the drill bit's

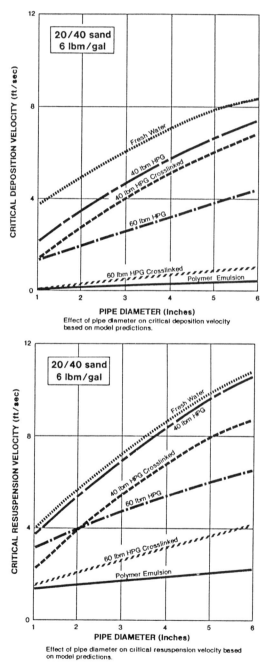

Figure 3-9. Critical deposition and resuspension velocities shown as a function of carrier fluid type and casing diameter. (Shah and Lord, 1989).

Completions 91

attitude. When the well profile is drawn to scale, the deviations can look minor over the length of the 2,000 ft wellbore, but during the production phase these seemingly slight deviations may have a significant impact on the well's performance (see Figure 3-10).

The dips and rises in the profile will act as downhole separators for the various phases of the produced fluids. In an oil well, oil and water will collect in the low spots and gas in the high areas. This will result in unstable flow patterns with irregular slugs of gas which will significantly decrease pump efficiency. In gas wells, water and condensate will collect in the low areas and unload in slugs. Depending on the flowing pressures of the well, these slugs may be big enough to load up and kill the well. Landing the tubing at the low point in the lateral will remove fluids as they are produced and minimize slugging.

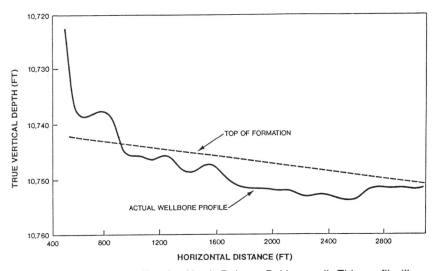

Figure 3-10. Actual profile of a North Dakota, Bakken well. This profile illustrates the many areas where liquid holdup can occur and impede the well's flow performance.

Tubing Stresses

Even for short radius wells, conventional $2^7/_8$-inch tubing can be run around the bend without yielding. However, with larger diameter pipe, especially liners, this can be a concern. Tubular stresses due to bending can be calculated using Equation 2-20 from Chapter 2.

If the allowable yield stress of the pipe is exceeded, then higher strength, or smaller diameter pipe should be used.

Another concern is buckling of the tubing when running or working in the horizontal section. The critical buckling load can be calculated as follows, using the equations developed by Dawson and Paislay (1984).

$$F_{crit} = 2A(EIW'\sin\theta/\Delta d)^{1/2} \tag{3-2}$$

where A = constant = 0.93 for steel
E = Young's modulus (psi)
W' = buoyed pipe weight (lb/ft)
θ = hole inclination from vertical (°)
Δd = radial clearance between tubing and casing (inches)
I = moment of inertia = 0.04909 $(D^2 - d^2)$
where D = pipe OD (in.)
d = pipe ID (in.)

F_{crit} calculates the critical tubing load that will put the pipe into a buckling mode. Care should be taken to stay below this load during any milling or other operations that put tubing in the lateral in compression. This equation can also be used to calculate the maximum weight that can be pushed into the lateral with coiled tubing.

To perform a complete analysis, the force due to friction of the tubing or coiled tubing in the horizontal section needs to be accounted for. The friction coefficient for unlubricated metals is 0.3. As the well will normally be filled with a completion fluid, which should act as a lubricant, the coefficient should be below 0.3.

PERFORATING

Reasons for perforating horizontal wells and the methodology used to place the perforations are as varied as the reservoirs the wells are placed in. Perforating guns can be conveyed in many different ways, but the following concerns unique to horizontal wells need to be addressed, regardless of the method employed:

Conveying Guns Around the Radius

Transporting guns around the radius is not a concern in long or medium radius wells. Casing guns exceeding 1,000 ft are routinely run in these wells (Weirich et al., 1987). Articulated casing guns are being developed to negotiate the tight corner in short radius wells.

Completions 93

Logging and Correlating Guns on Depth

In most horizontal wells this is not a concern, as long intervals are normally perforated and locating the guns within a few feet of the required interval is all that is required. However, it can be a concern if a short specific interval is to be shot for a stimulation or cement squeeze. The lateral is usually in one lithology type and the gamma ray will be quite undistinctive in character. The best way to accurately locate an interval is to run a marker joint in the vicinity of the selected interval. Prior to perforating, run a gamma ray/collar locator, then run a mechanical collar locator with the tubing-conveyed guns.

Gun Orientation

In consolidated formations, full-size guns with complete phasing are recommended. In unconsolidated formations, the perforations are normally limited to the lower 180° to 120° of the well, as shown in Figure 3-11. This is to avoid sloughing of formation from the high side of the hole during production.

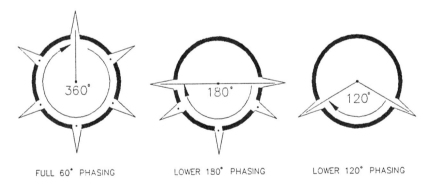

Figure 3-11. Perforation configuration in consolidated (360°) and unconsolidated formations (180° and 120°).

As well as minimizing sand production, this facilitates more efficient gravel packs, as gravel will settle to the low side during packing procedures. It also optimizes hole size, as gun standoff is minimized due to the gun resting on the low side of the hole.

When perforating the low side of the well, the entire gun section must be oriented. Steering tools and other methods can be used to orient the top of the gun. However, with these systems the length of the perforating assembly

94 Horizontal Wells

is normally limited to 150 ft. With guns hundreds of feet long there will be enough static friction between the gun and the liner to allow the gun to spiral when the top of the gun is being oriented. A commonly used method to overcome misorientation employs swivel subs and orientation fins. The fins are welded to the top side of the gun, lowering the gun's center of gravity, as shown in Figure 3-12. The gun's eccentric center of gravity prevents the gun from riding up the sides of the hole and keeps the perforating charges correctly positioned.

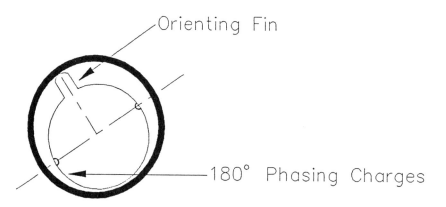

Figure 3-12. Orienting fin used to prevent "spiralling" of long guns during running and positioning.

Clearance Effects on Gun Performance

Both the perforation entry hole size and depth of penetration are affected by gun clearance or standoff. Amoco (King, 1989) generated plots showing the effects of standoff on entry hole size and depth of penetration for both deep and big hole penetrating charges (see Figure 3-13).

By nature of their size, large guns overcome standoff concerns. In consolidated formations, the largest gun that can be run and retrieved leaving room for gun swell and surge solids should be used.

In unconsolidated formations, the risk of surging significant sand into the casing requires that both the degree of underbalance and gun size be reduced. The gun size must be small enough, in unconsolidated formations, to be washed over if it becomes stuck. This is important as perforating guns are not millable.

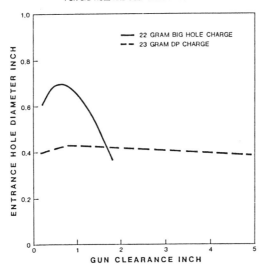

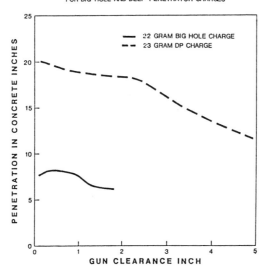

Figure 3-13. Plots showing the effects of standoff on entry hole size, and depth of penetration for big hole and deep penetrating charges (King, 1989).

96   *Horizontal Wells*

Perforation Density

In unstimulated vertical wells, shot density is maximized to minimize flow convergence effects at the wellbore. With so much zone exposed in a horizontal well, a lower shot density can be used. The shot density needs to be determined for each individual well, based on reservoir fluid viscosity, permeability, and reservoir heterogeneity.

Perforating can be a very significant percentage of the well cost and perforation density is a major factor in that cost.

To control perforating costs, the Alberta Energy Company utilized bullet guns which could fire 50 bullets, one at a time (Freeborn et al., 1989). The bullets had better penetration than jet charges in the low compressive strength sandstone they were working with, and allowed large and varied spacing between shots.

Gun Conveyance Methods. Perforating guns can be run in two different ways:

1. Tubing conveyed with or without packers
2. Coiled tubing conveyed

Tubing Conveyed Perforating Guns. With the few differences mentioned earlier, tubing conveyed perforating (TCP) in a horizontal well is very similar to perforating in a vertical well. Perforating can be carried out in a hydrostatically underbalanced or overbalanced wellbore, with or without a packer.

The TCP systems in vertical wells are normally actuated by a drop bar. In a horizontal well, hydraulic pressure is used as the drop bar is unable to reach the firing head. Hydraulic pressure actuation presents a problem if an underbalance is desired during perforating.

When a packer is run, an underbalance can be created by using a crossover kit or a hydraulic bypass in the packer. The guns are detonated by applying pressure to the annulus, as shown in Figure 3-14.

The tubing in this arrangement is run with a precalculated fluid column to create the desired underbalance. Annular pressure is applied to the firing head via the crossover at the packer. While the gun sees the annular pressure, the formation is exposed to only the tubing hydrostatic. With this system the packer has to be run reasonably close to the guns to accommodate the pressure bypass tools.

In an overpressured reservoir, a simpler way to underbalance the formation is with a time-delayed firing head (TDF) below a packer. The firing head is actuated by direct applied pressure and detonates the guns 5–7 minutes after being actuated. The TDF is actuated by pressuring up the tubing

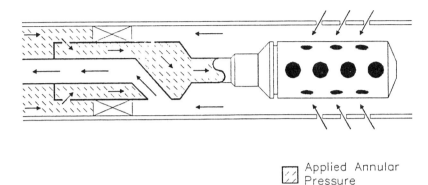

Figure 3-14. Diagram of crossover used to transmit annular pressure to activate pressure-sensitive detonators on TCP guns while maintaining an underbalance on the formation. (Weirich, Zaleski, and Mulcahy, 1989).

with water or a low density fluid, and bleeding the surface pressure off to create the desired underbalance prior to detonating the guns.

Tubing in a normally pressured reservoir can be run dry or swabbed down after the packer is set. The TDF is then actuated with nitrogen gas. The nitrogen is bled off to effect the desired underbalance prior to gun detonation.

If a packer is not run, the guns can still be hydraulically actuated while maintaining an underbalance by using a bypass assembly with a check valve in the pump seat of the bypass assembly. Nitrogen is injected into the tubing/casing annulus which displaces the annular fluid into the tubing. The nitrogen in the annulus is then bled off to create the desired underbalance. Pressuring up on the fluid-filled tubing actuates the differential firing head. The guns detonate with the formation exposed to only the underbalanced annular pressure.

All these systems use either differential or direct pressure operated firing heads. Multiple zones can be perforated with these systems by using multiple firing heads pinned to different actuation pressures.

Coiled Tubing Perforating. Coiled tubing with a conductive cable threaded through it can be used for both logging and perforating. Coiled tubing can be used to perforate multiple intervals using short switched guns. However, its limitations are quickly reached when longer guns are run. Coiled tubing will buckle when frictional forces associated with the guns in the horizontal hole exceed the coiled tubing critical buckling load.

The advantages of coiled tubing are the speed of running the guns, the ability to log perforating guns on depth, and the ability to make multiple runs.

Perforating Techniques. Perforated wells can be broken into two major classes.

Cemented Laterals. In this situation, the lateral has been cemented for the following reasons:

1. To allow effective isolation during stimulation
2. To allow segregation of wellbore intervals to control early water or gas breakthrough
3. To limit sand production in unconsolidated formations and facilitate gravel packing.

The most effective way to perforate a cemented lateral is with conventional perforating equipment. There has been discussion of using sand jetting to accommodate greater wellbore/fracture contact area during hydraulic fractures, but the problems associated with removing sand from the lateral during jetting operations has made this a little-used alternative.

Uncemented Laterals. The majority of horizontal wells are drilled into formations that, if left in an undamaged state, have sufficient permeability to produce without stimulation. A well that does not need to be stimulated or does not have sand production problems will probably be completed with uncemented casing or a liner. This opens the door to a number of innovative perforating approaches.

In wells requiring some stimulation, the lateral is normally set up to allow treatment at selected intervals. One of the more common approaches is to run external casing packers (ECP) in conjunction with preperforated pipe or tubing-operated port collars, as shown in Figure 3-1. The use of tubing-operated port collars allows intervals to be stimulated, evaluated, and selectively isolated at a later time if required.

A zone not requiring stimulation will often be cased to maintain wellbore integrity. Slots or wire-wrapped screens will be used in an unconsolidated formation. Due to the low flow velocities associated with the long wellbore/formation contact area, this is normally all that is required to control sand production.

In a consolidated formation, preperforated pipe works well if the liner or casing does not need to be circulated into place. If there is a need to circulate the pipe while running in, the casing cannot be perforated until after it's landed, unless an inner tubing string is run, as shown in Figure 3-3. Although somewhat awkward to run, this approach allows both well control and full length circulation while running in.

An alternative to this is a system currently being used in the Williston Basin in North Dakota and shown in Figure 3-4. To allow the casing to be circulated into place, the uncemented portion of the $5^{1}/_{2}$-inch casing is pre-

Completions 99

drilled with 1-inch holes, which are tapped and plugged with hollow aluminum plugs. Once the upper part of the string has been cemented, the plugs are milled to allow the well to produce. The hollow portion of the plugs allows entrance of formation fluids into the casing. Apart from being simple and relatively foolproof, this system is very economical. The entire 2,500 ft section can be put in place and opened up for under $20,000.

CASED HOLE LOGGING

Practically any log can be run in a horizontal well. As in perforating, the major consideration in logging is conveyance of the tools through the horizontal section. Wireline logging tools will drop via gravity to borehole angles up to 60°. From there they need to be pushed or pumped the rest of the way. The two main conveyance methods currently in use are pump down and coiled tubing.

Pump Down Tools

By far, the simplest logs to run are slim hole, pump down logs, such as Gamma Ray, Neutron, and Pulse Neutron tools. With tubing run to the end of the lateral, these tools can be pumped to bottom and then logged conventionally using the conductive cable to pull the tools. The logs will be sensing through the tubing and liner, but apart from minor signal attenuation this does not limit their usefulness.

Elf Aquitane has taken pump down logs a step further by attaching the logging tools to a rigid stinger, which is pumped out the end of the tubing. Using this approach, up to 1,600 ft of horizontal hole has been logged (see Figure 3-15).

The stinger is driven by a "locomotive" which pumps or moves the tools hydraulically into the horizontal section. A perforated joint is placed at the bottom of the string to prevent pumping the locomotive out the end of the tubing. This allows the tubing pressure to bleed off when the locomotive reaches the end of the tubing. Logging is initiated by applying tension to the conductive cable attached to the locomotive and logging in a conventional manner. This assembly can be used to production log if the annulus is open to surface or a second tubing string has been run to allow flow while logging. In early work, Elf Aquitane discovered that although the cable speed was constant at the surface, the tool moved in an irregular manner due to friction (Millich et al., 1988).

This assembly can be used to run perforating guns as well.

Horizontal Wells

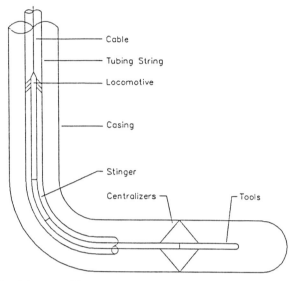

Figure 3-15. Stinger and locomotive used to push logging tools through the tubing and into the lateral. (Joly, 1987).

Coiled Tubing Logging

The tools in coiled tubing logging are run on the end of the tubing with the conductive line prewired inside. Most tools can be run, and the coiled tubing has the versatility of being run in casing/liner or through tubing. Other advantages of coiled tubing logging include faster running time, continuous logging capability, and total well control at all times.

As with perforating, the major limitation of coiled tubing is its inability to push logging tools over long horizontal distances without buckling. The heavier full-size logging tools can only be pushed 600–700 ft with 1 1/4-inch coiled tubing (Cooper, 1988), but can be transported further by stepping up to 1 1/2-inch coiled tubing. Production logging tools that are lighter than full-size standard logging tools have been pushed as far as 2,000 ft.

Tool Length

Not all full sized logging tools can be run in a cased horizontal hole. It is important to check to ensure the tool will travel around the radius. In short radius and some mid-radius wells this is critical. This check can be performed by using Equation 3-1.

Completions 101

This equation is for rigid tools only. Any tool with flex, such as a hollow steel carrier casing gun, can be run in longer lengths despite having a larger "OD" than this calculation would allow. Production logging tools, which are thin and flexible, can be run in greater lengths as well.

Logging Tools

Cased hole logs play a significant role in horizontal wells. In many cases it is either too expensive, or too risky to run openhole logs. This puts the onus on the cased hole logs to identify the quality of the reservoir penetrated, and its characteristics. Other than additional attention to centralization and conveyance methods, running of logs in horizontal wells differs little from operating in a vertical environment.

The one exception to this is production logs. One of the biggest problems in production logging is measurement and interpretation of multiphase flow. During multiphase flow in horizontal wells, as many as six different flow patterns can exist (Figure 3-16).

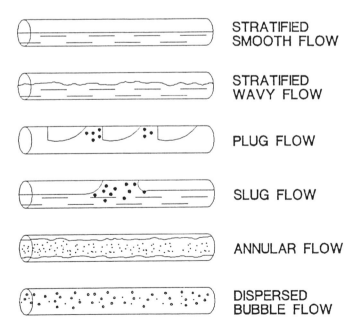

Figure 3-16. The six multiphase flow patterns that can exist in a horizontal well. (Cooper, 1988).

102 Horizontal Wells

Even under constant flow conditions, flow patterns can vary along the length of the horizontal well. Various flow patterns will significantly affect the flowmeter, the gradiometer, and the temperature profile.

The greatest effect will be felt by the flowmeter. If at one point the wellbore fluids are in the stratified flow regimes, the reduced cross-sectional area available to fluid flow will result in higher fluid velocities being recorded. The tool will interpret these as significantly higher fluid rates at this point in the well than are really present. To measure the flow across the entire cross-sectional area of the wellbore, variable diameter petal basket flowmeters are run just upstream from the flowmeter spinner (Figure 3-17). This tool supplies accurate readings of the flow profile across the horizontal well profile.

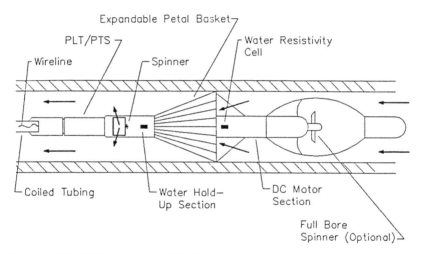

Figure 3-17. Full bore and variable diameter petal basket flowmeters run during production logging. (Cooper, 1988).

STIMULATION OF HORIZONTAL WELLS

Ideally, a horizontal well is drilled to replace a stimulated vertical well. The lateral is considered a replacement for the fracture in the vertical well. However, a horizontal well often needs to be stimulated to improve on its productivity.

The three main reasons why horizontal wells are stimulated are to remove damage, to overcome reservoir permeability anisotrophy, and to stimulate low permeability reservoirs.

Damage Removal

There are several ways a horizontal well can be damaged. The exposure time of the formation to drilling fluids is extended by as much as an order of magnitude in these wells. Not only is the exposure time greater, but the mud densities may be higher to deal with formation fluid influx or with hole stability problems in fractured reservoirs. The combination of long exposure times and high densities can result in very deep invasion that will require a large stimulation to correct.

Because of the exposure time and the fact many horizontal wells are expected to produce in a natural state, "nondamaging" mud systems are often substituted for conventional systems. Some systems can cause unexpected problems. An oil-based system used on a gas bearing, water sensitive, low permeability sandstone changed the wettability of the wellbore area from water to oil. Due to the small pore throat size of the fine grain sandstone, the thicker oil film on the pore throat reduced the cross-sectional area available to flow, resulting in a very high skin.

On another well a water based polymer mud with acid soluble calcium carbonate was used as a bridging agent. Even though the polymer mud and the bridging agent worked very well to minimize fluid loss, significant damage was measured during the production test. The use of the water based mud in this well resulted in a change in water saturation around the wellbore. The damage resulted in a relative permeability effect that corrected itself with time.

Two other types of damage that can impact a well's productivity are mechanical damage from the drillstring rotating on the low side of the hole grinding drill solids and mud into the formation; and cementing damage from either whole cement or filtrate invasion.

Regardless of the source of damage, it can have a significant impact on the productivity of a horizontal well. Figure 3-18 is a graph of expected flow rates for two lengths of horizontal wells. The flow rates are graphed against the skin effect for four different permeabilities. Other pertinent variables used for this calculation are $p - pwf = 1,000$ psi, $\beta = 1.1$ res bbl/stb, $\mu = 1$ cp, $h = 50$ ft, $ro = 745$ ft (40-acre spacing), and $rw = .326$ ft. For all values plotted, the kv is considered to be one-tenth of kh, which is a common reservoir anisotrophy.

The production of the well is nearly halved in all cases when a skin of $+5$ is imposed. Based on our experience this may be a conservative skin value. When comparing a skin of $+20$ to the undamaged case, a fourfold reduction in productivity is seen. A potential of up to a fourfold increase in productivity makes stimulation a viable economic option for all but the lowest permeability wells. The low permeability wells (0.1 md) see their productiv-

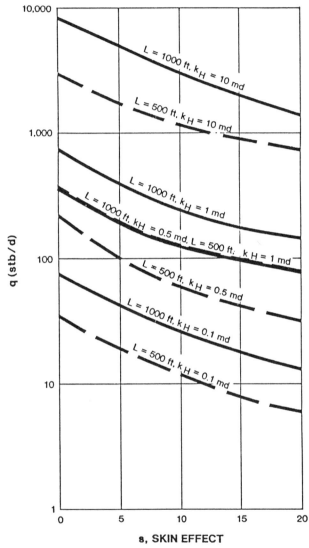

Figure 3-18. Pseudo steady-state flow rates of vertical and horizontal wells of various lengths in reservoirs with various permeabilities. (Economides, Ben-Naceur, and Klem, 1989).

ity increase from 8–40 bbl/day in the 500 ft lateral case and 18–80 bbl/day with the 1,000 ft lateral when the well is taken from a +20 to a zero skin. This probably will not pay out the stimulation cost if anything more than a matrix acid job is done.

Completions 105

Overcoming Reservoir Permeability Anisotrophy

Most reservoirs exhibit some degree of permeability anisotrophy. In sandstone reservoirs, the anisotrophy can be severe because the sand is often interbedded with shale. To quantify the degree of anisotrophy, a B factor is calculated using the equation $\beta = \sqrt{kH/KV}$. Most reservoirs exhibit some degree of anisotrophy and a value of 3 is a commonly used average value (Mukherjee and Economides, 1988). A β factor of 1 indicates complete isotrophy, and a factor of 0.25 shows highly favorable vertical permeability.

Figure 3-19 presents a comparison of productivity index ratios for the three situations (β = 3, 1, 0.25) comparing unfractured vertical well performance to unfractured horizontal well performance. For this illustration the drainage radius is 745 ft (40-acre spacing) and the well radius is 0.326 ft.

Two trends are evident from this graph. The higher the reservoir anisotrophy factor, the poorer the performance of unstimulated horizontal wells relative to vertical wells. The thicker the reservoir section, the more sensitive the horizontal wells are to permeability anisotrophy.

The β = 0.25 case corresponds to a naturally fractured reservoir. The high productivity index ratios clearly suggest that naturally fractured reservoirs are good candidates for horizontal wells.

The productivity index ratios for the thicker, conventional permeability reservoir situations (β = 3) indicate that unstimulated horizontal wells are not an obvious economic alternative to vertical wells. However, the reservoir anisotrophy can be overcome by hydraulic fracturing. One or more fractures, depending on fracture orientation, will significantly alter the reservoir permeability anisotrophy. In very low permeability reservoirs the β factor may be reduced to 1 or less (Mukherjee and Economides, 1988).

Stimulation of Low Permeability Reservoirs

Horizontal wells drilled in fields where vertical wells require extensive stimulation (more than to just remove near wellbore damage) are candidates for hydraulic fracturing. After all, a 2,500 ft horizontal well with a transverse fracture at each end would be very similar to two fractured vertical wells drilled on half-section spacing. The cost should be the same or less.

The spacing and number of fractures to optimally drain the reservoir will be dependent on the formation permeability and the cost of the stimulations, and will have to be determined using a coupled fracture/reservoir simulator.

106 Horizontal Wells

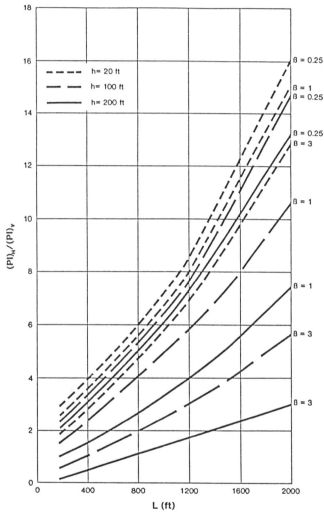

Figure 3-19. Comparison of productivity index ratios for three cases: normal anisotrophy (B = 3), complete isotrophy (B = 1), and highly favorable vertical anisotrophy (B = 0.25). (Economides et al., 1989)

Figure 3-20 is a plot of the productivity increase of a North Sea horizontal well drilled in a gas bearing chalk formation. The number of 100 ft radial fractures on the x-axis is plotted relative to the corresponding productivity increase on the y-axis.

A fourfold productivity gain is seen for 5 equidistant fractures per 1,000 ft of wellbore. Incremental gains for fracture densities greater than 5/1,000 become progressively less due to increased interference between the fractures.

Completions 107

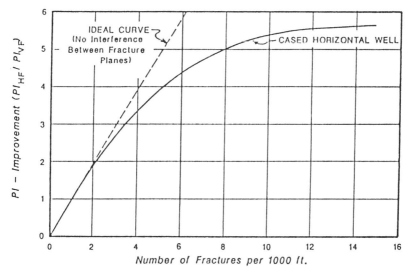

Figure 3-20. Plot of the productivity increase of a North Sea horizontal well with an increasing number of 100' radial longitudinal fractures. (Anderson, Hansen, and Fjeldgaard, 1988).

Stimulation Options

The stimulation approach will be dependent on the degree of stimulation necessary to bring the well to its economic potential. A well in a high permeability fractured gas reservoir needs a substantially smaller stimulation than a well in a low permeability oil reservoir. Three stimulation options are as follows:

Perforating. Perforating is not often thought of as a stimulation option. However, in horizontal wells it is one that should not be overlooked. In medium to high permeability wells, the wellbore will have sufficient exposed permeability to flow at an economic rate if drilling induced damage can be overcome. Drilling or cementing damage that extends less than a foot into the formation will be within the reach of deep penetrating perforating charges. If the charges are shot at high density and are successful in penetrating beyond the damage zone, they will extend the effective wellbore radius. This will result in a negative skin. In the worst case, where the charges do not reach beyond the damage, perforating becomes a necessary intermediate step preceding subsequent stimulations.

A near wellbore treatment is all that can be undertaken if the well has been drilled to overcome water or gas coning problems. In medium to high permeability wells, drilling damage may impair production and render an otherwise successful well uneconomic. In both cases, a near wellbore stimulation such as perforating may correct the problem.

108 Horizontal Wells

Matrix Acidizing. Horizontal wells drilled in medium to high permeability reservoirs (k > 1 md) with low permeability anisotrophy and limited reservoir thickness are candidates for matrix stimulation treatments. Figure 3-18 shows well production can be increased 50% in a 1,000 ft well by reducing the skin from + 10 to + 5. In a carbonate reservoir it would be possible to totally eliminate the skin, or even make it negative. However, the wellbore needs to have been previously set up to accommodate a matrix stimulation.

The chief advantages of matrix acidizing over sand or acid fracturing are cost and the ability to control the depth of penetration. This is an attractive option in a well drilled to overcome gas/water coning problems.

Fracturing. A well drilled in a formation with little or no vertical permeability may have to be fractured to be economical. Wells with low horizontal permeability or medium to high permeability with significant formation damage are also fracture candidates.

Fractures in a horizontal well may initiate in two modes. Longitudinal fractures initiate axially or along the wellbore. In this case the wellbore is oriented in a direction perpendicular to the least principal stress, and the fracture propagates along the wellbore connecting up perforations much as it does in a vertical well. In special instances, the well does not need to be drilled perpendicular to the least principal stress to induce a longitudinal fracture. Figure 3-21 is representative of a longitudinal fracture.

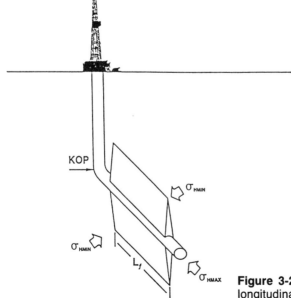

Figure 3-21. Horizontal well with a longitudinal fracture.

This type of fracture will effectively stimulate the reservoir in the near wellbore vicinity, and is chiefly used to extend the wellbore drainage radius to the upper and lower reservoir boundaries. Its principle uses are in overcoming deep wellbore damage, correcting reservoir anisotrophy, and reducing flow convergence around the wellbore.

The second mode of fracture initiation is transverse or perpendicular to the wellbore. The well is drilled in a direction parallel to the least principal stress. The fracture will initiate from one discrete set of perforations and propagate in a radial fashion from the wellbore. Multiple fractures may be initiated from the wellbore either simultaneously or sequentially. Figure 3-22 is representative of a transversely fractured horizontal well.

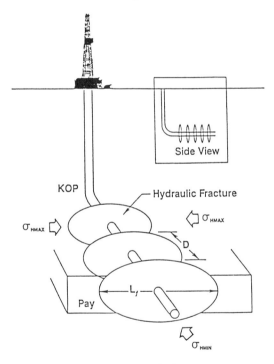

Figure 3-22. Horizontal well with multiple fracture transverse fractures.

This type of fracture is used chiefly to stimulate low permeability reservoirs. The fractures are designed more or less conventionally to optimally drain the reservoir.

Well Location. Prior to going into the design procedures for each type of fracture, two issues central to both fracture types need to be addressed: wellbore orientation and vertical placement of the wellbore in the reservoir.

Well Orientation. The intersection angle of the fracture with the wellbore is dependent on the wellbore orientation and the least principle stress direction. Once it has been established that the well will be fractured, and whether a longitudinal or transverse fracture will be used, the direction of the principle stresses in the reservoir need to be determined.

There are two methods to measure stress magnitude and orientation. Microfracturing, as described by Daneshy et al. (1986), may be used to directly measure least principal stress and orientation. Strain relaxation may also be used to estimate both magnitude and orientation (El Rabaa and Meadows, 1986). Because the openhole microfracture technique is a direct measurement of stress magnitude and orientation, it is the recommended technique in new reservoirs (Soliman et al., 1988).

To perform a microfracture, the vertical pilot hole is first drilled through the zone of interest. A DST packer is set above the interval and a small fracture induced by pressuring up the drill pipe. An orientated core is then cut through the interval and the orientation of the induced fracture recorded when the core is retrieved.

After the core has been recovered and the openhole evaluated, the well is plugged back and kicked off in the direction determined from the microfracture test. This allows measurement in the actual target formation of the magnitude and direction of the least principle stress. The measurements occur as close as practically possible to the location in which fracturing treatments will be performed.

Vertical Location. The vertical placement of the well should be chosen to optimize fracture length across the pay zone. This process will depend on the stress profile across the pay zone, pay zone thickness, and planned job size. Figure 3-23 shows a hypothetical formation where stress varies across the zone.

This figure shows that for a small stimulation, locating the horizontal well at the high stress part of the zone will most likely ensure coverage of the pay zone. On the other hand, if a large treatment is planned, locating the horizontal section in the low stress part of the zone is a wise choice (Soliman et al., 1989).

To accommodate the fracture design, the completion engineer needs to be involved in deciding which direction the well is to be drilled. If the two horizontal stresses are nearly equal this becomes a moot point, as the fracture direction cannot be predicted. However, if there is a variation between the two horizontal stresses, the orientation of the wellbore will determine the intersecting angle of the fracture with the wellbore.

In many cases, a well is drilled to intersect natural fractures or to overcome reservoir heterogeneity. It is only after the wellbore has been drilled and the predicted reservoir quality not encountered that the stimulation is planned. The completion engineer can only make the best of this situation.

Completions 111

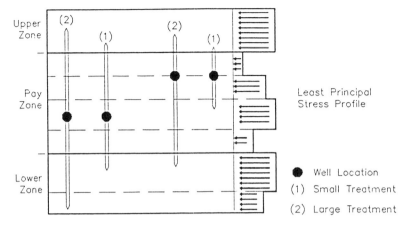

Figure 3-23. Prediction of frac height for fracs initiated from different depths. (Soliman, Hunt, and El Rabaa, 1989).

Fracture Design

In the ideal situation it is known before the well is spudded if the completion will include a fracture stimulation and the completion engineer can examine the pros and cons of designing a longitudinal or a transverse fracture. The decision of which fracture type to utilize depends on the ability of each fracture type to stimulate the reservoir. This in turn depends on the permeability of the reservoir and the achievable fracture conductivity contrast (Fcd). Figure 3-24 is a performance comparison plot of the productivity index ratio between a longitudinally fractured horizontal well and a fractured vertical well. The Fcd values refer to the fractured vertical well.

The comparison is for a horizontal well with a fully penetrating longitudinal fracture (i.e., 2 xf = L) and a fractured vertical well with a fracture length equal to xf.

The graph shows several trends. The longitudinally fractured well significantly outperforms the fractured vertical well at low Fcd values, and at high L/h ratios. Low Fcd values are seen in higher permeability reservoirs where it is difficult to get a high conductivity contrast between the fracture and the reservoir. High L/h values are seen in thinner reservoirs. Current drilling technology makes 2,500 ft laterals easily achievable. Using 2,500 ft as the L value, the definition of a thin reservoir becomes one with less than 75 ft of pay. The thinner the formation, the more attractive the longitudinally fractured horizontal well will become. Also, the higher the permeability the more attractive this type of completion becomes.

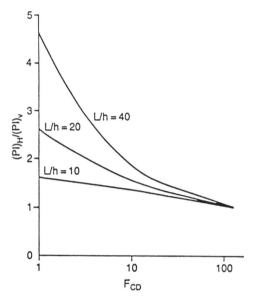

Figure 3-24. Productivity index ratios of vertical well/vertical fracture, and horizontal well with a longitudinal (parallel) fracture. (Economides et al., 1989).

In the region of low L/h values and high Fcd values, we enter the domain of the transversely fractured horizontal well. Figure 3-24 shows that the longitudinally fractured horizontal well only marginally outperforms the fractured vertical well. However, each transverse fracture in a horizontal well can be regarded as a discretely fractured vertical well, with several limitations. The first limitation is that multiple transverse fractures will interfere with each other's drainage radius during production, relative to fracture spacing and reservoir conductivity. The fracture spacing can be arranged to minimize early time pressure interference using a coupled fracture/reservoir simulator. The second limitation is the ability of the transverse fracture to adequately stimulate the reservoir. In a fractured vertical well, the fracture contacts the wellbore for the height of the perforations, which is usually the thickness of the formation net pay. This results in bilinear flow as the reservoir produces to the fracture, which then produces to the wellbore.

In a horizontal well with transverse fractures, the contact area of the fracture with the wellbore is limited by the diameter of the wellbore. This limited contact area results in severe flow convergence as the reservoir fluids flow from the reservoir to the fracture, and along the fracture to the wellbore. The flow regime can best be described as radial-linear. Figure 3-25 presents the two flow regimes.

The severity of the flow convergence will be proportional to the Fcd and the contact area of the transverse fracture with the wellbore. The contact area can be calculated using the equation:

Completions 113

$$L_w/f = D_w/ (\sin A \times \sin B) \tag{3-3}$$

where L_w/f = length of the intersection of the wellbore and fracture
D_w = effective wellbore diameter
A = angle between the fracture and the wellbore azimuth
B = angle of wellbore deviation from the vertical plane

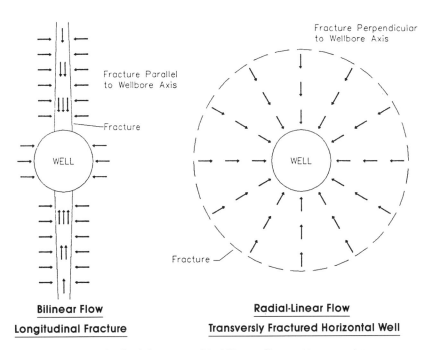

Figure 3-25. Longitudinal fracture with bilinear flow pattern, and transverse fracture with converging radial-linear flow pattern.

For intersection angles greater than 20 degrees, the wellbore/fracture contact length will be less than 3 times the effective wellbore diameter.

In Figure 3-26, performance of a vertical fracture intercepting a vertical well is compared to that of a vertical transverse fracture intercepting a horizontal well.

For $F_{cd} = 10$, the horizontal well (radial-linear flow) shows a significantly higher pressure drop for the same production rate. The higher pressure drop is due to the converging radial flow inside the fracture as fluid moves toward the wellbore. A fracture conductivity (F_{cd}) of 50 or more would produce a pressure drop similar to that of a vertical well intercepting a vertical fracture with $F_{cd} = 10$ (Soliman et al., 1988).

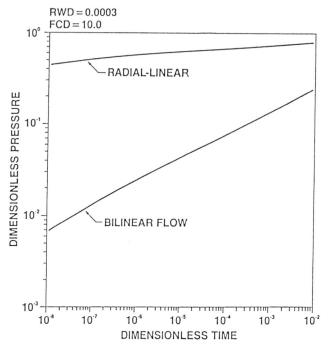

Figure 3-26. Performance of a vertical fracture intercepting a vertical well (bilinear flow) and a vertical transverse fracture intercepting a horizontal well (radial linear flow). (Soliman, Hunt, and El Rabaa, 1988).

This flow convergence can be characterized as a skin effect. The smallest contact theoretically occurs when a fracture and wellbore are exactly perpendicular. This configuration results in the largest skin effect and is given by (Economides et al., 1989):

$$(Sch)c = (kh/kfw) [\ln(h/2rw) - \pi/2] \qquad (3\text{-}4)$$

The skin is additive to the dimensionless pressure Pd for each individual fracture (Economides et al., 1989). When the well is vertical or horizontal, and colinear with the fracture(s), the effect is taken to be zero. The skin effect is directly proportional to formation permeability: it is small when the permeability is small and the corresponding Fcd is large.

Transverse fractures generally are viewed as viable in lower permeability reservoirs where Fcd > 10 can be achieved in the vertical well. Higher permeability reservoirs with lower achievable Fcd values (generally less than 2) are candidates for longitudinal fractures. Fracture type for reservoirs with achievable Fcd values between 2 and 10 need to be assessed on an individual basis.

Transverse fractures. The initial step in designing transverse fractures for a horizontal well is to determine the number of fractures and their spacing. This is done using an iterative method and the concept of Net Present Value (NPV) (Balen et al., 1988; Meng and Brown, 1987).

The wellbore length is divided equally for each transverse fracture and a fracture size for the reservoir area associated with that segment of wellbore is optimized using NPV. The fracture's performance over time is "choked," using Equation 3-4. To calculate the skin effect, some value of the kfw product must be initially assumed and later rechecked, after optimization of the fracture design.

Of particular importance is the maximum NPV of the individual fractures. The sum of the NPVs must be greater than the NPV of the longitudinal fracture case, or the vertical well with a vertical fracture. An optimization of the number of fractures versus total NPV can then be performed. There will be a different optimum fracture size for each number of transverse fractures. This optimum fracture size will always be smaller than for the vertical well case (Economides et al., 1989).

Multiple or Sequential Fracturing. After the optimum number of fractures have been determined, the designer must decide whether to place the fractures simultaneously, using limited entry technology; sequentially; or using some combination of the two methods. Due to the cost of sequential fracturing, the first fractured horizontal wells were approached with limited entry technology. Stimulations with as many as ten simultaneous fractures were designed. However, execution problems and inherent operational problems with this technology in horizontal wells have all but eliminated its use.

The major problems with limited entry stimulations have to do with fluid proportioning, sand transport, and high treatment pressures. Even with a limited number of perforated intervals there is strong likelihood that only one fracture will grow, until the pressure reaches a level where others will open. Consequently, the pad fluid may be largely diverted into a single fracture.

Sand transport problems are experienced when fluid velocity decreases as the fracture fluid progresses down the pipe to fractures at the far end of the casing. Some of the early treatments were designed at rates as high as 100 bbl/min to stimulate 10 intervals at once. In $5^{1}/_{2}$-inch casing, the 100bbl/min rate translates to a slurry velocity of 78 ft per second (fps). As the slurry progresses into the horizontal part of the hole and encounters the first set of perforations, a portion of it is diverted. The sand, with its higher density and proportionally greater momentum has problems making the right angle turn with as much efficiency as the fluid. This creates a centrifuge effect which tends to concentrate the sand in the slurry as it progresses down the

116   Horizontal Wells

pipe into areas of more reasonable velocity. At the far end of the pipe, the concern becomes one of sand transport as now the fluid velocity slows to the point where it is difficult to maintain the sand in suspension. Unlike vertical wells where gravity aids in keeping the sand moving in slower velocity areas, here gravity works against the process by pulling sand to the low side of the hole where it has a tendency to dune. This example is an extreme case, but these problems exist to some degree in all limited entry stimulations in horizontal wells.

The limited number of perforations is not the only reason for high treating pressures on limited entry stimulations. The formation breakdown pressure during stimulation is dependent on the wellbore orientation as shown in Figure 3-27 (El Rabaa, 1989). A well is rarely drilled exactly parallel to the least principle stress.

In this situation the breakdown pressures vary by 5,000 psi, as the orientation of the wellbore changes from parallel to perpendicular to the least principle stress. A vertical well in this stress field will be hydraulically fractured

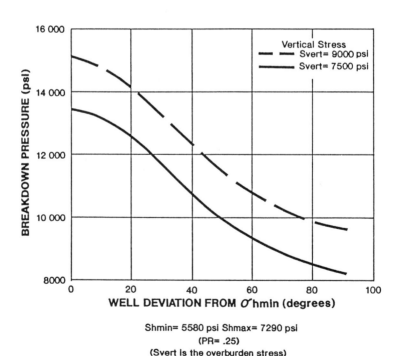

Figure 3-27. Fracture breakdown pressure as a function of wellbore orientation relative to the least principle stress direction. $\sigma = 0$ orients the well parallel to the least principle stress. (El Rabaa, 1989).

at a breakdown pressure of 9,950 psi. This plot is for a well drilled in the horizontal plane. As the wellbore is rotated through the vertical plane the breakdown pressures will change also.

A second major influence on treating pressures during limited entry stimulations is the impact that simultaneously propagating fractures have on one another. This interaction is not significant until the fracture's height or length approaches its spacing. This interaction is in proportion to the fracture width (Economides et al., 1989). Figure 3-28 is a graph of two simultaneously propagating fractures.

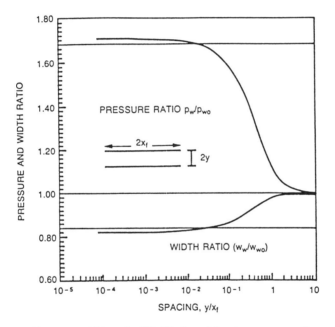

Figure 3-28. Fracture width ratio (W_w/W_{wo}) and fracture propagation ratio (P_w/P_{wo}) plotted as a function of the spacing of two transverse fractures propagated simultaneously. (Economides, 1989).

The fractures' interaction as they grow will increase treating pressures, decrease fracture widths, and cause possible deflection of the fractures away from each other. In a sequential fracturing, situation a previously placed fracture will influence a propagating fracture in proportion to its propped width (Jeffrey et al., 1987).

In summary, the likelihood of additional treating pressure due to the limited number of perforations, wellbore orientation effects, and fracture growth interaction makes it very difficult to design and successfully execute a limited entry treatment in a horizontal well. When the pressure concern is considered in combination with the fluid proportioning and sand transport concerns, it quickly becomes evident why few limited entry treatments are being performed on horizontal wells today.

Treating Pressures. Whether fractures are performed simultaneously or sequentially, attention needs to be paid to the anticipated treating pressure. As discussed, the treating pressure is a function of the wellbore orientation, fracture interaction, and perforation configuration. All of these need to be considered in some detail prior to initiating the fracture design.

As presented in Figure 3-27, the wellbore orientation has a major impact on treating pressures. It also has a major impact on fracture execution and subsequent productivity. If the wellbore is drilled in a direction other than parallel or perpendicular to the least principle stress, it is highly probable that an induced fracture will initiate perpendicular to the wellbore, and at an angle to its preferred orientation. As the fracture grows away from the wellbore it will reorient itself to be perpendicular to the least principle stress. During the reorientation process, the fracture is extended under both shearing and tensile modes, resulting in a rough surface on the fracture walls. The fracture widths are reduced in this region as well. The smaller fracture widths and rough surfaces significantly increase friction pressure. The distance over which the reorientation occurs is a function of the injection rate (Yew and Li, 1988) and the ratio between the horizontal principle stresses (Viola and Piva, 1984). The geometry of the main induced fracture is nonplanar (S-shaped) with nonidentical wings. The wing that extends at an acute angle from the wellbore will be shorter (El Rabaa, 1989).

The reduced fracture width in the wellbore vicinity not only impacts the treating pressure, but increases the likelihood of screen-out due to proppant bridging. Propping agents will bridge if the width of the fracture is less than 2.0–2.6 times the maximum proppant particle diameter (van der Vlis et al., 1975).

Perforations. Perforation configuration has a major impact on not only the perforation friction pressure, but the fracture initiation mode as well. It is possible to initiate multiple fractures from the same perforated or openhole interval during a stimulation treatment. Laboratory modeling work has revealed that multiple fractures can be created when the length of the perforated interval exceeds four wellbore diameters (P1 > 4D) (El Rabaa, 1989). In each case where multiple fractures were generated from a short perforated interval, one fracture eventually predominated. However, initial extension pressures for those cases were clearly higher than for cases where only one fracture was initiated. The same study revealed that the initial fracture connects all perforations spaced within one wellbore radius of each other, and that for small perforated intervals (<4D) only one fracture was likely to be induced.

To facilitate fracture initiation, the perforated interval for each fracture should be made as small as possible: certainly less than four times the well diameter. At the same time, the interval should be perforated at as high a density as possible. This will reduce friction pressure during the stimulation and minimize the choking effect during production. To be effective, the perforations need to be as large as possible without sacrificing penetration. They should be placed with a minimum 90° phasing and shot with as large a gun as possible to minimize standoff effects.

At this point the major variables in the transverse fracture design have been identified and dealt with. Using the method previously described, the number of fractures have been identified using maximum NPV as the criteria. Unless the heterogeneity of the formation has come into play, the fractures are spaced uniformly along the wellbore. The orientation of both the wellbore and the fracture are known and the possibility of the fracture initiating in a shear mode due to the fracture not being perpendicular to the wellbore has been considered. The accompanying high treating pressures have been anticipated and the tubulars are designed to accommodate them. The perforations for each fracture have been set in discrete short intervals and shot with as large a gun as possible.

The two remaining variables are fluid type and proppant selection.

Fracturing Fluids. Horizontal well conditions favor the use of gelled fluid systems that are complexed with a highly reactive metal. These fluids possess high short-term viscosity and cross-link density and will minimize sand fallout in the wellbore and fracture entrance area. If shear degradation can be minimized, fast cross-linking fluids will enable the creation of relatively wide fractures and transportation of proppant to the fracture extremities. When these preferred fluid systems cannot be used, the use of high injection rates may be necessary to improve proppant transport characteristics.

Gel break times should be delayed, if possible, to maintain proppant suspension properties until fracture closure is attained. This objective needs to be compatible with the goal of minimizing gel residue to maximize fracture conductivity (Cramer, 1989).

Proppant Selection. Narrow fracture widths near the wellbore and the need for high conductivity in this area are the two criteria that govern proppant selection. Depending on the mode of fracture initiation, fracture widths can be narrow in the near wellbore area, and if the fracture has been initiated in a shear mode the closure stresses can be higher than normal. To ensure placement, the proppant size should be as small as possible while ensuring adequate conductivity in this region. Conductivity needs to be maximized due to flow convergence effects. Figure 3-29 shows the importance of increasing the conductivity of the tail-in proppant.

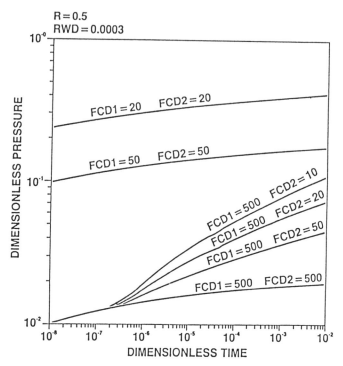

Figure 3-29. Transversely fractured wells with single and dual proppant types (and FCD values). The fractures with dual proppant types are propped with equal types of each proppant. (Soliman, Hunt, and El Rabaa, 1988).

This figure shows two fractures with consistent Fcd values of 20 and 50. It also shows four fractures with dual Fcd values. The near wellbore half of the fracture has a Fcd value of 500, while the other half of the fracture has varying Fcd values. There is a significant decrease in pressure drop when the tail-in conductivity is increased to 500. The effect of decreasing conductivity away from the wellbore is also demonstrated by the increase in pressure drop when the lead-in proppant conductivity ratios are decreased.

The proppant should be as small in diameter as possible to allow passage through the near wellbore area. Once transported across this area, it will encounter normal fracture widths. It may be advisable to choose proppants that are smaller than would normally be used in the early stages of the treatments, while the fracture in this area both grows and erodes. Larger proppants can be attempted as a tail-in to obtain maximum near wellbore conductivity.

Flowback of frac sand can be a major problem over the life of the well. There is a greater tendency for sand to flow back due to the high velocities associated with converging radial flow in the near wellbore areas. Once in the wellbore, the sand becomes a problem as it lies on the low side of the hole. Due to the difficulty and high cost of clean-out procedures it is advisable to run curable resin-coated sand as a tail-in. This will increase both the conductivity and the strength of the proppant near the wellbore, and minimize proppant flow back over the life of the well.

Longitudinal Fractures. Longitudinal fractures propagate along the axis of the horizontal wellbore. They naturally occur in wellbores drilled perpendicular to the least principle stress and can be artificially induced in wellbores drilled at other orientations. The major use of longitudinal fractures is in overcoming near wellbore damage and in correcting reservoir vertical permeability anisotrophy in medium to high permeability wells. They will impart a negative skin to the wellbore, but are a poor substitute for transverse fractures in low permeability reservoirs.

The concept of inducing a longitudinal fracture in a wellbore drilled perpendicular to the least principle stress is one that is easily grasped. More difficult to visualize is the process that occurs for longitudinal fractures induced in wellbores placed at other angles.

In damaged, medium-to-high-permeability wells, it may be more effective and economical to stimulate the reservoir with one or more longitudinal fractures than with numerous transverse fractures. The better permeability wells are often drilled to intercept natural fractures or are oriented for geologic considerations. The orientation of the wellbore with respect to area stresses is often a secondary consideration. When one of these wells requires stimulation to overcome damage or reservoir permeability anisotrophy, the stimulation designer is faced with the choice of using a transverse fracture in an environment where it will perform poorly, or trying to induce a quasi-longitudinal fracture.

Much research has been done on fracture initiation in both perforated and uncased horizontal wells. This research has confirmed simplified elastic solutions which suggest that, regardless of the orientation and azimuth of the wellbore, there will be a general tendency in openhole completions for the fracture to initiate along the wellbore (Economides et al., 1989). The fracture reorients to a more favorable direction once out of the influence of the borehole. Two independent laboratory studies (El Rabaa, 1989; Veeken et al., 1989) observed that longitudinal fractures did initiate in triaxially stressed blocks with modeled wellbores drilled at various angles to the minimum principle stress direction. The induced fractures initiated in a plane parallel to the wellbore axis, or interconnected the perforations if the wellbore was cased. El Rabaa observed that the initial fracture will connect all perforations helically spaced < D. A closer look at the fracture initiation process in the cased wellbore revealed that fractures initiate at an angle from a number of perforations and quickly coalesce to initiate a quasilongitudinal fracture. The fracture in both the perforated and openhole cases tends to have reduced fracture widths in the wellbore vicinity, prior to reorienting to the far field stresses.

Theory indicates the fracture will trace the wellbore axially as long as it is exposed to the treating pressures. This is supported by lab testing, and in field cases with perforated intervals covering hundreds of feet. (Whether the fracture can be propagated along a 3,000 ft horizontal wellbore is unknown at this time.)

It is difficult to design a fracture treatment to take advantage of this phenomenon as the fracture forms and propagates in an nonclassical manner. However, this method has been used to stimulate horizontal wells. Canadian Hunter perforated 300 ft of wellbore at 6 shots per foot (spf) and successfully initiated a quasilongitudinal fracture in a wellbore drilled parallel to the least principle stress. This longitudinal fracture, in the short term, produced at rates 3 times greater than those of an offsetting fractured vertical well in the same well spacing unit.

To use this type of stimulation, the designer must remember the mode of initiation, the potentially narrow fracture widths, and the very large fracture surface area created within the formation early in the stimulation. To initiate this type of fracture, the interval must be either openhole, or perforated at a minimum of 2 spf. As the fracture is initiating in a nonpreferred direction, it will require higher bottomhole treating pressures to propagate. At the same time, the potentially narrow fracture widths will create higher fluid friction pressure and make it more difficult to place proppant. Sand placement problems should be minimized with small, high-strength proppants. The conductivity of the fracture does not need to be as high as with the transverse fracture, or even as high as in a fractured vertical well, due to the large contact area of the fracture with the wellbore.

There are two schools of thought as to the best type of fluid to use. Crosslinked fluids can be used to maximize fracture widths, or a thin fluid can be used to ensure that if one or two transverse fractures initiate, they will have limited extension and won't prevent the initiation of the longitudinal fracture. Regardless of the fluid used, the early part of the sand schedule should be at low concentrations to give an indication of the fracture widths created.

Conventional Longitudinal Fractures. In this situation, the wellbore orientation is perpendicular to the least principle stress direction. The created fracture traces the wellbore along its length and does not change direction on leaving the influence of the wellbore. Of all the fracture types discussed, the conventional longitudinal fracture initiates at the lowest pressure and achieves its maximum width at the wellbore.

Fracture Design. The reservoir properties will determine the maximum length of fracture to be used. The leakoff will be very high because these fractures are normally used on the better permeability zones, and the fracture quickly establishes itself along the wellbore. The biggest fracture that can be effectively placed, coupled with the length of wellbore to be treated will decide the number of fractures necessary. The fractures will not interact with each other until their tips are extremely close (Van Damme, 1990).

Fluid selection should center around fairly efficient, cross-linked fluids that can deal with the potentially high leakoff situation. Pads can be designed using conventional fracture simulators once the proper reservoir configuration has been modeled. Conventional proppants can be used because fracture widths will be conventional and closure pressures lower than in similarly fractured vertical wells (Figure 3-27). Flow velocities will be low because of the long contact area between the wellbore and the fracture. However, curable resin-coated sands should be considered, due to the low closure stresses.

NOTATION

A	constant
d	pipe I.D.
D	pipe O.D.
Dw	diameter of wellbore
Ad	diametral clearance between tubing or tool and casing
E	Young's modulus
h	reservoir height
I	moment of inertia
Kf	fracture conductivity
kh	reservoir in situ permeability × reservoir height
L	tool length
Lw/f	contact length of fracture with wellbore
r_w	radius of wellbore
R	curve radius
w	fracture width
W'	buoyed pipe weight
o	hole inclination from vertical

REFERENCES

1. Hsiao, C., "A Study of Horizontal Wellbore Failure," SPE Paper 16927 presented at the 62nd Annual SPE Technical Conference and Exhibition held in Dallas, TX, September 27–30, 1987.
2. Haas, R. C. and Stokley, C. O., "Drilling and Completing a Horizontal Well in Fractured Carbonate," *World Oil*, October 1989, pp. 39–45.
3. Economides, M. J., Ben-Naceur, K., and Klem, R. C., "Matrix Stimulation Method for Horizontal Wells," SPE paper 19719 presented at the 64th Annual SPE Technical Conference and Exhibition held in San Antonio, TX, October 3–11, 1989.
4. Yost II, A. B. and Overbey, W. K., Jr., "Production and Stimulation Analysis of Multiple Hydraulic Fracturing of a 2,000-ft Horizontal Well," SPE paper 19090 presented at the SPE Gas Technology Symposium held in Dallas, TX, June 7–9, 1989.
5. Cooper, R. E. and Troncoso, J. C., "An Overview of Horizontal Well Completion Technology," SPE paper 17582 presented at the SPE International Meeting held in Tianjin, China, November 1–4, 1988.
6. Millich, E. et al., "New Technologies for the Exploration and Exploitation of Oil and Gas Resources," Volume 2, Proceeding of the 3rd E.C. Symposium held in Luxemburg, March 22–24, 1988.
7. Isvan, A., Wheatley, L. D., and Barry, P. M., "High Angle Development Wells in the Northwest Java Sea Bima Field," presented at the 1986 Indonesian Petroleum Association 15th Annual Convention, Jakarta, Indonesia, October 1986.
8. Zimmerman, J. C. et al, "Selection of Tools for Stimulation in Horizontal Cased Hole," SPE paper 18995 presented at SPE Joint Rocky Mountain Regional/Low Permeability Reservoirs Symposium, Denver, Colorado, March 6–8, 1989.
9. Brown, K. E., and Beggs, H. D., "Gas Lifting Directionally Drilled Wells," *Petroleum Engineer*, July 1977.
10. Anon., Oilfield Improvements Brochure, Tulsa, OK, 1984.
11. Zaleski, T. E., Jr., and Spatz, E., "Horizontal Completions Challenge for Industry," *Oil and Gas Journal*, May 2, 1988, pp. 59–70.
12. Motley, T. and Hollanby, R., "Novel Milling Fluid Saves Time, Cuts Costs," *World Oil*, March 1987.
13. Shah, S. N. and Lord, D. L., "Hydraulic Fracturing Slurry Transport in Horizontal Pipes," SPE paper 18994 presented at the SPE Joint Rocky Mountain Regional/Low Permeability Reservoirs Symposium and Exhibition, Denver, Colorado, March 6–8, 1989.
14. Dawson, R. and Paslay, P. R., "Drillpipe Buckling in Inclined Holes," *JPT*, October 1984, pp. 1,734–38.

15. Weirich, J. B., Zaleski, T. E., and Mulcahy, P. M., "Perforating the Horizontal Well: Designs and Techniques Prove Successful," SPE paper 16929 presented at the 62nd Annual Technical Conference, Dallas, TX, September 27–30, 1987.
16. King, G. E., "Perforating the Horizontal Well," JPT, July 1989, pp. 671–672.
17. Markle, R. D., "Drilling Engineering Considerations in Designing a Shallow, Horizontal Well at Norman Wells, N.W.T., Canada," SPE/IADC paper 16148, SPE/IADC Drilling Conference, New Orleans, LA, March 15–18, 1987.
18. Freeborn, R. W., Russell, B., and MacDonald, A. J., "South Jenner Horizontal Wells, A Water Coning Study," presented at the Horizontal Well Seminar, held by Heavy Oil Special Interest Group of the Petroleum Society of the CIM in Calgary, Alberta, Canada, April 20, 1989.
19. Joly, E. L. et al., "New Production Logging Technique for Horizontal Wells," SPE paper 14463 presented at SPE Annual Technical Conference and Exhibition held at Las Vegas, Nevada, September 22–25, 1987.
20. Cooper, R. E., "Coiled Tubing in Horizontal Wells," SPE paper 17581 presented at SPE International Meeting in Tianjin, China, November 1–4, 1988.
21. Mukherjee, H. and Economides, M. J., "A Parametric Comparison of Horizontal and Vertical Well Performance," SPE paper 18303 presented at the 63rd Annual SPE Technical Conference and Exhibition, Houston, TX, October 2–5, 1988.
22. Economides, M. J. et al., "Horizontal Wells," Dowell Schlumberger Technical Publication, 1989.
23. Andersen, S. A., Hansen, S. A., and Fjeldgaard, K., "Horizontal Drilling and Completion: Denmark," SPE paper 18349 presented at the SPE European Petroleum Conference, London, UK, October 16–19, 1988.
24. Austin, C. et al., "Halliburton Horizontal Completions Seminar," held in Calgary, Alberta, November 30–December 1, 1989.
25. Daneshy, A. A. et al., "In-Situ Stress Measurements During Drilling," August 1986, pp. 891–897.
26. El Rabaa, A. W. M. and Meadows, D. L., "Laboratory and Field Application of the Strain Relaxation Method," SPE paper 15072 presented at the 56th SPE California Regional Meeting, Oakland, CA, April 2–4, 1986.
27. Soliman, M. Y., and Hunt, J. L., El Rabaa, A. W. M., "On Fracturing Horizontal Wells," SPE paper 18542 presented at the SPE Eastern Regional Meeting Charleston, WV, November 1–4, 1988.
28. Soliman, M. Y. et al., "Planning Hydraulically Fractured Horizontal Completions," World Oil, September 1989, pp. 54–58.

29. Economides, M. et al., "Fracturing of Highly Deviated and Horizontal Wells," CIM paper 89-40-39 presented at the 40th CIM Annual Technical Meeting, Banff, AB, May 28–31, 1989.
30. Balen, R. M., Meng, H-Z., and Economides, M. J., "Application for the Net Present Value (NPV) in the Optimization of Hydraulic Fractures," SPE paper 18541, 1988.
31. Meng, H-Z. and Brown, K. E., "Coupling of Production Forecasting, Fracture Geometry Requirements and Treatment Scheduling in the Optimum Hydraulic Fracture Design," SPE paper 16435, 1987.
32. El Rabaa, A. W. M., "Experimental Study of Hydraulic Fracture Geometry Initiated from Horizontal Wells," SPE paper 19720 presented at the 64th Annual SPE Technical Conference and Exhibition, San Antonio, TX, October 9–11, 1989.
33. Jeffery, R. G., Vandamme, L., and Roegiers, J-C., "Mechanical Interactions in Branched or Subparallel Hydraulic Fractures," SPE/DOE paper 16422, 1987.
34. Yew, C. H. and Li, Y., "On Fracture Design of Deviated Wells," SPE *Production Engineering*, pp. 428, November 1988.
35. Viola, E. and Piva, A., "Crack Paths in Sheets of Brittle Material," *Engineering Fracture Mechanics*, Vol. 19, No. 6, pp. 1,069–1,084, 1984.
36. van der Vlis, A. et al., "Criteria for Proppant Placement and Fracture Conductivity," SPE paper 5637 presented at the 50th Annual SPE Technical Conference, Dallas, TX, September 28–October 1, 1975.
37. Cramer, D. D., "Guides Exist for Fracture Treatment in Horizontal Wells," *OGJ Special*, pp. 41–52, March 27, 1989.
38. Veeken, C. A. M., Davies, D. R., and Walters, J. V., "Limited Communication Between Hydraulic Fracture and (Deviated) Wellbore," SPE paper 18982 presented at the SPE Joint Rocky Mountain Regional/Low Permeability Reservoirs Symposium and Exhibition, Denver, Colorado, March 6–8, 1989.
39. Personal Communication, W. L. Van Damme, Noranda Research, January 1990.

4
Well Logging

Technology dealing with logging of horizontal wells has advanced very rapidly. It is not unreasonable to anticipate good quality results in horizontal wells when running spontaneous potential, natural and spectral gamma ray, dual induction, high resolution dipmeter with up to six hands, formation microscanner, shallow focused resistivity, dual induction, sonic (including long spacing), neutron, and density logs. Many of these tools are rated to over 400°F and more than 20,000 psi.

PIPE CONVEYED LOGGING SYSTEMS

In these systems, the tools are transported downhole using the drillpipe. The tools are accurately positioned at the bottom of the zone to be logged and then the collection of logging data can be started (Halliburton, 1990A).

At all times the logging tools remain attached to the transport assembly, allowing easy retrieval after the logging has been completed.

In most instances, the tools are protected from damage by special housings as shown in Figure 4-1, which displays a typical logging string and the protective house assembly. Notice that the logging string in this schematic includes, from top to bottom, a subsurface telemetry unit, and natural gamma ray, compensated neutron, density, dual induction, and shallow focused resistivity tools.

The protective housing assemblies presented in Figure 4-1 consist of the top subassembly and assemblies A, B, and C.

The top subassembly permits 360° rotation of the housing and provides a crossover to the drillpipe and circulating ports. This rotational feature per-

128 Horizontal Wells

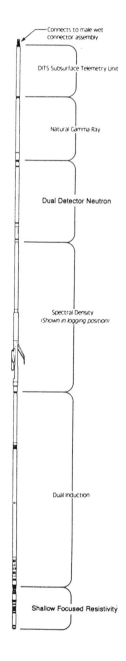
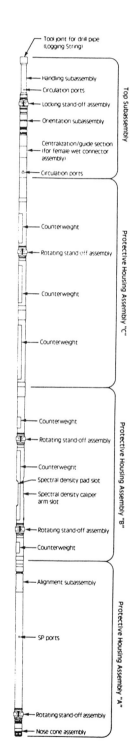

Figure 4-1. Typical toolpusher logging string and protective housing assembly. (Courtesy of Halliburton, 1990).

mits the six-arm dipmeter to swivel freely during logging operations and allows the density pad to be positioned at the low side of the wellbore.

The protective housing assembly A serves to transport the shallow focused resistivity and dual induction logging tools. Assembly B serves as transporter for the density logging tool. Assembly C serves as the transporter for the compensated neutron, gamma ray, subsurface telemetry unit, and male wet connector assembly (Halliburton, 1990A).

In some cases, such as holes having severe dogleg problems and small-diameter holes, the logging tools can be run without the protective housings. Special flex joints can be used to permit the tool string to pass through small radius deviations under these circumstances.

Figure 4-2 shows schematics of the surface unit, logging cable, side entry subassembly, female wet connector assembly, and male wet connector assembly. When the logging string is positioned just above the interval to be logged, the wireline cable is inserted into the drillstring through a side entry subassembly (Halliburton, 1990A). Both the cable and the subassembly can be released in case of an emergency.

The female wet connector is attached to the lower end of the cable, pumped down the drillpipe and latched onto the male wet connector assembly, which is attached to the upper end of the tool string. Following the latching, the subsurface telemetry unit is activated to permit simultaneous transmission of commands downlink and logging information uplink.

Drillpipe is added to log down and removed to log up. Once logging is completed and the side entry subassembly reaches the surface, the female wet connector is unlatched from the male wet connector and the wireline is removed from the drillstring. The remainder of the drillpipe and logging assembly are removed using standard tripping out procedures.

Advantages

The pipe-conveyed logging system (PCLS) presents several advantages over the more conventional wireline-conveyed methods which utilize pump-down techniques and gravity.

One of the key advantages is that the PCLS allows circulation at almost any time during the operation (Halliburton, 1990A; Western Atlas, 1990A). This in turn permits hole conditioning whenever it is required and allows continuous control of the well. The probabilities of a successful latch are increased by circulating just prior to latching the wet connectors, as this cleans the connectors' mating assemblies.

A downhole navigational system provides wellbore trajectory and monitors the position of the logging tools in the borehole.

130   *Horizontal Wells*

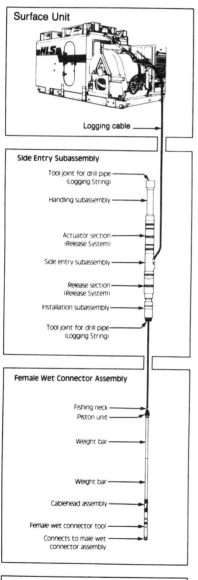

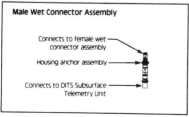

Figure 4-2. Toolpusher side entry subassembly and wet connector assemblies. (Courtesy of Halliburton, 1990).

Problems associated with doglegs, key seating, depth correlations, and high deviation angles are largely eliminated. Severe doglegs can be traversed by using special flex joints. Multiple logs can be made in a single run and standard full-size tools can be employed during logging operations of horizontal wells.

Stresses applied to the logging tools and pipe drag can be monitored through a load cell. This cell is very useful to indicate if the tools are working under extreme tension or compression. This is particularly important when the logging tools are run without the protective housings.

When logging openhole intervals that are larger than the cased section of the well, a multiple latching feature becomes very useful, as the wet connector assemblies can be latched and unlatched as many times as necessary without removing the logging tools and the protective assembly from the wellbore.

Operational Procedures

A typical pipe-conveyed logging job might involve the following operations (Halliburton, 1990A):

1. Insert logging cable into side entry subassembly; connect female wet connector assembly to logging cable.
2. Place protective housing assemblies in rotary table; insert logging tools and male wet connector assembly into protective housing assemblies.
3. Connect test cable to male wet connector assembly; pick up complete assembly until it clears rotary table.
4. Remove tool catcher; install shallow focused resistivity tool and nose cone assembly.
5. Test logging tools and male/female wet connector assemblies; lower complete assembly into rotary table and remove test cable.
6. Connect top subassembly to protective housing assemblies.
7. Add drillpipe until logging string is positioned just above zone to be logged.
8. Insert female wet connector assembly and logging cable into drillpipe; connect side entry subassembly to drillpipe.
9. Connect rig kelly onto side entry subassembly; pump to displace female wet connector assembly and logging cable down drillpipe onto male wet connector assembly at top of logging string. Remove rig kelly.
10. Log down by adding drillpipe; log up by removing drillpipe.

A tripping speed in open hole of 1 stand per 2 minutes is recommended to minimize the probabilities of damaging the logging tools (Western Atlas, 1990A).

Special Precautions

The following precautions are recommended by Halliburton (1990A) when carrying out pipe-conveyed operations:

1. Prevent foreign debris from entering drillpipe or contaminating drilling fluid.
2. Avoid logging cable damage when placing rotary slips around drillpipe.
3. Do not lower side entry subassembly below casing.
4. Maintain proper tension on logging cable.
5. Resolve any depth discrepancies between logging cable measurements and drillpipe measurements.
6. Avoid excessive downward force on logging assembly.
7. Avoid rotation of drillpipe.
8. Maintain constant communication between driller and wireline hoist operator.

COILED OR REELED-TUBING-CONVEYED WIRELINE SYSTEMS

In this approach standard wireline tools are attached at the surface to reeled-tubing through which a wireline cable has been run (Halliburton, 1990A).

No side entry subassemblies or wet connectors are required (as in the case of the pipe-conveyed logging systems) since the electrical connection between the logging tools and the cable is made previous to going into the hole.

Figure 4-3 shows a schematic of a reeled-tubing-conveyed downhole assembly. Included, from top to bottom, are the coiled tubing with logging cable, the cable head assembly, a tension/compression sub, a swivel sub, a flex-joint sub, a centralizer sub, a logging tool, another flex-joint sub, another centralizer sub, another logging tool, and a guide sub.

The injector assembly of the reeled-tubing unit pushes the tubing and tool string into the wellbore and later retrieves them. Electrical power and commands are sent downhole through the wireline cable, and the logging data is transmitted uphole.

Well Logging 133

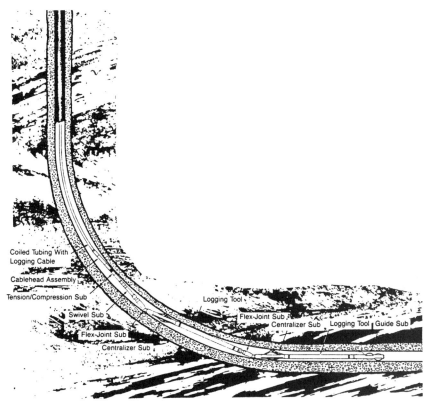

Figure 4-3. Reeled-tubing-conveyed downhole assembly (Courtesy of Halliburton, 1990).

Advantages

Many logging, perforating and auxiliary services can be performed, some of which are outlined in Table 4-1. Full-size standard logging tools can be employed in the operation.

The special flex-joints between logging tools and the flexibility of the reeled-tubing allows the assembly to traverse small radius deviations. Furthermore, rollers in the tools facilitate transport.

Because the tools are attached to reeled-tubing which is pushed and pulled by a tubing injector attached to the wellhead, no rig is required. Logging can be carried out in new or existing wells.

Centralization of the tools is possible. The tools can be oriented with respect to the low side of the wellbore, or they can be permitted to ride the low

Table 4-1
Reeled-Tubing-Conveyed Wireline Services

Logging	Perforating
Casing Collar Locator Natural Gamma Ray Spectral Natural Gamma Ray Single Detector Neutron Dual Detector Neutron Pulsed Neutron Capture Multiple Radioactive Tracer Fluid Travel Temperature Pressure Spinner Fluid Density Fluid Capacitance Conventional Cement Bond Ultrasonic Cement Bond Casing Inspection Caliper Depth Determination	Strip Guns Hollow Carrier Through-Tubing Guns Hollow Carrier Casing Guns Tubing Punchers Tubing Cutters Casing Cutters Drillpipe Cutters Severing Tools **Auxiliary Services** Junk Baskets Gauge Rings Bridge Plugs Free Point/Pipe Recovery

Source: Halliburton, 1990A.

side of the borehole. This radial positioning is very important when orientation of perforating guns is desired.

Circulation can be maintained during wireline operations. This allows lower borehole temperatures and cleaner fluids, which lead, in turn, to more reliable tool performance.

Should the operation require it, the tool string can be quickly released from the tubing and cable. This would leave a standard fishing neck exposed at the top of the tool string.

System Specifications

Some reeled-tubing and wireline specifications are presented in Table 4-2, including data for $1^{1}/_{4}$-inch and $1^{1}/_{2}$-inch OD reeled-tubing, and $5/_{16}$-inch and $7/_{16}$-inch OD electric lines (Halliburton, 1990B).

Factors such as well deviation, dogleg severity, length of the interval to be logged, borehole size, and rugosity must be carefully considered when designing a reeled-tubing-conveyed job. This will allow determination of critical operational factors, from which it is possible to anticipate the maximum forces that can be safely imparted to the tubing while logging the well.

Well Logging   135

Table 4-2
Reeled-Tubing-Conveyed System Specifications

	Reeled Tubing			
	1¼ in. OD		1½ in. OD	
Reel Capacity	15,000 ft	4572 m	15,000 ft	4572 m
Outside Diameter	1.25 in.	31.75 mm	1.50 in.	38.10 mm
Inside Diameter	1.084 in.	27.53 mm	1.282 in.	32.56 mm
Weight (in air)	1.034 lbm/ft	1.539 kg/m	1.619 lbm/ft	2.409 kg/m
Yield Load Rating	21,300 lbf	94,700 N	33,320 lbf	148,200 N
Burst Pressure Rating	9,960 psi	68,700 kPa (700 kg/cm²)	12,430 psi	85,700 kPa (874 kg/cm²)
Collapse Pressure Rating	8,800 psi	60,700 kPa (619 kg/cm²)	9,430 psi	65,000 kPa (663 kg/cm²)
Volume Capacity (w/o logging cable)	47.94 gal/1000 ft	595.4 L/km	67.06 gal/1000 ft	832.8 L/km
Running Speed				
Into hole	100–220 ft/min	30.5–67.1 m/min	100–200 ft/min	30.5–67.1 m/min
Out of hole	20–220 ft/min	6.1–67.1 m/min	20–220 ft/min	6.1–67.1 m/min
Injector Capability				
Push	24,000 lbf	106,800 N	24,000 lbf	106,800 N
Pull	24,000 lbf	106,800 N	24,000 lbf	106,800 N
	Electric Line			
	5/16 in. OD		7/16 in. OD	
Outside Diameter	5/16 in.	7.94 mm	7/16 in.	11.11 mm
Temperature Rating	350°F	177°C	400°F	204°C
Number of Conductors	One	One	Seven	Seven
Weight (in air)	189 lbm/100 ft	281 kg/km	305 lbm/100 ft	454 kg/km

Source: Halliburton, 1990A.

MEASUREMENT WHILE DRILLING (MWD)

These systems provide measurements of downhole data from sensors mounted in collars directly above the drill bit. The data are transmitted to the surface using a mud pulse telemetry system.

MWD data are very useful while drilling horizontal wells because they allow efficient steering of downhole motors, provide early information related to basic geological and reservoir engineering parameters, and give critical information conditions such as overpressures (Halliburton, 1990B).

MWD sensors are placed in nonmagnetic collars. These include the gamma-directional and the resistivity-gamma-directional combinations. The directional sensors measure toolface orientation azimuth and inclination.

Other sensors commonly available in MWD systems include conventional and focused gamma ray, resistivity, porosity, temperature, and pressure sensors.

136 Horizontal Wells

Lateral and bit resistivity measurements can give an indication with respect to overpressuring, and when combined with porosity give estimates of water saturation from conventional equations.

An interval temperature sensor is used to correct wellbore directional measurements for temperature. An annulus sensor is used to correct resistivity measurements for temperature. Local geothermal gradients and potentially overpressured zones are indicated by the temperature sensors.

Downhole pressure is measured directly by sensitive pressure transducers. This allows determination of borehole-assembly and swab-surge pressures, pressure drop across the bit, and equivalent circulating density.

LOG INTERPRETATION

Evaluation of horizontal wells is different from that of deviated or vertical wells (Struyk and Poon, 1990; Gianzero et al., 1990). The problem arises from the design of conventional logging tools which assumes symmetry in the formation around the wellbore.

This assumption is reasonable in most vertical wells. In horizontal wells, however, this is not necessarily the case because the formation above, below, and on the sides of the wellbore can change considerably. Thus, there is lack of symmetry around the wellbore and the use of logging tools that are radially averaged, such as neutron, gamma ray, laterolog, and induction requires special considerations when evaluating horizontal wells.

On the other hand, directionally focused logging devices such as sidewall neutron, density, and dipmeter are very useful for evaluating horizontal wells if the location of the pads is known (Struyk and Poon, 1990).

Differences Between Horizontal and Vertical Wells

The discussion presented in this section follows very closely the original work of Struyk and Poon (1990). Subjects covered include borehole rugosity, mud cake, anisotropy, invasion, bed boundaries intersecting the wellbore and away from the wellbore, true vertical depth, and natural fractures.

Borehole Rugosity. In all probability, the ceilings of horizontal wells are more rugose than the floor or the sides. Consequently, better results can be obtained if pad devices are directed to the floor and preferentially to the sides. Single pad devices tend to read along the low side of the wellbore due to their weight distribution. Thus, a single caliper registers hole diameters between the floor and ceiling. The borehole rugosity shown by this caliper consequently does not provide a valid check on the quality of the pad reading.

Well Logging   137

Mud Cake. Solids settled on the floor of the wellbore, also referred to as a "cuttings bed," can be easily confused with mud cake along the floor of the hole. Under these circumstances, it is difficult to obtain good qualitative indications of permeability from microlog tools. In principle, this problem could be solved by directing the pads in such a way that they read along the sides of the borehole. This is reasonable because the sides are not as affected by rugosity or the cuttings bed.

Anisotropy. The effects of anisotropy on the logging responses can be unexpected and difficult to interpret because the readings are directly influenced by the vertical components of the formation parameters.

Invasion. Historically, the invasion profile of a vertical well has been modeled by assuming a cylinder that surrounds the wellbore. This does not necessarily apply to horizontal wells.

In single-porosity conventional reservoirs without natural fractures, the value of horizontal permeability is generally larger than the vertical permeability. For this situation the invasion profile in the horizontal well is probably elliptical about the axis of the borehole.

In the case of bed boundaries and changes in permeability, complex invasion profiles are probably created due to the anisotropic nature of the formation around the wellbore. This is illustrated in Figure 4-4, which presents the perceived invasion profile in a horizontal well at different times. The invasion is probably elliptical and increases when in contact with an aquifer.

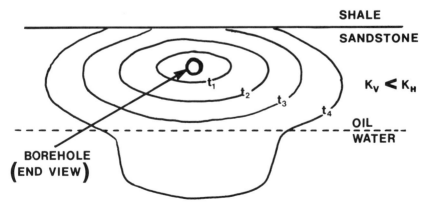

Figure 4-4. Perceived invasion profile in a horizontal well at time t_1, t_2, t_3, t_4. Source: Struyk and Poon (1990).

138 Horizontal Wells

Bed Boundaries. It is important to properly determine the location of the bed boundaries to qualify the success of the horizontal well with respect to the actual production area.

Bed Boundaries Intersecting the Borehole. Logging measurements in a horizontal well will not adequately reveal low angle boundaries and fluid contacts. The boundary indicated in a log depends on the tool's depth of investigation, measurement bias, measurement resolution, and direction of the tool reading. Thus, logs might give the impression of being off depth with each other. For proper log correlation it is important to run a base log such as a gamma ray on each logging run.

A tool that seems to accurately locate bed boundaries in horizontal wells at this writing is the four-or six-arm dipmeter.

Bed Boundaries Away From the Wellbore. Having good information about bed boundaries that do not intersect the wellbore is very important in horizontal wells. A general and oversimplified assumption is that the formation tops form a straight line between two vertical wells. This is not necessarily correct.

A horizontal well permits, in some instances, keeping track of these bed boundaries, allowing determination of their position with respect to nearby fluid contacts or shale zones. This can probably be accomplished with a dipmeter tool by taking oriented resistivity measurements with various depths of investigation.

True Vertical Depth. It is common practice to redisplay well logs of deviated wells to true vertical depth to allow sound correlation with other wells and other types of geological data. This does not appear to be very meaningful in horizontal wells. What does appear worthwhile is redisplaying horizontal well logs into true horizontal well logs, because the horizontal well is in many cases sinusoidal. With this type of work, the lateral extent of the penetrated zone can be more easily determined.

Natural Fractures. The formation microscanner (FMS) and the borehole televiewer (CBIL) have been used successfully in some cases to locate natural fractures intersected by a horizontal wellbore. Most open natural fractures are vertical or of high inclination, and show up in the microscanner and borehole televiewer as low angle or horizontal events (Bourke et al., 1989).

The response of sonic and waveform logs is different in vertical and horizontal wells. Usually, in vertical wells, the shear waves are more attenuated than the compressional waves. In horizontal wells, however, both waves are significantly attenuated, because the acoustic energy is more exposed to the vertical and high angle fractures.

Well Logging 139

If the compressional wave is sufficiently diminished, cycle skipping might develop. It must be kept in mind, however, that there are other potential reasons for cycle skipping, for example, the presence of gas or road noise. On the other hand, natural fractures might be present even if cycle skipping does not develop. This might be the result of solid to solid contact across the natural fracture, in which case the degree of acoustic discontinuity is diminished. Furthermore, modern software processing is designed to mask cycle skipping.

Types of Tool Measurements

Logging tools carry out two types of measurements: radially averaged measurement and directionally focused measurement.

The radially averaged measurement takes an average reading over a plane perpendicular to and radiating out from the borehole, as illustrated in Figure 4-5. The induction log, for example, provides this type of measurement. Radial averaging provides excellent readings as long as the formation is homogeneous and isotropic. This probably happens in most vertical wells because the plane over which an averaged reading is taken is parallel to the bedding planes.

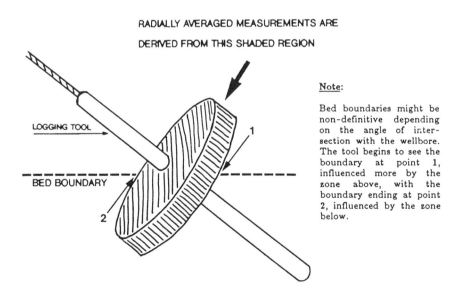

Figure 4-5. Illustration of a radially averaged logging tool. Source: Struyk and Poon (1990).

140 Horizontal Wells

In horizontal wells, however, radially averaged measurements are in a plane perpendicular to the bedding planes. The measurements are therefore carried out in a nonhomogenous, nonisotropic medium, and the radial averaging might provide unreasonable results.

On the other hand, a directionally focused measurement is normally recorded by a pad-type tool that takes readings in a particular side of the borehole, as illustrated in the schematic in Figure 4-6. The density tool, for example, provides directionally focused measurements readings in one side of the borehole, while the dipmeter takes four separate directionally focused measurements, each from a different side of the borehole. Because of their characteristics, directionally focused tools identify bed boundaries in horizontal wells much more distinctly than radially averaged measurements.

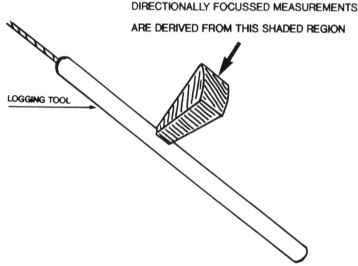

Figure 4-6. Illustration of a directionally focused logging tool. Source: Struyk and Poon (1990).

Radially Averaged Measurements. Induction and laterologs provide radially averaged measurements for which some resistivity modeling has been done in horizontal wells (Gianzero et al., 1990). This modeling indicates that the laterolog and induction tools can be used in some instances to identify nearby beds that are parallel to the borehole. However, only beds within one or two feet of the borehole can be identified. Furthermore, it is not possible to determine if the bed is located above or below the borehole.

The gamma ray, pulsed neutron, compensated neutron (mandril-type neutrons) and MWD tools are examples of radially averaged logging tools. Radially averaged measurements can be used to give an idea of the angle between the borehole and the intersecting bed, although a dipmeter would be the preferred tool for this purpose.

If a boundary is reasonably sharp on a radially averaged tool, one can assume the boundary intersects the borehole at approximately right angles. The more gradual the bed boundary appears on radially averaged logs in horizontal wells, the more parallel are the bed boundary and the borehole.

Some of the radially averaged tools tend to ride along the floor of the borehole. In the case of large diameter holes, these tools will be more influenced by the formation below the wellbore. It must also be kept in mind that laterolog readings are shifted to a higher value when run on the end of the drillpipe. This is caused by the drillpipe acting as a current return. Correction charts should be generated to account for this shift.

Directionally Focused Measurements. The density and epithermal neutron (pad-type) tools are directionally focused. The pads of these tools are heavy, and in many cases cause the tool to read along the floor of the horizontal hole. The cutting bed probably does not affect the readings because the plow-shaped design of the pad pushes the cuttings and mud cake aside.

Normally, the exact direction in which the pad is reading is not known, but it is assumed to be downwards. A directional device run in combination with these tools could be used to indicate the exact pad orientation. It may also prove useful to design a directionally focused gamma ray to read in the same direction as the density or pad-type neutron tools. Under these circumstances it would be possible to correct the density or pad-type neutron more accurately for shale effects.

One type of dipmeter consists of four directionally focused resistivity readings each at ninety degrees to the adjoining one. Six-arm dipmeters are also available. These resistivity readings can be made to approximate a shallow resistivity with a similar depth of investigation to that of a spherically focused log. Under this approximation, a dipmeter could provide resistivity information along the top, bottom, and sides of the well.

This directional resistivity log could give an indication of the proximity and location of bed boundaries. The curves along the horizontal axis could be considered as the "normal" curves, while the curves from the high and low sides of the well could be compared to these normal curves. It would be important to record tool rotations with the corresponding resistivity readings. Bed boundaries intersecting the wellbore could thus be identified by the dipmeter. Other important information from the dipmeter would include a directional survey and dip information.

The dipmeter could prove to be a very valuable resistivity tool in horizontal wells. It could prove to be even more useful if the resistivity readings could have various depths of investigation.

SIMULATION OF RESISTIVITY BEHAVIOR

As mentioned in the previous section, the induction and laterologs provide radially averaged measurements.

Gianzero et al. (1990) have modeled the resistivity behavior of induction and dual laterolog near a single horizontal bed boundary.

Dual Induction

Figure 4-7 shows the computed response of the dual induction in a horizontal well near a bed boundary. If the dual induction is far below the boundary, it simply reads the resistivity R_{LOWER} and is not affected by the upper resistivity. As the sonde approaches the formation boundary, both the medium and deep induction measurements sense the boundary long before it is reached. Obviously, at several feet above the interface, the sonde will read a resistivity equal to R_{UPPER}.

Quantifying the distance at which the sonde senses a resistive formation is complicated because the progressive resistivity transition presents "hornlike" features, as illustrated in Figure 4-7. The error in the deep induction reading is in excess of 10% when the horizontal well is within 5 ft of the resistive bed, and approximately 25% when the nearby shoulder is conduc-

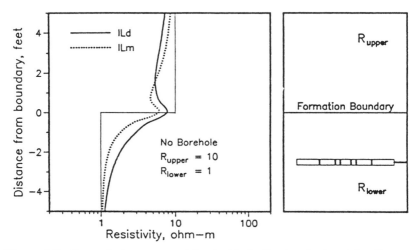

Figure 4-7. Computed response of the dual induction in a horizontal well near a bed boundary. Source: Gianzero et al. (1990).

Well Logging 143

tive. Figure 4-7 shows that these effects are less pronounced in the medium induction curve. A significant change in resistivity is a good indication of the presence of a nearby bed.

Dual Laterolog

Figure 4-8 shows the computed response of the dual laterolog in a horizontal well near a bed boundary. The simulation was carried out using a three-dimensional finite element program. The simulation indicates that a resistive bed is not identified by the dual laterolog until the sonde has actually penetrated the bed. On the other hand, a conductive bed is seen at about 0.5 ft by the shallow laterolog and at 1 ft by the deep laterolog.

A comparison of induction and laterolog responses in a horizontal well is presented in Figure 4-9. It is obvious that the resistivity measurements complement each other.

Shoulder bed corrections for the deep and medium induction logs in a horizontal well are presented in Figures 4-10 and 4-11 as crossplots of true resistivity, R_t, vs bed thickness, in feet. The values of R_s represent the shoulder resistivities. Examination of these figures indicates that the correction is an approximate monotonic function of bed thickness as long as it exceeds 4 ft. Charts for correcting the dual laterolog have also been presented by Gianzero et al. (1990).

POROSITY AND WATER SATURATION

As of this writing, all conventional equations developed throughout the years for calculating porosities and water saturations appear to be useful for evaluating reservoirs penetrated by horizontal wells. The changes appear to be only in the corrections to the input data.

CASE HISTORIES

Well logs have been instrumental for interpreting the geology of many reservoirs where horizontal wells have been drilled, as shown in the following case histories.

text continues on page 147

144 Horizontal Wells

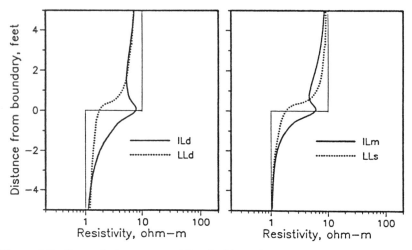

Figure 4-8. Computed response of the dual laterolog in a horizontal well near a bed boundary. Source: Gianzero et al. (1990).

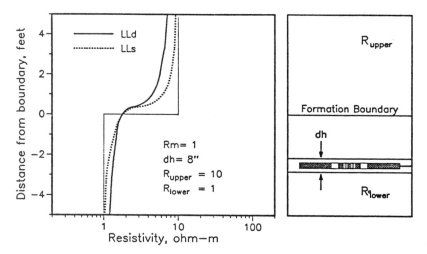

Figure 4-9. Comparison of induction and laterolog in a horizontal well. Source: Gianzero et al. (1990).

Horizontal Wells

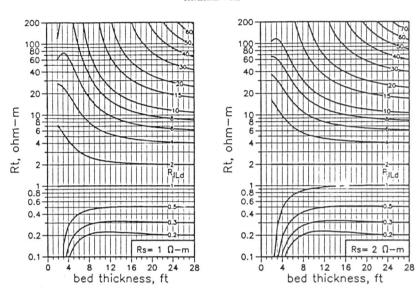

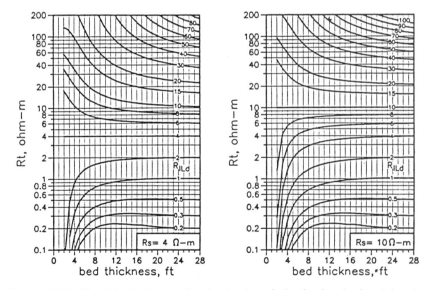

Figure 4-10. Shoulder bed correction for the deep induction in a horizontal well. Source: Gianzero et al. (1990).

146 Horizontal Wells

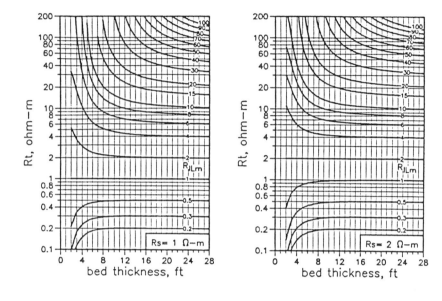
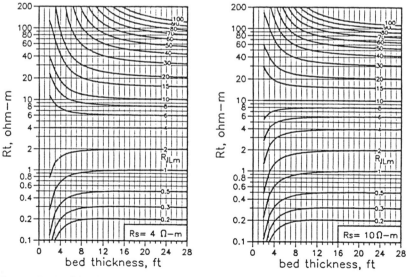

Figure 4-11. Shoulder bed correction for the medium induction in a horizontal well. Source: Gianzero et al. (1990).

Well Logging   147

text continued from page 143

Rospo Mare Field, Italy

This is a Karst-type reservoir with an intense degree of natural fracturing located offshore. Total mud losses occurred during drilling of horizontal wells and no cuttings were recovered (de Montigny et al., 1988).

Some of the logs run in this field included the sonic, natural gamma ray, dual laterolog, microspherical focused, compensated neutron, lithodensity, an auxiliary measurement tool (AMS), general purpose inclination tool (GPIT), the measurement-while-drilling (MWD) gamma ray tool, and the formation microscanner.

The AMS detected theft zones by measuring temperature and checked the advancement of the sonde by measuring the tension/compression exerted on the logging tools. The GPIT oriented the pads of the tool in the lower part of the hole. All the logs mentioned above, with the exception of the formation microscanner, can be recorded during the same run.

Production logs including full bore spinner and sensors for temperature and pressure measurements were also run successfully.

In the end, the most useful logs run in the horizontal wells to interpret the geology of Rospo Mare were as follows (de Montigny et al., 1988):

1. The formation microscanner (FMS) gave a good indication of natural fractures, vugs, and caves.
2. The compensated neutron and the lithodensity logs also pinpointed wide natural fractures and large caves.
3. The natural gamma ray, after calibration with cores, distinguished the natural fractures filled with clays from overlying shales containing thorium, uranium, and potassium from those fractures filled with weathered clays that contained only thorium.

Figure 8-4 shows a correlation of geology and well logs based on data from well Rospo Mare 6 d. The production logging information suggests that the fluid entry corresponds to Karstic voids and high inclination natural fractures identified by the well logs (de Montigny et al., 1988).

Bima Field, Indonesia

Atlantic Richfield reported on the drilling of 48 offshore horizontal wells from seven platforms in the northwest Java Sea, including 16 horizontal (90°) and high angle (85°) wells through the productive zone (Barry et al., 1988) of the Batu Raja limestone.

Openhole well logs were obtained using drillpipe conveyed logging technology. All logs were obtained in a single run, ensuring that the hole would be open for the shortest possible time prior to running a 7-inch liner.

After obtaining openhole logs in several wells, the conclusion was reached that the gamma ray and resistivity measurements obtained with MWD technology were enough and that no additional openhole well logs were required. This came as a result of the observation that the logs' signatures were repetitive in the Batu Raja limestone.

Elimination of the conventional logs resulted in a cost-saving opportunity, although no porosity logging information was obtained. It must be noted that now it is possible to obtain good porosity values with MWD systems.

Spraberry Field, Texas

After horizontal drilling operations by ARCO (Edlund, 1987) in the Spraberry field that were designed to intersect as many vertical natural fractures as possible, it became desirable to pinpoint the exact formation top and bottom.

The angle build portion of the well was logged while going in the hole at 75 ft/hr and rotating at 60 rpm. The logging rate was increased to 120 ft/hr once the horizontal portion was reached.

The logging tool incorporated into the MWD equipment provided a gamma ray which was correlated with the gamma ray obtained in a $8^{3}/_{4}$-inch vertical open hole. There was a good correlation between the two tops which indicated basically no dip. At a vertical section equal to 560 ft, the formation began to dip upwards approximately 1%. The conclusion was reached that the base of the Lower Spraberry was 12 ft higher than in the vertical borehole.

Austin Chalk, Texas

Schlumberger's formation microscanner has been reported to give good indications of natural fractures in the Austin chalk (Fett and Henderson, 1990; Stang, 1989). If there are open fractures and they are filled with salty water, they are obviously conductive and show up in the microscanner as a black feature. If they are mineralized and healed they show up in the microscanner as white. High inclination natural fractures show up as black sine waves. Multiple sine waves are indicative of multiple fractures.

Figure 4-12 shows an example of a formation microscanner in the Austin chalk in south central Texas. The log was made by attaching the logging tool to the bottom of the drillpipe and operating from a wet connector. High inclination natural fractures in the horizontal borehole appear as very dark, steeply inclined lines. A large natural fracture appears just left of X958 ft in Figure 4-12.

Well Logging 149

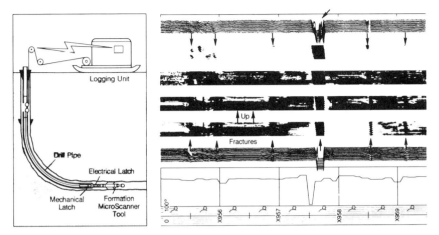

Figure 4-12. Formation microscanner of horizontal well in Austin chalk. Source: Bourke et al. (1989).

Another tool that has been reported to provide good results in horizontal wells drilled in the Austin chalk is the Western Atlas circumferential borehole imaging log (CBIL) (Western Atlas, 1990B). The tool is pipe-conveyed and a combination of centralizers and knuckle joints prevent eccentric positioning of the CBIL instrument.

Figure 4-13 shows a combination of CBIL, gamma ray, and spectralog. The spectralog was recorded while going into the hole. The CBIL and gamma ray measurements were made coming out of the hole through the entire 1,500 ft of horizontal section. Logging operations were carried out at a standard speed of 600 ft/hr.

The left side of Figure 4-13 shows a natural fracture intercepted by the horizontal wellbore. This is confirmed by the spectralog which displays an increase in uranium counts.

The right side of Figure 4-13 shows vee pattern in the CBIL image indicating that the wellbore exited the chalk and drifted into a marl. The spectralog verifies once again the presence of the chalk-marl interface.

Other Examples

Struyk and Poon (1990) have reported some interesting case histories without giving details as to the location of the horizontal wells.

Figure 4-14 shows the response of various logs in the presence of a boundary. The density log (a directionally focused tool) sharply identifies the bed boundary at A. On the other hand, the laterolog, gamma ray, and pulse neutron log-sigma (radially averaged tools) are nondefinitive in locating the

text continued on page 152

150   Horizontal Wells

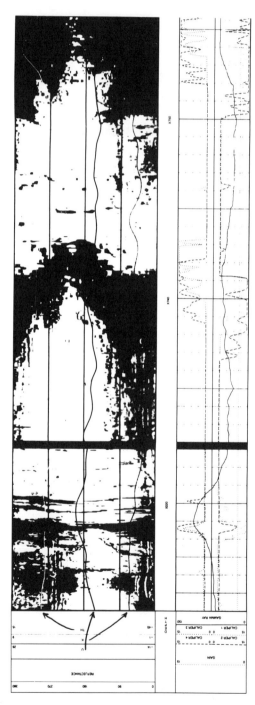

Figure 4-13. Circumferential borehole imaging log (CBIL) of horizontal well in Austin chalk. (Courtesy of Western Atlas, 1990).

Well Logging 151

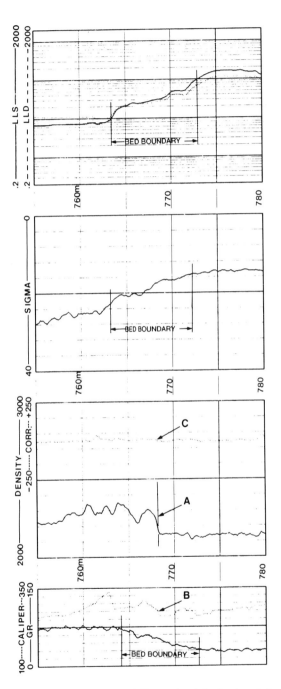

Example #1

1. Note the sharp bed boundary at 'A' on the density (directionally focussed tool) and how non-definitive this same bed boundary appears on the sigma, laterolog, and gamma ray (radially averaged tools). This confirms that the borehole and bed boundary intersect each other at a very low angle.

2. Note that rugose hole is indicated by the density caliper (B) while the density correction (C) indicates good pad contact with the wall.

Figure 4-14. Logs of Example 1 in horizontal well. Notice bed boundaries. Source: Struyk and Poon (1990).

152   Horizontal Wells

text continuted from page 149

same bed boundary. The caliper shows a rugose borehole (B), and the density correction curve suggests good pad contact with the wall.

The upper half of Figure 4-15 shows a schematic of a sinusoidal horizontal well. The formation of interest is bounded by shales above and below. A water-oil contact is also present.

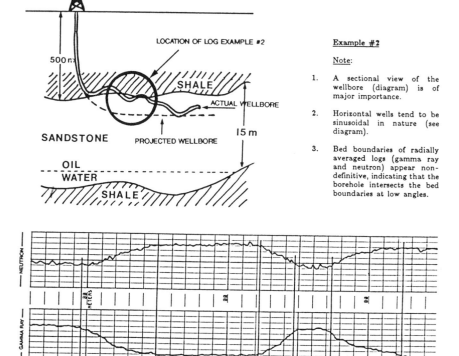

Figure 4-15. Gamma ray and neutron logs of Example 2 in horizontal well. Source: Struyk and Poon (1990).

Gamma ray and neutron logs are shown in the lower half of Figure 4-15, together with indications of the bed boundaries. Because both tools provide radially averaged logs, the bed boundaries appear nondefinitive, suggesting that the borehole intercepts the bed boundaries at low angles. The places where the wellbore drifts upwards into the overlying shale, however, are clearly delineated.

Figure 4-16 shows an example of laterolog, density, caliper, and gamma ray run in a horizontal well drilled into an oil-filled sand. The borehole and

Well Logging   153

Example #3

- Example of a horizontal well drilled into an oil-filled sand.
- Note a low angle of intersection between the borehole and bed boundary at 'A' and a high angle at 'B.'
- The anomaly on the density at 'C' was interpreted as a cemented zone.

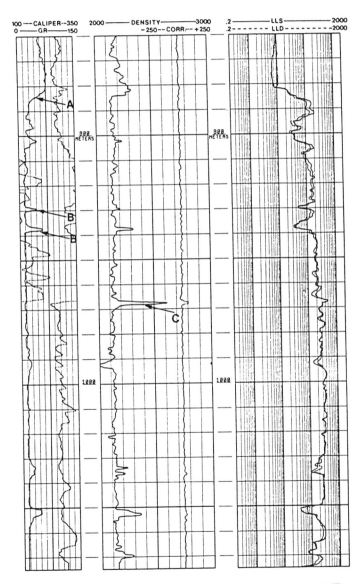

Figure 4-16. Logs of a horizontal well drilled into an oil-filled sand (Example 3). Source: Struyk and Poon (1990).

bed boundary intersect at a low angle at point A and at a high angle at point B. A similar type of conclusion could be reached by examining the density and laterologs. The anomaly on the density log at point C was interpreted as a cemented zone.

LOOKING AT THE FUTURE

In this writer's opinion, horizontal and highly deviated wells are here to stay. They can be best evaluated using directionally focused logging devices of known orientation (Struyk and Poon, 1990). The formation microscanner (Fett and Henderson, 1990) and circumferential borehole imaging logs (Western Atlas, 1990B) are excellent tools for qualitative analysis of horizontal wells and efforts are currently under way to use them for quantitative purposes.

As pointed out by Struyk and Poon (1990), horizontal well log interpretation is new and requires further study. More resistivity modeling is needed to supplement the efforts of Gianzero et al. (1990). Correction charts are required for the dual laterolog when used in conjunction with drillpipe-conveyed logging. Mechanical devices will help to orient directionally focused tools to the sides of the borehole. A four (or more) pad resistivity tool with different depths of investigation should prove useful. Invasion profiles will become better understood (Struyk and Poon, 1990).

In summary, we are just getting started, and much remains to be learned regarding logging devices and their proper interpretation in horizontal wells.

REFERENCES

1. "Tool Pusher Pipe-Conveyed Logging Systems," *Halliburton Horizontal Completions Seminar Manual* (1990A).
2. "Pipe-Conveyed Logging System," a Western Atlas Videotape (1990A).
3. *Extended-Reach and Horizontal Well Services*, a Western Atlas publication (1990B).
4. "Measurement While Drilling," *Halliburton Horizontal Completions Seminar Manual* (1990B).
5. Struyk, C. and Poon, A., "Log Interpretation in Horizontal Wells," *Horizontal Wells, Evaluation and Completion*, A Schlumberger Well Services Publication (January 1990).

Well Logging 155

6. Gianzero, S., Chemali, R. and Su, S. M., "Induction, Resistivity, and MWD Tools in Horizontal Wells," *The Log Analyst* (May-June, 1990) 158-171.
7. Bourke, L., et al., "Using Formation Microscanner Images," *The Schlumberger Technical Review* (January 1989) 16-40.
8. de Montigny, O., et al., "Horizontal-Well Drilling Data Enhance Reservoir Appraisal," *Oil and Gas Journal* (July 4, 1988) 40-48.
9. Barry, P. M., et al., "High Angle Development Wells in the N.W. Java Sea Bima Field: An Update," OSEA paper 88130 presented at the 7th Offshore Southeast Asia conference held in Singapore (February 2-5, 1988).
10. Edlund, P. A., "Application of Recently Developed Medium-Curvature Horizontal Drilling Technology in the Spraberry Trend Area," paper 16170 presented at the SPE/IADC Drilling Conference held in New Orleans, LA (March 15-18, 1987).
11. Fett, T. and Henderson, J., "Fracture Evaluation in Horizontal Wells Using the Formation Microscanner (FMS) Imaging Device," a Schlumberger presentation (1990).
12. Stang, C. W., "Alternative Electronic Logging Technique Locates Fractures in Austin Chalk Horizontal Well," *Oil and Gas Journal*, (November 6, 1989) 42-45.

5
Well Test Analysis

INTRODUCTION

The mechanical process of drilling highly accurate horizontal wells is now a common occurrence throughout the world. While there will continue to be many innovations in the realm of drilling technology, it is apparent that our practical understanding of horizontal wells in petroleum disciplines such as completion, production forecasting, and well test interpretation is far behind. The scope of this chapter will be to review the current theoretical understanding associated with pressure transient analysis in horizontal wells and to assess the practical application of these analytic solutions.

The material presented is based on a review of the available literature. There have been many groups of authors that have published articles in the area of pressure transient analysis for horizontal wells. When one compares the mathematical models introduced by these authors, it is apparent that the analytic techniques necessary for pressure transient analysis are still in the development stage. In preparing this chapter it was elected to focus primarily on the solutions developed by Goode (1985), and Kuchuk and Thambynayagam et al. (1988A & 1988B).

The majority of onshore horizontal wells are being drilled in unconventional and complex reservoirs and there is a growing awareness of the necessity to complete these wells effectively. One of the primary goals of pressure transient testing is to provide a method of differentiating between completion success and in situ reservoir quality. The following question is often asked after a horizontal well has been drilled successfully, but completed for a production rate below what was anticipated: Is the reservoir poorer than anticipated or has the wellbore been damaged?

The value of properly assessing the condition of the wellbore in relation to the reservoir through simple pressure tests can be very significant. The cost of stimulating a horizontal well is often economically prohibitive and in most cases should be avoided, if possible.

ANALYSIS PROCEDURE

It is much more difficult to interpret well test data from a horizontal well than from a vertical counterpart. The flow geometry is three-dimensional and radial symmetry no longer exists. Several flow regimes must be considered to analyze the data. Wellbore storage effects are much more significant, and horizontal wells will commonly exhibit partial penetration and end effects that make interpretation very difficult.

In the vertical realm, we are accustomed to dealing with single dimensional variables such as average permeability, net vertical thickness, and skin. With horizontal wells, this simplification is no longer available to us. Not only is vertical thickness important, but the horizontal dimensions of the reservoir, relative to the horizontal wellbore, need to be known.

It is suggested that there are three steps necessary to effectively evaluate pressure transient data from a horizontal well. The first step requires correctly identifying the specific flow regimes associated with the test data. Typically, when engineers evaluate conventional data from a vertical wellbore, the techniques they employ generally center on a single flow regime known as the middle time region. A pressure transient in a horizontal well can involve four major and distinct regimes which need to identified. These regimes may or may not exist and may or may not be obscured by wellbore storage effects, end effects, or transition effects.

The second step is to correctly apply the right analytic and graphical procedures to the data. Each flow regime has a corresponding mathematical solution which may be used to estimate various reservoir parameters. At best, only parameter groups can be determined directly from the equations. In most cases, when solving for specific parameters, the application of these analytic expressions involves a complex iterative process requiring the use of a computer.

Finally, the third step involves evaluating the uniqueness and sensitivity of the solution. One may question the wisdom of even bothering with this step, but experience suggests that the solution of horizontal well test derivations will seldom provide unique answers. These derivations, which require the aid of computers to be used effectively for practical applications, provide a foundation for establishing the key characteristics and variables associated with the test. They also help to establish the current limitations involved with horizontal well pressure transient analysis.

The purpose of this chapter is to provide the reader with an introduction to the concepts associated with horizontal well tests.

HORIZONTAL WELL MODEL

Figure 5-1 illustrates the horizontal well configuration associated with the analytic solutions that will be presented.

The physical model consists of an infinite conductivity horizontal wellbore of radius r_w and length L located in a homogeneous yet anisotropic reservoir of width h_x, length h_y, and uniform thickness h_z. A fluid of slight but constant compressibility is produced through the horizontal well. The fluid properties are independent of pressure, and gravity effects are neglected. It is also assumed that the length of the horizontal wellbore is much larger than the thickness of the formation. The pressure in the reservoir prior to producing the well is uniform and equal to P_i. The well is produced at a constant rate of q.

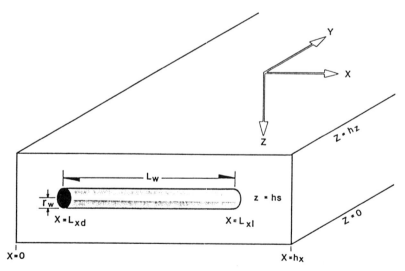

Figure 5-1. Horizontal well configuration. Source: Goode and Thambynayagam (December 1987)

FLOW REGIMES

There are four primary flow regimes that are theoretically possible with a drawdown or buildup test in a horizontal well. In many instances, these primary patterns will be distorted or eliminated because of reservoir heterogeneities, end effects, wellbore storage, or boundary effects. As the horizontal well is first produced, the pressure transient will move out perpendicular to the wellbore as illustrated in Figure 5-2, and radial flow is established. This is very similar to a vertical well in an infinite acting environment. The duration of the regime is normally very short unless the reservoir thickness is large or the vertical permeability is very low.

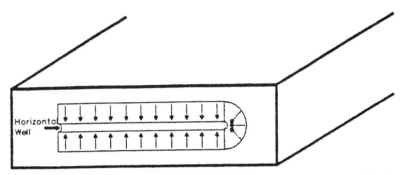

Figure 5-2. Early time radial flow: The pressure transient is moving radially from the wellbore and has not encountered any boundaries.

Once the nearest boundary is contacted by the pressure transient, a flow pattern which is hemicylindrical (Figure 5-3) is established. This type of flow is only significant when the well is close to a no-flow boundary.

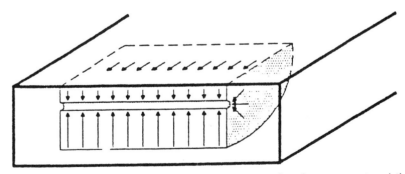

Figure 5-3. Hemicylindrical flow: The pressure transient has encountered the nearest horizontal boundary (this is not a major flow regime).

In most cases, the length of the horizontal well is much greater than the thickness of the reservoir. When this is the case, a linear flow regime will be established once the pressure transient has reached both the upper and lower boundaries of the reservoir as represented in Figure 5-4. This is the second

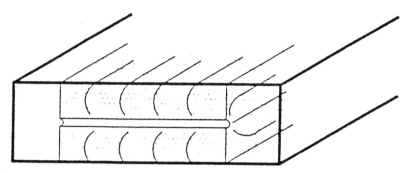

Figure 5-4. Intermediate time linear flow: The duration of this second major flow regime is directly related to the effective length of the horizontal well.

primary flow regime, which has an effective duration directly related to the onset of the end effects. If the horizontal well length is not long compared to the formation thickness, then this flow regime will not develop. Instead, a lengthy transition zone will develop prior to the next identifiable flow period.

If a constant pressure source is not present and there are no boundaries to horizontal flow over a reasonable distance, flow towards the horizontal wellbore becomes effectively radial in nature after a long enough time. This is depicted by Figure 5-5.

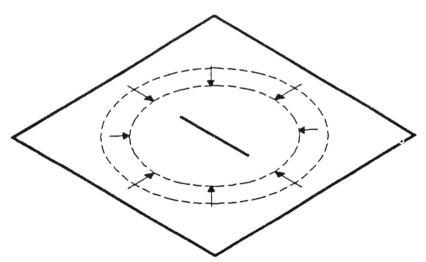

Figure 5-5. Late time radial flow: Pressure transient becomes effectively radial in nature after a long enough time.

In a semi-infinite reservoir, once the parallel boundaries have been contacted (Figure 5-6), a linear flow regime will develop.

A regime that might be established when a horizontal well is drawn down, but which is not considered common, is steady state. This will only develop when there is a constant pressure source such as an aquifer or gas cap.

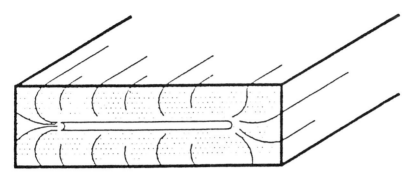

Figure 5-6. Late time linear flow: although this is the last major flow regime, it is not commonly seen in tests.

In summary, there are four primary flow regimes that can theoretically develop when a horizontal well is drawn down or built up. In chronological order of development, they are:

- Early time radial flow
- Intermediate time linear flow
- Late time radial flow
- Late time linear flow (pseudo steady state)

The identification of these flow regimes is critical to the proper interpretation of a horizontal well test. Unfortunately, even though all the flow patterns presented are mathematically possible, in many circumstances these regimes will either not exist or will be obscured.

DURATION OF FLOW REGIMES

Prior to initiating a horizontal well test or analyzing the results of one, it is suggested that the time criteria should be estimated for the various flow regimes. Goode and Thambynayagam (1987) empirically derived the following equations for estimating these time criteria. It is suggested to the reader that more accurate criteria can be mathematically extracted from equations published by Kuchuk et al. (1988A).

Early Time Radial Flow

The early time radial flow period ends at approximately:

$$t_{erf1} = \frac{190.0 \, h_s^{2.095} \, r_w^{-0.095} \, \phi\mu c_t}{k_z} \quad (5\text{-}1)$$

Intermediate Time Linear Flow

Intermediate time linear flow is estimated to end at:

$$t_{elf1} = \frac{20.8 \, \phi\mu c_t L_w^2}{k_x} \quad (5\text{-}2)$$

If the time calculated from Equation 5-2 is less than the time calculated for the early time radial flow to end (Equation 5-1), it may mean that the length of the horizontal well is not sufficient compared to the formation thickness. Thus, the intermediate time linear flow may not develop.

Late Time Radial Flow

If late time radial flow (also known as pseudo radial flow) develops, it will begin at approximately:

$$t_{brf2} = \frac{1{,}230.0 \, L_w^2 \, \phi\mu c_t}{k_x} \quad (5\text{-}3)$$

and for a reservoir of finite width, would end at:

$$t_{erf2} = 297.0 \, \frac{(L_{xl} + L_{xd})^{2.095} \, L_w^{-0.095} \, \phi\mu c_t}{k_x} \quad (5\text{-}4)$$

If the estimated time to the end of late time radial flow (Equation 5-4) is less than the calculated beginning (Equation 5-3), this may imply that the reservoir is smaller than anticipated and this radial flow period will not develop.

Example 1: Calculation of End of Early Time Radial Flow

$$t_{erf1} = \frac{190.0 \, hs^{2.095} \, r_w^{-0.095} \, \phi\mu c_t}{k_z}$$

$L_w = 2,000$ ft
$h_x = 20$ ft
$h_s = 10$ ft (horizontal well in center of reservoir)
$r_w = 0.35$ ft
$\phi = 0.10$
$c_t = 12 \times 10^{-6}$ psi^{-1}
$\mu = 0.4$ cp
$k_x = k_y = 15$ md
$k_z = 2$ md

$$t_{erf1} = \frac{190.0 \, (10)^{2.095} \, (0.35)^{-0.095} \, (0.10) \, (0.4) \, (12 \times 10^{-6})}{2}$$

$= 0.0063$ hrs.

This simple example illustrates that early time radial flow normally cannot be seen on a pressure test unless the reservoir is very thick, or the fluid very viscous, or the vertical permeability very poor.

Example 2: Calculation of End of Intermediate Time Linear Flow

$$t_{elf1} = \frac{20.8 \, \phi\mu c_t L_w^2}{k_x}$$

$L_w = 2,000$ ft
$h_x = 20$ ft
$h_s = 10$ ft
$r_w = 0.35$ ft
$\phi = 0.10$
$c_t = 12 \times 10^{-6}$ psi^{-1}
$\mu = 0.4$ cp
$k_x = k_y = 15$ md
$k_z = 2$ md

$$t_{elf1} = \frac{20.8 \, (0.10) \, (0.4) \, (12 \times 10^{-6}) \, (2,000)^2}{15}$$

$= 2.6624$ hrs.

Example 3: Calculation of Beginning of Pseudo Radial Flow

$$t_{brf2} = \frac{1,230.0 \, L_w^2 \, \phi\mu c_t}{k_x}$$

Horizontal Wells

$L_w = 2,000$ ft
$h_x = 20$ ft
$h_s = 10$ ft
$r_w = 0.35$ ft
$\phi = 0.10$
$c_t = 12 \times 10^{-6}$ psi^{-1}
$\mu = 0.4$ cp
$k_x = k_y = 15$ md
$k_z = 2$ md

$$t_{brf2} = \frac{1,230.0 \ (2,000)^2 \ (0.10) \ (0.40) \ (12 \times 10^{-6})}{15}$$

$$= 157.44 \text{ hrs.}$$

A comparison of the equations that govern the end of intermediate time linear flow (Equation 5-2) and the beginning of pseudo radial flow (Equation 5-3) indicates a simple multiplicity of 59.13. The beginning of pseudo radial flow will occur 59.13 times later than the end of intermediate linear flow.

Example 4: Well Test Planning Using the Flow Time Equations

An 8½-inch, 2,000 ft horizontal well has been drilled into a thin fractured source rock. The following data applies:

Reservoir thickness: 10 ft
Porosity: 3%
c_t: 50×10^{-6} (the fracture system creates a system with a very high compressibility)
μ: 0.2 cp
$k_x = k_y = k_z$: 1.0 md

• $t_{elf1} = \dfrac{190 \ (5)^{2.095} \ (0.35)^{-0.095} \ (0.03) \ (0.2) \ (50 \times 10^{-6})}{1.0}$

$= 0.0018$ hrs.

• $t_{elf1} = \dfrac{20.8 \ (0.03) \ (0.2) \ (50 \times 10^{-6}) \ (2,000)^2}{1.0}$

$= 24.96$ hrs.

Well Test Analysis   165

- $t_{brf2} = \dfrac{1{,}230.0\ (0.03)\ (0.2)\ (50 \times 10^{-6})\ (2{,}000)^2}{1.0}$

  $= 1{,}476$ hrs.

In this situation, a two-day drawdown or buildup test using a bottomhole shut-off device and accurate bottomhole gauges would be recommended. There is little advantage to a longer test.

Example 5: Well Test Planning Using the Flow Time Equations

An 8½-inch, 2,500 ft horizontal well is to be drilled into a low permeability oil reservoir that could not be economically developed with vertical wells. Analysis of data for the area suggests the following:

Reservoir thickness:	25–40 ft
Porosity:	8–12%
Environment:	consolidated sandstone (c_t approximately 8×10^{-6} psi^{-1})
Permeability:	approximately 0.5 md vertically and less than 3.0 md horizontally
Oil viscosity:	approximately 0.5 cp

Early time radial flow will end approximately 0.11 hr after the start of flow, based on Equation 5-1. The maximum values of reservoir thickness and porosity were assumed. It was also assumed that the well would be drilled in the center of the reservoir. 0.11 hr represents a maximum value.

Intermediate time linear flow will end approximately 14 hr after the start of flow (Equation 5-2). In this case, the minimum value of porosity was used because this flow regime is not immediately followed by any distinct flow regime. Rather, a long period of transition occurs prior to the next regime, which is late time radial flow. From Equation 5-3, it is estimated that late time radial flow will begin approximately 820 hr after the end of intermediate linear flow.

In this situation, an effective and economic well test may be difficult to conduct. It is highly probable that storage effects will obscure both the early time radial flow and the intermediate time linear flow periods. Because late time radial flow will not develop for at least 820 hr, it is also unlikely that a drawdown or buildup test would be conducted for that length of time.

Example 6: Well Test Planning Using the Flow Time Equations

A 1,500 ft horizontal well is drilled through the center of a thick, highly fractured reservoir. There is no effective matrix.

$h_x = 120$ ft
$\phi = 2\%$
$c_t = 20 \times 10^{-6}$ psi^{-1}
$\mu = 0.6$ cp
$k_x = k_z = 190$ md
$r_w = 0.35$ ft

- $t_{erf1} = \dfrac{190 \, (60)^{2.095} \, (0.35)^{-0.095} \, (0.02) \, (0.6) \, (20 \times 10^{-6})}{190}$

  $= 0.0014$ hrs.

- $t_{elf1} = \dfrac{20.8 \, (0.02) \, (0.6) \, (20 \times 10^{-6}) \, (1{,}500)^2}{190}$

  $= 0.0591$ hrs.

- $t_{brf2} = \dfrac{1{,}230 \, (0.02) \, (0.6) \, (20 \times 10^{-6}) \, (1{,}500)^2}{190}$

  $= 3.5$ hrs.

A short test would readily develop into radial flow, and a test of 1–7 days could evaluate reservoir size and boundaries.

Other authors have also developed similar analytic expressions for estimating the time criteria associated with these flow periods. Odeh and Babu (1989) provided the following:
Early Time Radial Flow ends at the minimum of

$$t_{erf1} = \frac{1{,}800 \, d_z^2 \, \phi\mu c_t}{k_z} \qquad (5\text{-}5)$$

or

$$t_{erf1} = \frac{125 \, L_w^2 \, \phi\mu c_t}{k_x}$$

Intermediate Time Linear Flow starts at

$$t_{blf1} = \frac{1{,}800 \, D_z \, \phi\mu c_t}{k_z} \qquad (5\text{-}6)$$

and ends at

$$t_{elf1} = \frac{160 \ L_w^2 \ \phi\mu c_t}{k_x} \quad (5\text{-}7)$$

Late Time Radial Flow starts at

$$t_{brf2} = \frac{1,480 \ L_w^2 \ \phi\mu c_t}{k_x} \quad (5\text{-}8)$$

and ends at the minimum of

$$t_{erf2} = \frac{2,000 \ \phi\mu c_t \left(d_x + \frac{L_w}{4}\right)^2}{k_x} \quad (5\text{-}9)$$

or

$$t_{erf2} = \frac{1,650 \ \phi\mu c_t \ d_y^2}{k_y}$$

Late Time Pseudo Steady State (Linear) ends at the larger of

$$t_{blf2} = \frac{4,800 \ \phi\mu c_t \ D_x^2}{k_x} \quad (5\text{-}10)$$

or

$$t_{blf2} = \frac{1,800 \ D_z^2 \ \phi\mu c_t}{k_z}$$

PRESSURE RESPONSE FUNCTIONS

It is important to estimate the times relating to each of the flow regimes because each flow regime has a unique mathematical solution relating to it, and certain reservoir parameters can only be approximated during particular flow regimes. Goode and Thambynayagam developed generalized mathematical response functions at the horizontal wellbore for conditions of both pressure drawdown and pressure buildup. These general equations, published in 1985, assumed an effective pressure point along the horizontal wellbore. Later work by Kuchuk et al. (1988A & 1988B) was based on pressure averaging.

168 Horizontal Wells

Under conditions of pressure drawdown, the following analytic expressions apply. These equations provide an excellent summary of the key variables associated with each flow regime.

Analytic Solutions Under Pressure Drawdown
(Goode and Thambynayagam)

General Solution

$$p_i - p_{wf} = \frac{282.4 q \mu B_o r_w'}{h_x h_z k_y} \left[\sqrt{\pi t_D} + \frac{h_x^2}{\pi^2 v_x} \sum_{n=1}^{\infty} \frac{1}{n} \right.$$

$$\times \operatorname{erf}(v_x \pi n \sqrt{t_D}) \, \Xi_n^2 + \frac{h_x h_z}{L_w' v_z \pi} \sum_{m=1}^{\infty} \frac{1}{m}$$

$$\left. \times \operatorname{erf}(v_z \pi m \sqrt{t_D}) \, \Xi_m \cos(m\pi z_e) + \frac{h_z h_x \sqrt{k_y/k_z}}{2 r_w' L_w} S_m \right] \quad (5\text{-}11)$$

where $\Xi_n = \dfrac{1}{nL_w} \left[\sin\left(\dfrac{n\pi L_{xl}}{h_x}\right) - \sin\left(\dfrac{n\pi L_{xd}}{h_x}\right) \right]$

$$\Xi_m = \frac{1}{4 m r_w'} \left\{ \sin\left[\frac{m\pi}{h_z}(h_s + 2r_w')\right] - \sin\left[\frac{m\pi}{h_z}(h_s - 2r_w')\right] \right\}$$

$$z_e = \frac{1}{h_z}(h_s + 1.47 r_w')$$

$$t_D = \frac{0.000264 k_y t}{\phi \mu c_t r_w'^2}$$

$$v_x = \frac{r_w'}{h_x} \sqrt{\frac{k_x}{k_y}}$$

$$v_z = \frac{r_w'}{h_z} \sqrt{\frac{k_z}{k_y}}$$

$$L_w = (L_{xl} - L_{xd})$$

$$r'_w = r_w \left(\frac{k_z}{k_y}\right)^{1/4}$$

$$\Delta t = t - t_0$$

Early Time Radial Flow

$$p_i - p_{wf} = \frac{162.6q\mu B_o}{\sqrt{k_zk_y}\, L_w} \left[\log_{10}\left(\frac{\sqrt{k_yk_z}\, t}{\phi\mu c_t r_w^2}\right) - 3.227 + 0.868 S_m\right] \quad (5\text{-}12)$$

Intermediate Time Linear Flow

$$p_i - p_{wf} = \frac{8.128qB_o}{L_wh_z}\sqrt{\frac{\mu t}{k_y\phi c_t}} + \frac{141.2q\mu B_o}{L_w \sqrt{k_yk_z}}(S_z + S_m) \quad (5\text{-}13)$$

Late Time Radial Flow

$$p_i - p_{wf} = \frac{162.6q\mu B_o}{\sqrt{k_xk_y}h_z}\left[\log_{10}\left(\frac{k_x t}{\phi\mu c_t L_w^2}\right) - 2.023\right]$$

$$+ \frac{141.2q\mu B_o}{L_w \sqrt{k_yk_z}}(S_z + S_m) \quad (5\text{-}14)$$

Late Time Pseudo Steady State

$$p_i - p_{wf} = \frac{8.128qB_o}{h_xh_z}\sqrt{\frac{\mu t}{k_y\phi c_t}} + \frac{141.2q\mu B_o}{L_w \sqrt{k_yk_z}}(S_x + S_z + S_m) \quad (5\text{-}15)$$

In the case of a pressure buildup test, similar equations were developed for two scenarios. The first, and probably most common, is where the reservoir is effectively infinite acting during the course of the test. The second is where the pressure transient has reached the reservoir boundaries, and the late time pseudo steady state was established before the well was shut in.

Horizontal Wells

Analytic Solutions Under Pressure Buildup
(Goode and Thambynayagam)

General Solution

$$p_i - p_{wf} = \frac{282.4 q \mu B_o r'_w}{h_x h_z k_y} \left\{ \sqrt{\pi} \left(\sqrt{t_D} - \sqrt{\Delta t_D} \right) + \frac{h_x^2}{\pi^2 v_x} \sum_{n=1}^{\infty} \frac{1}{n} \right.$$

$$\times \left[\text{erf} \left(v_x \pi n \sqrt{t_D} \right) - \text{erf} \left(v_x \pi n \sqrt{\Delta t_D} \right) \right] \Xi_n^2 + \frac{h_x h_z}{L_w v_z \pi} \sum_{m=1}^{\infty} \frac{1}{m}$$

$$\left. \times \left[\text{erf} \left(v_z \pi m \sqrt{t_D} \right) - \text{erf} \left(v_z \pi m \sqrt{\Delta t_D} \right) \right] \Xi_m \cos \left(m \pi z_e \right) \right\}$$

(5-16)

where $\Xi_n = \dfrac{1}{nL_w} \left[\sin \left(\dfrac{n \pi L_{xl}}{h_x} \right) - \sin \left(\dfrac{n \pi L_{xd}}{h_x} \right) \right]$

$\Xi_m = \dfrac{1}{4 m r'_w} \left\{ \sin \left[\dfrac{m \pi}{h_z} (h_s + 2 r'_w) \right] - \sin \left[\dfrac{m \pi}{h_z} (h_s - 2 r'_w) \right] \right\}$

$z_e = \dfrac{1}{h_z} (h_s + 1.47 r'_w)$

$t_D = \dfrac{0.000264 k_y t}{\phi \mu c_t r'^2_w}$

$v_x = \dfrac{r'_w}{h_x} \sqrt{\dfrac{k_x}{k_y}}$

$v_z = \dfrac{r'_w}{h_z} \sqrt{\dfrac{k_z}{k_y}}$

$L_w = (L_{xl} - L_{xd})$

$r'_w = r_w \left(\dfrac{k_z}{k_y} \right)^{1/4}$

$\Delta t = t - t_0$

Early Time Radial Flow

infinite reservoir

$$p_i - p_{wf} = \frac{162.6q\mu B_o}{\sqrt{k_z k_y}\, L_w} \left[\log_{10}\left(\frac{t_0 + \Delta t}{\Delta t}\right) + \gamma 1 \right] \tag{5-17}$$

where $\gamma 1 = \dfrac{L_w}{h_z}\sqrt{\dfrac{k_z}{k_x}}\left[\log_{10}\left(\dfrac{k_x t}{\phi \mu c_t L_w^2}\right) - 2.023\right] - \log_{10}(t)$

$$- \log_{10}\left(\frac{\sqrt{k_z k_y}}{\phi \mu c_t r_w^2}\right) + 3.227 + 0.868 S_z$$

bounded reservoir

$$p_i - p_{wf} = \frac{162.6q\mu B_o}{\sqrt{k_y k_z}\, L_w} \left[\log_{10}\left(\frac{t_0 + \Delta t}{\Delta t}\right) + \gamma 2 \right] \tag{5-18}$$

where $\gamma 2 = \dfrac{0.05\, L_w}{h_z h_x}\sqrt{\dfrac{k_z t}{\phi \mu c_t}} - \log_{10}\left(\dfrac{\sqrt{k_y k_z t}}{\phi \mu c_t r_w^2}\right)$

$$+ 3.227 + 0.868\, (S_x + S_z)$$

Intermediate Time Linear Flow

infinite reservoir

$$p_i - p_{wf} = \frac{8.128 q B_o}{h_z L_w}\sqrt{\frac{\mu \Delta t}{k_y \phi c_t}} + \gamma 3 \tag{5-19}$$

where $\gamma 3 = \dfrac{162.6 q \mu B_o}{h_z \sqrt{k_x k_y}}\left[\log_{10}\left(\dfrac{k_x t}{\phi \mu c_t L_w^2}\right) - 2.023\right]$

172   Horizontal Wells

bounded reservoir

$$p_i - p_{ws} = \frac{8.128qB_o}{h_z L_w}\sqrt{\frac{\mu}{k_y \phi c_t}}\left(\sqrt{\Delta t} - \frac{L_w \sqrt{t}}{h_x}\right) + \frac{141.2q\mu B_o}{L_w \sqrt{k_y k_x}} S_x \quad (5\text{-}20)$$

Late Time Radial Flow

infinite reservoir

$$p_i - p_{ws} = \frac{162.6q\mu B_o}{h_z \sqrt{k_x k_y}}\left[\log_{10}\left(\frac{t_0 + \Delta t}{\Delta t}\right)\right] \quad (5\text{-}21)$$

bounded reservoir

$$p_i - p_{ws} = \frac{162.6q\mu B_o}{h_z \sqrt{k_x k_y}}\left[\log_{10}\left(\frac{t_0 + \Delta t}{\Delta t}\right) + \gamma 4\right] \quad (5\text{-}22)$$

where $\gamma 4 = \dfrac{0.05}{h_x}\sqrt{\dfrac{k_x t}{\phi \mu c_t}} - \log_{10}\left(\dfrac{k_x t}{\phi \mu c_t L_w^2}\right) + 2.023 + 0.868 S_x$

Late Time Pseudo Steady State

bounded reservoir

$$p_i - p_{ws} = \frac{8.128qB_o}{h_z h_x}\sqrt{\frac{\mu}{k_y \phi c_t}}(\sqrt{t} - \sqrt{\Delta t}) \quad (5\text{-}23)$$

It is apparent from the analytic solutions that it is very difficult to solve directly for specific reservoir parameters. At best, we are able to determine only the product of variables. Tables 5-1 and 5-2 summarize the methods necessary to solve for these parameter sets for drawdown and buildup tests, and show during which flow regime these solutions exist.

In general, the pressure transient plotted as the wellbore pressure versus the logarithm of time will yield two straight lines corresponding to the early time and late time radial flow regimes. The pressure transient plotted as the wellbore pressure versus the square root of a time could also yield two straight lines. The first corresponds to intermediate time whereas the second

Well Test Analysis   173

Table 5-1
Pressure Drawdown

Flow Regime	Reservoir Parameter	Plot	Calculation
Early Time Radial	$\sqrt{k_z k_y}$	P_{wf} or $(P_i - P_{wf})$ vs log t	$\sqrt{k_z k_y} = \dfrac{162.6\, q\mu B_o}{m\, L_w}$
	S_m	P_{wf} or $(P_i - P_{wf})$ vs t	$S_m = 1.15 \left[\dfrac{(P_i - P_{wf})}{m} + 3.227 \right]$ where P_{wf} is at 1 hour
Intermediate Time Linear	$\dfrac{\phi c_i k_y}{\mu}$	P_{wf} or $(P_i - P_{wf})$ vs $\sqrt{t}$	$\sqrt{\dfrac{\phi c_i k_y}{\mu}} = \dfrac{8.128\, qB_o}{mL_w h_z}$
Late Time Radial	$\sqrt{k_x k_y}$	P_{wf} or $(P_i - P_{wf})$ vs log t	$\sqrt{k_x k_y} = \dfrac{162.6\, q\mu B_o}{mL_w}$
Late Time Linear (Pseudo Steady State)	$\dfrac{\phi c_i k_y}{\mu}$ or h_x	P_{wf} or $(P_i - P_{wf})$ vs $\sqrt{t}$	$\sqrt{\dfrac{\phi c_i k_y}{\mu}} = \dfrac{8.128\, qB_o}{mh_x h_z}$ or $h_x = \dfrac{8.12\, qB_o}{mh_z}\, \dfrac{\sqrt{\mu}}{\sqrt{\phi c_i k_y}}$

Table 5-2
Pressure Buildup

Flow Regime	Reservoir Parameter	Plot	Calculation
Early Time Radial	$\sqrt{k_z k_y}$	ΔP vs $\log \frac{(t_o + \Delta t)}{\Delta t}$	$\sqrt{k_z k_y} = \frac{162.6\, q\mu B_o}{mL_w}$
Intermediate Time Linear	$\sqrt{\dfrac{\mu}{k_y \phi c_t}}$	ΔP vs $\sqrt{\Delta t}$	$\sqrt{\dfrac{\mu}{k_y \phi c_t}} = \dfrac{8.128\, q B_o}{m h_z L_w}$
Late Time Radial	$\sqrt{k_z k_y}$	ΔP vs $\log \frac{(t_o + \Delta t)}{\Delta t}$	$\sqrt{k_z k_y} = \dfrac{162.6\, q\mu B_o}{m h_z}$

Generally, the initial part of this Horner plot will be a straight line. If the reservoir is of finite width, and the pressure transient has reached the final linear flow period before shut-in, and when the inequality $t_o + \Delta t$ becomes invalid, this Horner relationship will no longer produce a straight line.

Flow Regime	Reservoir Parameter	Plot	Calculation
Late Time Linear (only exists for a finite width reservoir)	$\sqrt{\dfrac{\phi c_t k_y}{\mu}}$ or h_x or P_i	P vs $\sqrt{t} - \sqrt{\Delta t}$ ΔP vs $\sqrt{t} - \sqrt{\Delta t}$	extrapolate to P_i $\sqrt{\dfrac{\phi c_t k_y}{\mu}} = \dfrac{8.128\, q B_o}{m h_z h_x}$

Well Test Analysis   175

would relate to late time pseudo steady state. Odeh and Babu (1989) correctly stressed that not all four flow regimes will automatically exist during every pressure test.

PRESSURE TRANSIENT CHARACTER

Experience to date suggests that horizontal well tests are best interpreted using a log-log representation of the pressure-time data in conjunction with the derivative curve. This representation provides the best distinction between the various flow regimes, which not only improves the ability to correctly identify the flow regimes, but also maximizes the chance for regressing to a more unique solution of the data.

Figure 5-7 illustrates the character of the log-log plot in relation to the two radial flow regimes under ideal conditions. Following the effects of

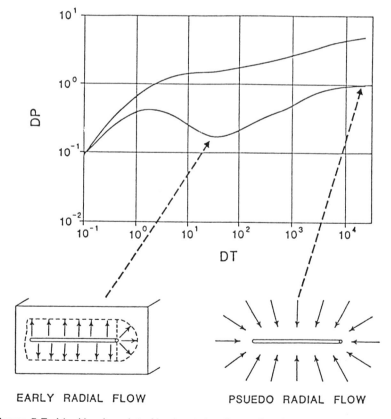

Figure 5-7. Ideal log-log plot of horizontal well test showing unique fingerprint of major flow regimes.

176 Horizontal Wells

wellbore damage, they are associated with the minimum and maximum flexures (i.e., zero slopes) on the derivative curve. The intermediate time linear flow period is theoretically located as a constant slope line between these two radial flexures. Each of the flow regimes, which ideally establish a unique fingerprint on the log-log plot, provides an opportunity for estimating particular reservoir parameters that are much more difficult (and often impossible) to determine during other flow regimes (Figure 5-8).

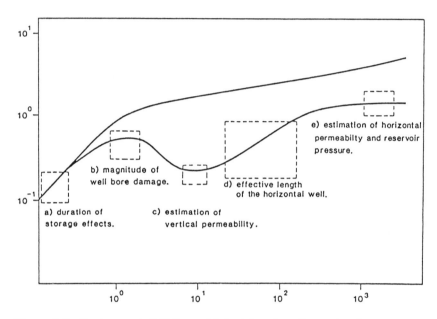

Figure 5-8. Daviau et al. (1985) established that the effects of wellbore storage often prevent early time radial flow from being seen with pressure transient data.

WELLBORE STORAGE EFFECTS

Typically, there is a substantial storage volume associated with a horizontal wellbore which can have serious consequences on the effectiveness of a pressure transient test, even when the measurement tool is located below a downhole shut-in device. In a paper published by Daviau et al. (1985), it was shown that the first semilog straight line associated with early time radial flow almost always disappears because of the effects of wellbore storage (Figure 5-9).

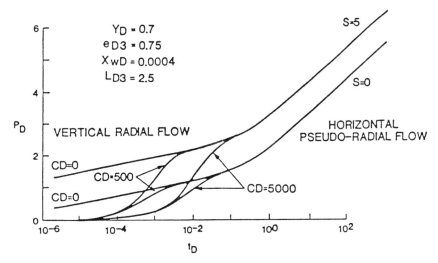

Figure 5-9. Log-log plot of horizontal well test showing radial flow regimes.

Kuchuk and Goode et al. (1988B) noted that the storage effect in a horizontal well typically lasts longer than that for a vertical well in the same formation, because of greater wellbore volume and because anisotropy reduces the effective permeability for a horizontal well.

It is significant that the first semilog straight line (early time radial flow) commonly does not appear due to wellbore storage because true wellbore damage S_m and horizontal permeability h_z can only be effectively estimated during that time. It has been suggested that this concern can be overcome by concurrently measuring downhole flow rates and pressures.

SKIN EFFECTS

It is becoming apparent that although the mechanical process of drilling a horizontal well has become reasonably accurate, the process all too often results in restrictive amounts of wellbore damage. The potential cost associated with overcoming wellbore damage in a horizontal well can be very high, therefore it becomes important to be able to correctly identify and quantify actual wellbore damage.

Mathematically, three components of skin are inherent to the analytic solutions published by Goode and Thambynayagam (1987). There is the mechanical component of skin S_m, attributable to wellbore damage, which typically results from the drilling and completion process. In the majority of circumstances, only this component is effected by stimulation and workover

178 Horizontal Wells

programs. The second component S_z, referred to as a pseudo skin, can be related to the effects of partial penetration in the vertical direction. The magnitude of this skin is a direct function of reservoir thickness and the geometric location of the horizontal well in the vertical plane. This pseudo skin is theoretically not present during early time radial flow. The third component S_x is also a pseudo skin caused by partial penetration effects, but in the x direction (or parallel to the wellbore), and is only present during the late time linear flow period.

For each of the flow regimes, the mechanical skin factor S_m can be calculated from a drawdown analysis, as follows:

Early Time Radial

$$S_m = 1.151\left[\frac{p_i - p_{wf(1hr)}}{m_{1r}} - \log_{10}\left(\frac{\sqrt{k_v k_z}}{\phi \mu c_t r_w^2}\right) + 3.227\right]$$

Intermediate Time Linear

$$S_m = \frac{0.058}{h_z}\sqrt{\frac{k_z}{\phi \mu c_t}}\left(\frac{p_i - p_{wf(0hr)}}{m_{1l}}\right) - S_z$$

Late Time Radial

$$S_m = \frac{1.151\, L_w}{h_z}\sqrt{\frac{k_z}{k_x}}\left[\frac{p_i - p_{wf(1hr)}}{m_{2r}} - \log_{10}\left(\frac{k_x}{\phi \mu c_t L_w^2}\right) + 2.023\right] - S_z$$

Late Time Linear

$$S_m = \frac{0.058\, L_w}{h_z h_x}\sqrt{\frac{k_z}{\phi \mu c_t}}\left(\frac{p_i - p_{wf(0hr)}}{m_{2l}}\right) - (S_x + S_z)$$

The equations also apply for a buildup analysis, once P_i is replaced by $P_{ws(1hr)}$ or $P_{ws(0hr)}$, and P_{wf} is measured immediately prior to shut in.

NOTATION

X horizontal coordinate in direction parallel to horizontal well (ft)	Y horizontal coordinate in direction perpendicular to horizontal well (ft)

Z	vertical coordinate (ft)	t_{brf1}	time to the end of the second radial flow (hr)
h_x	width of reservoir parallel to horizontal well (ft)	t_{erf2}	time to the end of the second radial flow (hr)
h_y	width of reservoir perpendicular to horizontal well (ft)	t	time (hr)
		t_o	shut-in time (hr)
h_z	effective thickness of reservoir (ft)	P_i	initial pressure (psia)
		P_{wf}	bottom hole flowing pressure (psia)
r_w	radius of wellbore (ft)		
L_w	effective length of horizontal well (ft)	P_{ws}	bottom hole shut-in pressure (psia)
L_{xl}	distance in x direction to beginning of horizontal wellbore	$r_w l$	effective wellbore radius (ft)
		B_o	formation volume factor (RB/STB)
L_{xd}	distance in x direction to end of horizontal wellbore	k_x	permeability in X direction (md)
h_s	distance from the upper reservoir boundary to the center of the horizontal well (ft)	k_y	permeability in Y direction (md)
		k_z	permeability in Z direction (md)
q	flow rate (STB/D)	m	Fourier transform variable
ϕ	porosity (factor)	n	Fourier transform variable
μ	viscosity (cp)	S_m	mechanical skin
c_t	total compressibility (psi^{-1})	S_z	pseudo skin in Z direction
t_{erf1}	time to the end of the first radial flow (hr)	S_x	pseudo skin in X direction
t_{elf1}	time to the end of the first linear flow (hr)		

REFERENCES

1. Daviau, F., Mouronval, G., Bourdarot, G., and Curutchet, P., "Pressure Analysis for Horizontal Wells," SPE 14251 presented at the 1985 SPE Annual Technical Conference and Exhibition, Las Vegas, Sept. 22-25, 1985.

2. Goode, P. A. and Thambynayagam, R. K. M., "Pressure Drawdown and Buildup Analysis of Horizontal Wells in Anisotropic Media," SPE 14250 presented at the SPE Annual Technical Conference and Exhibition, Las Vegas, Sept. 22-25, 1985.

3. Clonts, M. D. and Ramey Jr., H. J., "Pressure Transient Analysis for Wells with Horizontal Drainholes," SPE 15116 presented at the 56th California Regional Meeting in Oakland, CA, April 2–4, 1986.
4. Orkan, E., Raghavan, R., and Joshi, S. D., "Horizontal Well Pressure Analysis," SPE 16378 presented at the 1987 SPE California Regional Meeting in Ventura, CA, April 8–10, 1987.
5. Kuchuk, F. J., Goode, P. A., Wilkinson, D. J., and Thambynayagam, R. K. M., "Pressure Transient Behavior of Horizontal Wells With and Without Gas Cap and Aquifer," SPE 17413 presented at the 1988 California Regional Meeting in Long Beach, CA, March 23–25, 1988A.
6. Kuchuk, F. J., Goode, P. A., Brice, B. W., Sherrard, D. W., and Thambynayagam, R. K. M., "Pressure Transient Analysis and Inflow Performance for Horizontal Wells, SPE 18300 presented at the 1988 SPE Annual Technical Conference and Exhibition, Houston, Oct. 2–5, 1988B.
7. Odeh, A. S. and Babu, D. K., "Transient Flow Behavior of Horizontal Wells, Pressure Drawdown, and Buildup Analysis," SPE 18802 presented at the SPE California Regional Meeting held in Bakersfield, CA, Apr. 5–7, 1989.
8. Salamy, S. P., Locke, C. D., and Overbey Jr., W. K., "Four Pressure Buildup Analysis Techniques Applied to Horizontal and Vertical Wells with Field Examples," SPE 19101 presented at the SPE Gas Technology Symposium held in Dallas, TX, June 7–9, 1989.
9. Goode, P. A. and Thambynayagam, R. K. M., "Pressure Drawdown and Buildup Analysis of Horizontal Wells in Anisotropic Media," SPE Formation Evaluation, Dec. 1987. Copyright 1987 Society of Petroleum Engineers.

6
Well Performance and Productivity

Caution should be exercised, as not a lot is known about well production, and production and performance are in their infancy. These mathematical models are not based on actual experience but rather, on theoretical developments, and so care should be taken with the models.

MATHEMATICAL MODEL

A simple way to look at a horizontal well is to consider a conventional vertical well rotated at 90°, the effects of gravity excepted. For example, in the left-handed system, the xyz axis will become the $x'y'z'$ axis: the old x axis becomes the new x' axis; the old y axis becomes the new z' axis; the old z axis becomes the y' axis; and the horizontal well is parallel to the y' direction.

The usual assumptions apply here, i.e., we have a single phase flow of a slightly compressible fluid, a homogeneous reservoir with uniform thickness, and gravity and capillary effects are negligible with impermeable upper and lower boundaries for the reservoir. Porosity and absolute permeability are independent of position and pressure. We assume uniform flux along the wellbore. Initial condition is that throughout the reservoir pressure is uniform. We neglect wellbore storage and skin for the moment.

PRODUCTIVITY OF VERTICAL WELLS

The inflow equation at pseudo steady state for a conventional vertical well in the center of a circular drainage area is documented in the literature:

182   *Horizontal Wells*

$$J_v = \frac{7.08 \times 10^{-3} * h \sqrt{k_x k_y}}{\mu B \left[\ln \frac{0.565\sqrt{ab}}{r_w} - 0.75 \right]} \tag{6-1}$$

where: J_v = productivity index of a vertical well, STB/(day-psi)
       h = formation thickness, ft
       k_x = permeability in x direction, md
       k_y = permeability of y direction, md
       ab = drainage area, ft^2
       a = reservoir length, ft
       b = reservoir width, ft
       r_w = wellbore radius, ft
       μ = viscosity of oil, cp
       B = formation volume factor of oil, bbl/STB

The porosity ϕ is constant. All the boundaries of the drainage volume are sealed. The reservoir is box-shaped. At time zero, the well starts to flow at constant rate q, uniform flux. Here the subscript v stands for vertical well.

WELLBORE FLOWING PRESSURE OF HORIZONTAL WELLS, P_{wf}

Recently, numerous papers have been published which give analytical solutions on various horizontal well models. The results of many of these papers, in their limiting cases, can be considered as horizontal wells. For example:

1. Slanted well of Cinco-Ley (1974), when the angle of inclination from the vertical is equal to 90°
2. Horizontal fracture solution of Gringarten and Ramey (1974), when the fracture thickness becomes the thickness of the wellbore of the horizontal well
3. A partially penetrating vertical fracture solution should approach that of a horizontal well when the fracture height approaches zero (Raghaven et al., 1976)

Various solutions have been published recently for P_{wf} of horizontal wells under various assumptions such as infinite x direction, infinite y direction, infinite in both directions, and in a bounded reservoir, either single or double porosity. Many give P_{wf} in analytical closed forms, usually slowly convergent series like error functions, exponential integrals, Struve functions, or even Mathieu functions. Some give solutions in the form of Laplace transforms (Goode and Thambynayagam, 1985; Daviau et al., 1985; Clonts and

Ramey, 1986; Kuchuk et al., 1988 A and 1988 B; Carvalho and Rosa, 1988; Aguilera and Ng, 1988). Some researchers are particularly interested in the formulas for productivity index (Joshi, 1988; Babu and Odeh, 1988; Karcher et al., 1986).

GENERAL REVIEW

Goode and Thambynayagam (1985) presented an excellent paper on pressure drawdown and buildup analysis of horizontal wells in anisotropic media. Their well, finite in length, is perpendicular to the linear barriers that are no-flow boundaries, and these boundaries are parallel to the y-axis. The well lies on the x-axis and the y-axis is infinite.

They give specialized semilog analysis on early radial flow, intermediate linear flow, late radial flow, and late linear flow, with the different time periods defined by the various parameters. They present analytical expressions in closed form for the pressure drop for both the drawdown and the buildups.

One shortcoming of this model is that if a horizontal well lies parallel to the linear barriers, the result given in this paper must be modified. Otherwise, one must turn to another model.

Clonts and Ramey (1986) have pointed out that Cinco-Ley's solution for an angle of inclination equal to 75° gives a very good approximation to the solution of a horizontal well.

Another drawback of the Goode and Thambynayagam method is that it uses Fourier series and error functions and the full expression converges very slowly. A large number of terms is needed for acceptable accuracy, especially for early time in buildups.

A possible improvement would be to convert the infinite series to infinite integral whenever possible (Gringarten and Ramey, 1973) and use numerical integration instead. Another computational improvement would be to use numerical inversion like the Stehfest algorithm on their Equation A7, rather than analytical inversion (Goode and Thambynayagam, 1985) of the Laplace transform.

Aguilera and Ng extended the model to a double porosity reservoir. Based on the theory of this paper, a program has been written to do nonlinear regression of the estimation of the parameters, generation of the pressure data in one single simulated curve, and graphical semilog specialized analysis.

The constrained optimization method of Box (1965) has been adopted rather than the Levenberg-Marquardt method (Fletcher, 1971) so that wide range of the parameters can be input as initial guesses.

Daviau et al. (1985) analyzed horizontal wells of uniform flux and/or infinite conductivity in the wellbore, with wellbore storage and skin. The homogenous reservoir can be infinite or limited by impermeable or constant pressure boundaries.

Clonts and Ramey (1986) gave pressure transient analyses for wells with single horizontal drainholes, and also for a system of multiple drainholes. They presented a finite line source solution with the reservoir being infinite in the x and y directions.

If the production rate varies and can be measured, then, in an analytical solution such as

$$\Delta P (X,Y,T) = \frac{1}{\phi C} \int Q (T) \, S (X,Y,T) \, D \, T \qquad (6\text{-}2)$$

we simply apply the method of superposition

$$\Delta P = \frac{1}{\phi C} \left[\sum Q_i \int S \, D \, T \right] \qquad (6\text{-}3)$$

where q_i = step functions defined in T (which lies between corresponding time intervals) and are considered constant in the respective intervals
$s(x, y, t)$ = the line source function

Kuchuk et al. (1988 A) deals with horizontal wells with and without a gas cap on top or an aquifer in the bottom in the infinite x and y directions. Kuchuk et al. (1988 B) also discussed the inflow performance in terms of $<p> - P_{wf}$ for a horizontal well that is producing from a closed rectangular region. This also elaborates on the application of the method of nonlinear least squares regression in the estimation of the reservoir parameters. Here $<p>$ is the average pressure in the reservoir.

Carvalho and Rosa (1988) discussed transient pressure behavior in horizontal wells in naturally fractured reservoirs with infinite lateral extension. They quoted a general scheme for converting a solution for single porosity reservoirs to a solution for double porosity reservoirs. If the single porosity solutions are in Laplace space, the Laplace variable s is replaced by sf(s) to obtain the solution for the dual porosity system in terms of the pressure derivative. Here:

$$f(s) = \frac{\omega (1 - \omega) s + \lambda}{(1 - \omega) s + \lambda}$$

where λ and ω are defined in the Warren and Root (1963) model.

Well Performance and Productivity   185

λ = the interporosity flow coefficient and determines how rapidly the matrix starts contributing to the flow
ω = the ratio of the storativity of the fracture system to the total reservoir storativity

PRODUCTIVITY IN HORIZONTAL WELLS

A solution published by Joshi (1988) deals with flow in vertical planes perpendicular to the well axis at early time, and in horizontal planes at late time.

The three-dimensional problem is broken into a series of two-dimensional problems. The solution is sought by means of potential theory in fluid mechanics.

Babu and Odeh (1988), using Green's function and Newman's product method, applied the intersection of three infinite source planes in a bounded reservoir. This results in an instantaneous source point.

They first integrate the point y_0 between two limits, y_1 and y_2, to obtain a finite source line. They then integrate the finite source line between time zero and a dummy time parameter t and finally obtain a line source continuous in time.

For conventional vertical wells, it can be shown that the line source solution gives good approximation to a solid cylinder solution (Streltsova, 1988).

They derive the expression q_h = (P.I.) $(p_r - p_{wf})$ from the analytical solution of their mathematical model, and therefore, they need an independent method to evaluate p_r. Babu and Odeh use a material balance equation for this purpose. Here P.I. is the productivity index and p_r is the average reservoir pressure at pseudosteady state condition.

In Babu and Odeh's method, all boundaries are no-flow boundaries. They consider a uniform flux solution given by the equation (Figure 6-1):

$$q = \frac{7.08 \times 10^{-3} \, b \, \sqrt{k_x k_y} \, (\bar{p}_r - p_{wf})}{B\mu \left[\mathrm{Ln} \, \frac{A^{1/2}}{r_w} + \mathrm{Ln}C_H - 0.75 + S_R \right]} \quad (6\text{-}4)$$

where p_r = volumetric average pressure of reservoir, psi
p_{wf} = flowing bottomhole pressure, psi
B = formation volume factor, bbl/STB
μ = viscosity, cp
C_H = geometric factor
r_w = wellbore radius, feet

186   Horizontal Wells

S_R = pseudo skin factor due to fractional penetration, $S_R = 0$ if $L = b$
h = thickness of the formation, ft
a = length of reservoir in x direction, ft
b = width of reservoir in y direction, ft
L = well length = $y_2 - y_1$, ft
x_o = position of well, ft
y_o = position of well, ft
z_o = position of well, ft
q_h = constant rate of production, STB/d

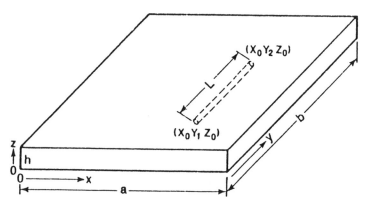

Figure 6-1. Physical model of horizontal well. Source: Babu and Odeh (1988).

The well is drilled in a box-shaped drainage volume parallel to the y direction, and therefore the drainage area is ah, while for a conventional vertical well, the drainage area is ab. The following equations follow very closely the original work of Babu and Odeh's geometric factor:

$$LnC_H = 6.28 \frac{a}{h} \sqrt{\frac{k_y}{k_x}} \left[\frac{1}{3} - \frac{x_o}{a} + \left(\frac{x_o}{a}\right)^2 \right] - Ln \sin$$

$$\left[\frac{180°z_o}{h} \right] - 0.5Ln \left[\frac{a}{h} \sqrt{\frac{k_z}{k_x}} \right] - 1.088 \tag{6-5}$$

$$J = \frac{7.08 \times 10^{-3} \, b\sqrt{k_y k_z}}{B\mu \left[Ln \left(\frac{C_H A^{1/2}}{r_w} \right) - 0.75 + s_R \right]} \tag{6-6}$$

Well Performance and Productivity   187

$$\frac{a}{\sqrt{k_x}} \geq \frac{0.75b}{\sqrt{k_y}} > \frac{0.75h}{\sqrt{k_z}} \tag{6-7}$$

Case 1

$$PXYZ = \left[\frac{b}{L} - 1\right]\left[Ln\frac{h}{r_w} + 0.25\,Ln\frac{k_x}{k_z} - 1.05\right] \tag{6-8}$$

$$PXY' = \frac{2b^2}{Lh}\sqrt{\frac{k_z}{k_y}}\left[F\left(\frac{L}{2b}\right) + 0.5\left[F\left(\frac{4y_o + L}{2b}\right) - F\left(\frac{4y_o - L}{2b}\right)\right]\right] \tag{6-9}$$

if $\dfrac{4y_o - L}{2b} \leq 1$ \hfill (6-10)

$$F\left(\frac{4y_o - L}{2b}\right) = -\left(\frac{4y_o - L}{2b}\right)\left[0.145 + Ln\left(\frac{4y_o - L}{2b}\right)\right.$$

$$\left. - 0.137\left(\frac{4y_o - L}{2b}\right)^2\right] \tag{6-11}$$

$$\left(\frac{4y_o + L}{2b}\right) > 1.0, \tag{6-12}$$

$$F\left(\frac{4y_o + L}{2b}\right) = \left(2 - \frac{4y_o + L}{2b}\right)\left[0.145 + Ln\left(2 - \frac{4y_o + L}{2b}\right)\right.$$

$$\left. - 0.137\left(2 - \frac{4y_o + L}{2b}\right)^2\right] \tag{6-13}$$

and, for $\left(\dfrac{4y_o - L}{2b}\right) > 1.0,$ \hfill (6-14)

$$F\left(\frac{4y_o - L}{2b}\right) = \left(2 - \frac{4y_o - L}{2b}\right)\left[0.145 + \text{Ln}\left(2 - \frac{4y_o - L}{2b}\right)\right.$$

$$\left. - 0.137\left(2 - \frac{4y_o - L}{2b}\right)^2\right] \tag{6-15}$$

Case 2

$$\frac{b}{\sqrt{k_y}} > \frac{1.33a}{\sqrt{k_x}} > \frac{h}{\sqrt{k_z}} \tag{6-16}$$

$$s_R = PXYZ + PY + PXY \tag{6-17}$$

$$PY = \frac{6.28b^2 \sqrt{k_x k_z}}{ah \; k_y}\left[\left(\frac{1}{3} - \frac{y_o}{b} + \frac{y_o^2}{b^2}\right) + \frac{L}{24b}\left(\frac{L}{b} - 3\right)\right] \tag{6-18}$$

$$PXY = \left(\frac{b}{L} - 1\right)\left[\frac{6.28a}{h}\sqrt{\frac{k_z}{k_x}}\right]\left[\frac{1}{3} - \frac{x_o}{a} + \frac{x_o^2}{a^2}\right] \tag{6-19}$$

for $[\text{Min}(x_o, a - x_o) \geq 0.25a]$ \hfill (6-20)

Example 1: How to Use Equation 6-5 (Babu and Odeh, 1988)

Data

A horizontal well 1,000 ft long is drilled in a box-shaped drainage volume 4,000 ft long, 2,000 ft wide, and 100 ft thick. The well lies between $y_1 = 750$ ft and $y_2 = 1,750$ ft. The x_o and z_o coordinates of the well are, respectively, 3,000 ft and 50 ft. The permeabilities in the x, y, and z directions are, respectively, 200 md, 200 md, and 50 md and the well radius = 0.25 ft. To calculate the productivity index we assume the following:

$$B_o = 1.3 \qquad \mu = 0.2$$
$$P_R = 29.4 \qquad P_{wf} = 14.7$$

Now

$$a = 4,000 \qquad b = 2,000$$
$$h = 100 \qquad x_o = 3,000$$
$$z_o = 50 \qquad k_x = 200$$
$$k_y = 200 \qquad k_z = 50$$
$$r_w = 0.25$$

From Equation 6-5

$$\text{LnC}_H = 6.28 * \frac{4,000}{100} \sqrt{\frac{50}{200}} \left[\frac{1}{3} - \frac{3,000}{4,000} + \left(\frac{3,000}{4,000}\right)^2\right]$$

$$- \text{Ln}\left[\sin\frac{180° \times 50}{100}\right]$$

$$- 0.5\, \text{Ln}\left(\frac{4,000}{100}\sqrt{\frac{50}{200}}\right) - 1.088 = 15.73$$

From Equation 6-8

$$\text{PXYZ} = (2 - 1)\left(\text{Ln}\frac{100}{0.25} + 0.25\, \text{Ln}\frac{200}{50} - 1.05\right) = 5.29$$

From Equation 6-9

$$\frac{L}{2b} = \frac{1,000}{4,000} = 0.25$$

$$y_o = \frac{750 + 1,750}{2} = 1,250$$

$$\frac{4y_o + L}{2b} = \frac{4 \times 1,250 + 1,000}{4,000} = 1.5$$

$$\frac{4y_o - L}{2b} = 1$$

$$F\left(\frac{L}{2b}\right) = F(0.25) = -0.25\,(0.145 + \ln 0.25 - 0.137$$
$$\times\, 0.25 \times 0.25) = 0.312$$

$$F\left(\frac{4y_o - L}{2b}\right) = F(1.0) = -1\,(0.145 + \ln 1 - 0.137)$$
$$= -0.008$$

$$F\left(\frac{4y_o + L}{2b}\right) = F(1.5) = (2 - 1.5)\,(0.145 + \ln(2 - 1.5)$$
$$- 0.137\,(2 - 1.5)(2 - 1.5)) = -0.291$$

$$PXY' = \frac{2 \times (2{,}000)\,(2{,}000)}{1{,}000 \times 100}\sqrt{\frac{50}{200}}\,(0.312 + 0.5$$
$$(-0.291 + 0.008)) = 6.82$$

$$S_R = PXYZ + PXY' = 5.29 + 6.82 = 12.1$$

$$L_N\,\frac{(a \times h)^{0.5}}{R_w} = L_N\,\frac{(4{,}000 \times 100)^{0.5}}{0.25} = 7.836$$

$$q_H = \frac{7.08 \times 10^{-3} \times 2{,}000 \times 100\,(p)}{B\,\mu \times 34.92} = 2{,}292.62$$

$$J_H = \frac{q}{p} = \frac{40.55}{B\,\mu} = 155.96$$

$$J_V = \frac{7.08 \times 10^{-3} \times 100 \times 200}{B\,\mu\left[\dfrac{L_N\,0.565\,(a \times b)^{1/2}}{R_W} - 0.75\right]}$$

$$\text{Area} = a \times b = 4{,}000 \times 2{,}000$$

$$J_V = \frac{7.08 \times 10^{-3} \times 20{,}000}{B\,\mu \times 8.013} = \frac{17.67}{B\,\mu} = 67.96$$

Well Performance and Productivity

$J_H/J_V = 155.96/67.96 = 2.295$

Thus, in this example the horizontal well flow rate will be about 2.3 times that of the vertical well.

Example 2

Data

A horizontal well 1,000 ft long is drilled in a drainage volume which is 2,000 ft long, 4,000 ft wide, and 200 ft thick. The well lies between $y_1 = 1,000$ ft and $y_2 = 2,000$ ft. The x_o and z_o coordinates are 1,000 ft and 150 ft respectively. The k_x, k_y, and k_z values are, respectively, 100 md, 100 md, and 20 md. The wellbore radius is 0.25 ft. Calculate the improvement in the productivity index over that of a vertical well in the same drainage volume.

$$a = 2,000, \; b = 4,000, \; h = 200$$

We again assume, as follows:

$$B_o = 1.3 \qquad \mu = 0.2$$
$$P_R = 29.4 \qquad P_{wf} = 14.7$$

In this case,

$$\ln C_H = 6.28 \left[\frac{2,000}{200} \sqrt{\frac{20}{100}} \right] \left(\frac{1}{3} - 0.5 + .25 \right)$$

$$- \ln \left[\left(\sin \frac{180° \times 150}{200} \right) \right] - 0.5 \ln \left(\frac{2,000}{200} \sqrt{\frac{20}{100}} \right)$$

$$- 1.088 = 0.85$$

$$PXYZ = \left(\frac{4,000}{1,000} - 1 \right) \left[\ln \frac{200}{.25} + 0.25 \ln \frac{100}{20} \right.$$

$$\left. - 1.05 \right] = 18.1$$

Horizontal Wells

$$PY = \frac{6.28 \times (4,000)^2}{200 \times 2,000} \frac{\sqrt{20 \times 100}}{100}$$

$$\left[\frac{1}{3} - \frac{1,500}{4,000} + \left(\frac{1,500}{4,000}\right)^2 + \frac{1,000}{24 \times 4,000}\left(\frac{1,000}{4,000} - 3\right)\right] = 7.9$$

Now, from Equation 6-21

$$PXY = \left(\frac{4,000}{1,000} - 1\right)\left(\frac{6.28 \times 2,000}{200}\sqrt{\frac{20}{100}}\right)$$

$$\left(\frac{1}{3} - \frac{1,000}{2,000} + \left(\frac{1,000}{2,000}\right)^2\right) = 7.0$$

Therefore,

$$S_R = PXYZ + PY + PXY$$
$$= 18.1 + 7.9 + 7.0 = 33.0$$

$$J_H = \frac{q}{\Delta p} = \frac{7.08 \times 10^{-3} \sqrt{k_y k_z}\, b}{B\mu \left(\text{Ln}\, \frac{A^{1/2}}{r_w} + \text{Ln} \times C_H - 0.75 + S_R\right)}$$

$$= \frac{7.08 \times 10^{-3} \sqrt{100 \times 20} \times 4,000}{B\mu \left(\text{Ln}\, \frac{(2,000 \times 200)^{1/2}}{r_w} + 0.85 - 0.75 + 33\right)} = \frac{30.9}{B\mu}$$

$$J_V = \frac{7.08 \times 10^{-3} \times 100 \times 200}{B\mu \left(\text{Ln}\, \frac{0.565 \times (4,000 \times 2,000)^{1/2}}{0.25} - 0.75\right)} = \frac{17.7}{B\mu}$$

$$\frac{J_H}{J_V} = \frac{30.9}{17.7} = 1.75$$

Hence, the improvement in the productivity index is about 1.75 times greater.

A flow chart to calculate J_H and J_v and their ratio appears in Figure 6-2. A readily available expression for $p_R - p_{wf}$ has been published by Kuchuk et al. (1988 B).

Well Performance and Productivity

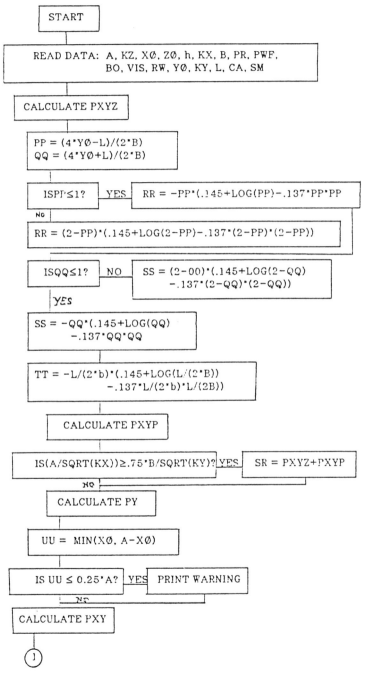

Figure 6-2 continued on next page

194   Horizontal Wells

Figure 6-2 continued from previous page

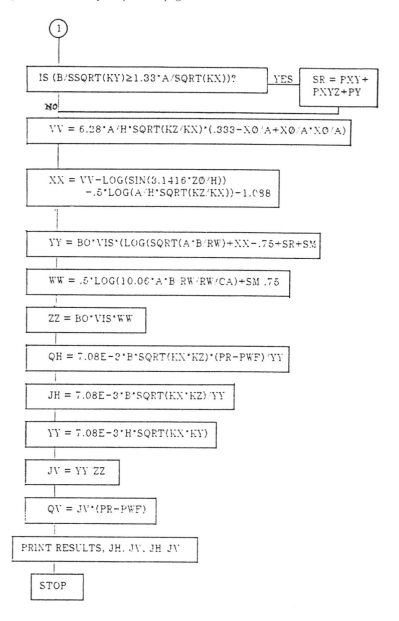

Figure 6-2. Flow chart for calculating horizontal and vertical productivity indices.

Well Performance and Productivity   195

To calculate p_R, Odeh and Babu used a program based on a material balance equation as a function of time.

Material Balance

The material balance of a slightly compressible fluid at pseudo steady state is:

$$N_p B_o = N B_{oi} \frac{(S_w C_w + S_o C_o + C_f) \Delta p}{1 - S_{wc}} \quad (6\text{-}21)$$

where $\Delta \bar{p} = p_r - p_{wf}$, psi   (6-22)

where N_p = cumulative production, STB
B_o = oil formation volume factor, bbl/STB
N = original hydrocarbon in place, STB
B_{oi} = initial oil formation volume factor, bbl/STB
S_w = saturation of water, fraction
C_w = compressibility of water, psi^{-1}
S_o = saturation of oil, fraction
C_o = compressibility of oil, psi $^{-1}$
C_f = rock compressibility, psi^{-1}
S_{wc} = connate water saturation, fraction

Equation 6-21 can be treated by a straight line passing through the origin (Havlena and Odeh, 1963) when $N_p B_o$ is plotted against Δp; by the method of linear least squares (Craft and Hawkins, 1959); or by the method of Monte Carlo (Murtha, 1987). Results of a set of sample data by the method of least squares are shown in Examples 3 and 4.

Example 3: Linear Least Squares (Havlena and Odeh, 1963)

Given:

$S_w C_w = 0.868E - 06$
$C_f = 4.95E - 06$
$p_i = 3,685$

With Aquifer

Number of samples = 12
Sum of samples = 0.130E + 05
Sum of squares of samples = 0.293E + 08
Variance of samples = 0.266E + 07
Standard deviation = 0.163E + 04
Mean of samples = 0.108E + 04

	$NB_{oi}C_e dp \times 10^{-6}$	$(N_p B_o + W_p) \times 10^3$
	1	2
1	0.9300E + 02	0.2684E + 02
2	0.1680E + 03	0.4554E + 02
3	0.3370E + 03	0.1029E + 03
4	0.3930E + 03	0.1335E + 03
5	0.8440E + 03	0.2827E + 03
6	0.1504E + 04	0.4782E + 03
7	0.2224E + 04	0.7127E + 03
8	0.3217E + 04	0.1106E + 04
9	0.4506E + 04	0.1675E + 04
10	0.6228E + 04	0.2230E + 04
11	0.7898E + 04	0.2806E + 04
12	0.9703E + 04	0.3400E + 04

The parameters of the regression equation are:

1. $-0.19710E + 02$
2. $0.35658E + 00$

Slope = NB_{oi} = 356.5×10^6 Res. bbl
Intercept = -19.710

column 1 = x variable = $N B_{oi} C_e dp \times 10^{-6}$
column 2 = y variable = $(N_p B_o + W_p) \times 10^3$
column 3 = y value based on regression equation = slope × x + intercept
column 4 = column 2 − column 3

Well Performance and Productivity 197

	1	2	3	4
1	0.9300E + 02	0.2684E + 02	0.1345E + 02	0.1339E + 02
2	0.1680E + 03	0.4554E + 02	0.4020E + 02	0.5340E + 01
3	0.3370E + 03	0.1029E + 03	0.1005E + 03	0.2490E + 01
4	0.3930E + 03	0.1335E + 03	0.1204E + 03	0.1304E + 02
5	0.8440E + 03	0.2827E + 03	0.2812E + 03	0.1489E + 01
6	0.1504E + 04	0.4782E + 03	0.5166E + 03	− 0.3837E + 02
7	0.2224E + 04	0.7127E + 03	0.7733E + 03	− 0.6067E + 02
8	0.3217E + 04	0.1106E + 04	0.1127E + 04	− 0.2177E + 02
9	0.4506E + 04	0.1675E + 04	0.1587E + 04	0.8767E + 02
10	0.6228E + 04	0.2230E + 04	0.2201E + 04	0.2875E + 02
11	0.7898E + 04	0.2806E + 04	0.2797E + 04	0.9149E + 01
12	0.9703E + 04	0.3400E + 04	0.3440E + 04	− 0.4051E + 02

Order of equation = 1
Number of samples = 12
Sums of squares of Ymea - Yave = 15198080.0000
Sums of squares of Ycal - Yave = 15181832.0000
Deviation of sums of squares = 16248.0000
Goodness of fit = 0.998931
Correlation coefficient = 0.999465

For a graphical solution, refer to Figure 6-3.

Example 4: Linear Least Squares (Havlena and Odeh, 1963)

$S_w C_w = 0.842E - 06$
$c_f = 5.5E - 06$
$p_i = 3,654$

No Aquifer

Number of samples = 10
Sum of samples = 0.446E + 04
Sum of squares of samples = 0.396E + 07
Variance of samples = 0.440E + 06
Standard deviation = 0.663E + 03
Mean of samples = 0.446E + 03

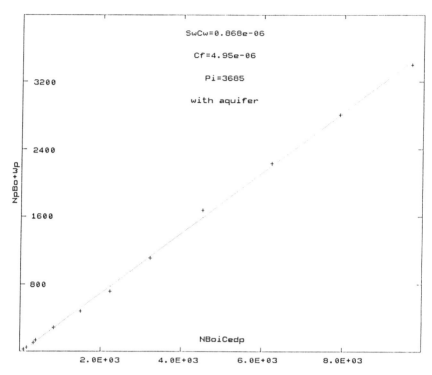

Figure 6-3. The material balance as an equation of a straight line including an aquifer; Example 3.

$$NB_{oi}C_e dp \times 10^{-6} \qquad N_p B_o \times 10^3$$

	1	2
1	0.1820E + 03	0.2153E + 02
2	0.3040E + 03	0.3072E + 02
3	0.6900E + 03	0.5361E + 02
4	0.1505E + 04	0.1167E + 03
5	0.2465E + 04	0.1964E + 03
6	0.3772E + 04	0.3398E + 03
7	0.5637E + 04	0.5454E + 03
8	0.8060E + 04	0.7663E + 03
9	0.1080E + 05	0.1046E + 04
10	0.1371E + 05	0.1345E + 04

The parameters of the regression equation:

1. $-0.18322E + 02$
2. $0.98540E - 01$

Slope = NB_{oi} = 98.3×10^6 Res. bbl
Intercept = -18.322

column 1 = x variable = $N\ B_{oi}\ C_e\ dp \times 10^{-6}$
column 2 = y variable = $N_p\ B_o \times 10^3$
column 3 = y value based on regression equation
          = slope × x + intercept
column 4 = column 2 − column 3

	1	2	3	4
1	$0.1820E + 03$	$0.2153E + 02$	$-0.3873E + 00$	$0.2192E + 02$
2	$0.3040E + 03$	$0.3072E + 02$	$0.1163E + 02$	$0.1909E + 02$
3	$0.6900E + 03$	$0.5361E + 02$	$0.4967E + 02$	$0.3935E + 01$
4	$0.1505E + 04$	$0.1167E + 03$	$0.1300E + 03$	$-0.1332E + 02$
5	$0.2465E + 04$	$0.1964E + 03$	$0.2246E + 03$	$-0.2823E + 02$
6	$0.3772E + 04$	$0.3398E + 03$	$0.3533E + 03$	$-0.1349E + 02$
7	$0.5637E + 04$	$0.5454E + 03$	$0.5372E + 03$	$0.8245E + 01$
8	$0.8060E + 04$	$0.7663E + 03$	$0.7759E + 03$	$-0.9581E + 01$
9	$0.1080E + 05$	$0.1046E + 04$	$0.1046E + 04$	$-0.6777E + 00$
10	$0.1371E + 05$	$0.1345E + 04$	$0.1333E + 04$	$0.1211E + 02$

Order of equation = 1
Number of samples = 10
Sums of squares of Ymea − Yave = 1969044.0000
Sums of squares of Ycal − Yave = 1966721.0000
Deviation of sums of squares = 2323.31450000
Goodness of fit = 0.998820
Correlation coefficient = 0.999410

A graphical solution is shown in Figure 6-4.

Horizontal Wells

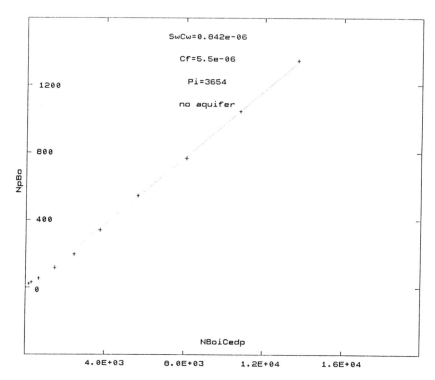

Figure 6-4. The material balance as an equation of a straight line with no aquifer; Example 4.

Alternative Methods of Calculating P_R

Mead's rectangular hyperbola method (Van Everdingen and Hurst, 1949), which is derived from the steady state Darcy's equation, gives very good results in calculating P_R even for horizontal wells. This leads to the speculation that in the absence of other available tools, an acceptable answer can be obtained using Matthews, Brons, and Hazebrook's theory of average reservoir pressure and Dietz shape factors developed for conventional vertical wells.

Alternative Methods of Calculating Productivity Index

Two other useful formulas for productivity index calculations have been published by Sherrard et al. (1987) and Reiss (1987).

From Sherrard, we get a productivity index given by:

Well Performance and Productivity   201

$$J_H = \frac{7.08 \times 10^{-3}k}{\mu_o B_o} \times \frac{1}{\frac{1}{h} \text{Ln}\left[1 + \frac{\sqrt{1 + (L/2r_d)^2}}{L/2r_d}\right] + \frac{1}{L} \text{Ln}\left(\frac{h}{2\pi r_w}\right)} \quad (6\text{-}23)$$

$$J_v = \frac{7.08 \times 10^{-3}kh}{\mu_o B_o \ln\left(\frac{r_d}{r_w}\right)} \quad (6\text{-}24)$$

L = length of well, ft
r_d = drainage radius, ft
h = thickness of formation, ft
J_H = productivity index of horizontal well, STB/(day-psi)
J_v = productivity index of vertical well, STB/(day-psi)

On the other hand, Reiss's formula is as follows:

$$\frac{J_H}{J_v} = \frac{\ln (r_{ev}/r_{wv})}{\ln \frac{1 + [1 + (L/2r_{eH})^2]^{1/2}}{(L/2r_{eH})} + \frac{h}{L} \ln (h/2\pi r_{wH})} \quad (6\text{-}25)$$

r_{ev} = radius of outer boundary for vertical well
r_{wv} = radius of wellbore for vertical well
r_{eH} = radius of outer boundary for horizontal well
h = thickness of formation
L = length of horizontal well

WELLBORE STORAGE AND SKIN

Once we have a solution of dimensionless wellbore pressure drop p_D for no wellbore storage and no skin, we can use the convolution integral to find dimensionless pressure p_{wd} with wellbore storage and skin for a horizontal well. The procedure is the same as that used in vertical wells (Streltsova, 1988; Van Everdingen and Hurst, 1949; Agarwal et al., 1970).

Then, we face two alternatives to treat the following dimensionless pressure equation:

$$p_{wD}(t_D) = \int_0^{t_D} \left[1 - \bar{C} \frac{dp_{wD}(t_D')}{dt_D'}\right] \left[\frac{dp_D(t_D - t_D')}{dt_D}\right]$$

$$dt_D' + S \left[1 - \bar{C} \frac{dp_{wD}(t_D)}{dt_D}\right] \quad (6\text{-}26)$$

202 Horizontal Wells

where t_D = dimensionless time
S = skin
C_D = dimensionless wellbore storage

Alternative 1: Finite Difference Methods

In this case, we can apply the findings of Cinco-Ley and Samaniego (1977) to Equation 6-26, treating the skin equal to zero. Ikoku and Ramey (1978) gave a finite difference formula that is easier for programming purposes and also equates the skin to zero. Kabik et al. (1986) presented a finite difference solution with a nonzero skin. Note that the skin effect can be added if we introduce an effective radius $r_{we} = r_w \exp(-S)$ in the analytical solution because the effective skin formulation is also good for negative skins (Daviau et al., 1985).

Alternative 2: Laplace Transform

If we take the Laplace transform of Equation 6-26, we obtain:

$$\bar{p}_{wD}(s) = \frac{s\,\bar{p}_D(s) + S}{s + C_D s^2\,[s\bar{p}_D(s) + S]} \qquad (6\text{-}27)$$

Where s is the Laplace parameter. As pointed out by Kuchuk et al. (1988A), the computation by the finite difference formula takes a lot of computer time. This is because the integral is of a convolution type. In other words, for a given time, p_{wD} is always computed starting from t_D = zero.

On the other hand, to apply Equation 6-27 one needs the Laplace transform of p_D available, namely, $\bar{p}_D$. The Laplace transform for a vertical well in a homogenous reservoir is elegant:

$$\bar{p}_D(s) = \frac{1}{s} K_o(\sqrt{s})$$

s = Laplace transform variable (6-28)

Where K_o is the modified Bessel function of zero order, the corresponding expression for a horizontal well can be lengthy, complicated, and sometimes horrible looking.

To invert p_D to real space by the Stehfest algorithm, $p_D(s)$ has to be a continuous function of s for the range of interval in question. For this reason one could not use a step function with jump discontinuities.

Roumboutsos and Stewart (1988) suggested using linear line segments in a numerical Laplace transform to transform p_D to $\bar{p}_D$, and then invert Equation 6-27 back to p_{wd} using the Stehfest algorithm, as follows:
Assuming a table of numerical functions:
In each interval $T_{i-1} < t < T_i$, the function is approximated by a linear line segment.

$$\begin{array}{ll} T_0 & f(T_0) = f_0 \\ T_1 & f(T_1) = f_1 \\ T_2 & f(T_2) = f_2 \\ \vdots & \vdots \\ T_i & f(T_i) = f_i \\ \vdots & \vdots \\ T_N & f(T_N) = f_N \end{array} \quad (6\text{-}29)$$

In each interval $T_i - 1 < t < T_i$, the function is approximated by a linear line segment.

$$f_i(t) = f_{i-1} + \dot{f}_{i-1}(t - T_{i-1}) \quad T_{i-1} \leq t \leq T_i \quad (6\text{-}30)$$

$$\dot{f}_{i-1} = \frac{f_i - f_{i-1}}{T_i - T_{i-1}} \quad (6\text{-}31)$$

$$f(t) = \begin{cases} f_1(t) & 0 \leq t \leq T_1 \\ f_2(t) & T_1 \leq t \leq T_2 \\ \vdots & \vdots \\ f_N(t) & T_{N-1} \leq t \leq \infty \end{cases} \quad (6\text{-}32)$$

$$\mathcal{L}(f(t)) = \bar{f}(s) = \int_0^{T_1} f_1(t) e^{-st} \, dt + \quad (6\text{-}33)$$

$$\int_{T_1}^{T_2} f_2(t) e^{-st} \, dt + \ldots \int_{T_{N-1}}^{\infty} f_N(t) e^{-st} dt \quad (6\text{-}34)$$

$$L(f(t)) = \tilde{f}(s) = \frac{\dot{f}_o}{s^2}(1 - e^{-sT_1})$$

$$+ \sum_{i=1}^{N-2} \frac{\dot{f}_i}{s^2}(e^{-sT_i} - e^{-sT_{i+1}}) + \frac{\dot{f}_N}{s^2} e^{-sT_{N-1}} \quad (6\text{-}35)$$

where T_i = time point
f_i = value of the function at T_i
$\dot{f}_i$ = slope of f_i at T_i

The problem with the above approach is that one needs to write a very accurate routine to handle the exponential terms with large absolute arguments. The ordinary, built-in, intrinsic routines in a personal computer for calculating exponential functions probably lose accuracy on this. We use the following method, found in *Handbook of Mathematical Functions* by Abramowitz and Stegun (1972).

Compute $\exp(-987.6543)$
Let $x = 987.6543$
Define $y = x/\ln 10 = 428.93282$
Write n = integral part of $(x/\ln 10) = 428$
Put $z = y - \text{float}(n) = 0.93282$

$$\exp(987.6543) = \exp\left(\frac{987.6543}{\ln 10} \cdot \ln 10\right)$$
$= \exp(428.93282 \ln 10)$
$= 10^{428} \exp(.93282 \ln 10)$
$= 10^{428} \exp(2.1479199)$
$\exp(-987.6543) = 1./\exp(987.6543)$
$= 10^{-428} \exp(-2.1479199) = 0.1167267 \times 10^{-428}$

For those interested in writing their own subroutine for calculating a negative exponential, expansion of $\exp(-x)$ in Chebyshev polynomials is, for $-1 \leq x \leq 1$, $\exp(-x) = 1.000045 - 1.000022\,x + 0.499199\,x^2 - 0.166488\,x^3 + 0.043794\,x^4 - 0.008687\,x^5$ and in terms of Pade approximations, defined as:

$f(x) = 1.0000000007 - 0.4759358618x + 0.0884921370x^2 - 0.0065658101x^3$
$g(x) = 1 + 0.5240642207x + 0.1125548636x^2 + 0.0106337905x^3$
$\exp(-x) = f(x)/g(x)$ error $= 7.34 \times 10^{-10}$

ANALYTICAL SOLUTION TO A GAS HORIZONTAL WELL

The physical model considered in this analysis consists of an infinite conductivity horizontal well located at the center of a semi-infinite homogeneous and isotropic medium of uniform thickness and width (Figure 6-5).

The formation fluid properties are independent of pressure. A compressible fluid is produced through the horizontal well at constant rate. The well lies on the x-axis and is parallel to the x-axis. The well is perpendicular to the linear barriers of no-flow boundary conditions which are parallel to the y-axis. Gravity effects are neglected. The well is only partially penetrating the formation.

Well Performance and Productivity 205

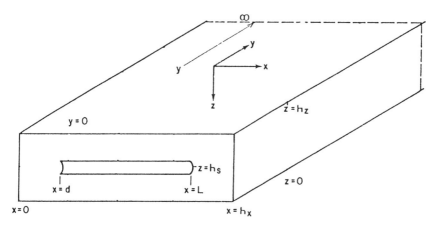

Figure 6-5. Transient pressure analysis model, horizontal well. Source: Babu and Odeh (1988).

The partial differential equation governing the isothermal flow is given by

$$\frac{\phi \mu c}{k_y} \frac{\partial \psi}{\partial t} = \frac{\partial^2 \psi}{\partial y^2} + \frac{k_z}{k_y} \frac{\partial^2 \psi}{\partial z^2} + \frac{k_x}{k_y} \frac{\partial^2 \psi}{\partial x^2} \tag{6-36}$$

$$\psi = 2 \int_o^P \frac{p'}{Z(p')\mu(p')} \, dp' \tag{6-37}$$

Where Z = gas deviation factor
C = compressibility of gas
ψ = pseudo pressure of gas
μ = viscosity

The initial and boundary conditions are:

1. The reservoir pressure is initially constant.
2. The reservoir is assumed to be semi-infinite. The pressure at infinity is not affected by the pressure disturbance at the wellbore.
3. The line source is replaced by a thin strip source of width (b − a) and length (L − d).
4. No fluid flow across the over and underburden.
5. No fluid flow across the lateral extremities of the reservoir, mathematically.

We have the following:

p = p_i as y tends to infinity for all x, z, and t   (6-38)

$$\lim_{y \to 0} (b - a)(L - d) \frac{\partial p}{\partial y} = \frac{-q\mu}{2k_y} \qquad (6\text{-}39)$$

Using the same approach as Goode and Thambynayagam (1985), we get an analytical solution to Equation 6-38, similar to Goode and Thambynayagam's, with pressure replaced by real gas pseudopressure and the unit conversion factor value changed.

NOTATION

a	length of reservoir in Equation 6-5	K_o	modified bessel function in Equation 6-28
a	vertical distance from the top reservoir boundary to the top edge of the horizontal strip source	L	horizontal well length
		l	lateral distance from the left reservoir boundary to the far end of the well
b	width of reservoir in Equation 6-4	N_p	cumulative production
		p	pressure
b	vertical distance from the top reservoir boundary to the bottom of the horizontal strip source	$<p>$	average reservoir pressure
		$\bar{p}_r$	average reservoir pressure in Equation 6-4
B	formation volume factor	$\bar{P}$	Laplace transform of p in Equation 6-28
C	compressibility in Equation 6-21	q	well flowing rate
		S	source function in Equation 6-2
C	wellbore storage in Equation 6-26	S	saturation in Equation 6-21
C_H	geometric factor in Equation 6-4	S	skin in Equation 6-26
		s	Laplace parameter in Equation 6-27
d	lateral distance from the left reservoir boundary to the bottom of the horizontal strip source	S_R	pseudo skin factor in Equation 6-4
$\dot{f}$	slope in Equation 6-30	x	space lateral coordinate
		y	space lateral coordinate
H	height	z	space coordinate
h	formation thickness	t	time
J	productivity	T	dummy variable
k	permeability		

Greek Symbols

Δ difference
ℒ Laplace transform in Equation 6-33
λ interporosity flow coefficient
φ porosity in Equation 6-2
ψ pseudo pressure in Equation 6-36
μ viscosity

Subscripts

d drainage radius
D dimensionless
e external radius of reservoir
f formation in Equation 6-21
H horizontal
o oil in Equation 6-21
M mechanical
R reservoir
V vertical
W wellbore
w water in Equation 6-21
i initial
wf well flowing

Functions

$$\frac{\sqrt{\pi}}{2} \text{ERF}(X) = \int \text{EXP}(-W \times W) \, DW$$

REFERENCES

1. Cinco-Ley, H., "Unsteady State Pressure Distributions Created by a Slanted Well or a Well with an Inclined Fracture," Ph.D. dissertation, Stanford University 1974, Department of Petroleum Engineering.
2. Gringarten, A.C. and Ramey, H.J., Jr., "Unsteady State Pressure Distribution Created by a Well with a Single Horizontal Fracture, Partial Penetration or Restricted Entry," A.I.M.E. Trans., August 1974.
3. Raghavan, R., Uraiet, A., and Thomas, G.W., "Vertical Fracture Height: Effect on Transient Flow Behaviour," Paper SPE 6016, presented at the SPE 51st Annual Technical Conference and Exhibition, New Orleans, Louisiana, October 3–6, 1976.
4. Goode, P.A. and Thambynayagam, R.K.M., "Pressure Drawdown and Buildup Analysis of Horizontal Wells in Anisotropic Media," SPE Paper 14250, September 1985.
5. Daviau, F., Mouronval, G., and Bourdarot, G., "Pressure Analysis for Horizontal Wells," SPE Paper 14251, September 1985.

6. Clonts, M.D. and Ramey, H.J., Jr., "Pressure Transient Analysis for Wells with Horizontal Drainholes," SPE Paper 15116, April 1986.
7. Kuchuk, F.J., Goode, P.A., Wilkinson, D.J., and Thambynayagam, R.K.M., "Pressure Transient Behaviour of Horizontal Wells with and Without Gas Cap or Aquifer," SPE Paper 17413, March 1988A.
8. Kuchuk, F.J., Goode, P.A., Brice, B.W., Sherrard, D.W., and Thambynayagam, R.K.M., "Pressure Transient Analysis and Inflow Performance for Horizontal Wells," SPE Paper 18300, October 1988B.
9. Carvalho, R.S. and Rosa, A.J., "Transient Pressure Behaviour for Horizontal Wells in Naturally Fractured Reservoirs," SPE Paper 18302, October 1988.
10. Aguilera, R. and Ng, M.C., "Transient Pressure Analysis of Horizontal Wells in Anisotropic Naturally Fractured Reservoirs," SPE Paper 19002, 1988.
11. Joshi, S.D.: "Augmentation of Well Productivity with Slant and Horizontal Wells," *Journal of Petroleum Technology*, June 1988.
12. Babu, D.K. and Odeh, A.S.,"Productivity of a Horizontal Well," SPE Paper 18298, October 1988.
13. Karcher, B.J., Giger, F.M., and Combe, J., "Some Practical Formulas to Predict Horizontal Well Behaviour," SPE Paper 15430, October 1986.
14. Gringarten, A.C. and Ramey H.J., Jr., "The Use of Source and Green's Functions in Solving Unsteady Flow Problems in Reservoirs," SPE Paper 3818, 1973.
15. Streltsova, T.D., "Well Testing in Heterogeneous Formations," John Wiley and Sons, First Edition, 1988.
16. Havlena, D. and Odeh, A.S., "The Material Balance as an Equation of a Straight Line," *Journal Of Petroleum Technology*, August 1963.
17. Craft, B.C. and Hawkins, M.F., "Applied Petroleum Reservoir Engineering," Prentice Hall, 1959.
18. Murtha, J.A., "Infill Drilling in the Clinton, Monte Carlo Techniques Applied to the Material Balance Equation," SPE Paper 17068, October 1987.
19. Sherrard, S.W., Brice, B.W., and MacDonald, D.G., "Application of Horizontal Wells at Prudhoe Bay," *Journal of Petroleum Technology*, November, 1987.
20. Reiss, L.H., "Production from Horizontal Wells After Five Years," *Journal of Petroleum Technology*, November 1987.
21. Van Everdingen, A.F. and Hurst, W., "The Application of the Laplace Transformation to Flow Problems in Reservoirs," *Petroleum Transactions, A.I.M.E.*, December 1949.
22. Agarwal, R.G., Al-Hussainy, R., and Ramey, H.J., Jr., "An Investigation of Wellbore Storage and Skin Effect in Unsteady Liquid Flow," *Society of Petroleum Engineers Journal*, September 1970.

23. Cinco-Ley, H. and Samaniego, F.V., "Effect of Wellbore Storage and Damage on the Transient Pressure Behaviour of Vertically Fractured Wells," SPE Paper 6752, October 1977.
24. Roumboutsos, A. and Stewart, G., "A Direct Deconvolution or Convolution Algorithm for Well Test Analysis," SPE Paper 18157, October 1988.
25. Abramowitz, M. and Stegun, A., "Handbook of Mathematical Functions," Dover, New York, 1972.
26. Kabik, C.S., Kuchuk, F.J., and Hasan, A.R., "Transient Analysis of Acoustically Derived Pressure and Flow Rate Data," SPE Paper 15481, October, 1986.
27. Mead, H.N., "Using Finite System Buildup Analysis to Investigate Fractured, Vugular, Stimulated and Horizontal Wells," *Journal of Petroleum Technology*, October 1988.
28. Ikoku, C.U. and Ramey, H.J., Jr., "An Investigation of Wellbore Storage and Skin Effects during the Transient Flow of Non-Newtonian Power-Law Fluids in Porous Media," SPE Paper 7449, October 1978.
29. Box, M. J., "A New Method of Constrained Optimization and a Comparison with Other Methods," *Computer Journal* (1965) *vol.* 8, p. 42.
30. Fletcher, R., "A Modified Marquardt Subroutine for Nonlinear Least Squares," *Atomic Energy Research Establishment Report*, 6799 (May 1971).
31. Warren, J. E. and Root, P.J., "The Behavior of Naturally Fractured Reservoirs," *Society of Petroleum Engineers Journal*, (Sept. 1963), pp. 245–255.

7
Thermal Recovery and Primary Production for Heavy Oils

INTRODUCTION

To employ thermal recovery methods with horizontal wells (HWs) frequently requires different, more specialized approaches than are typically followed in conventional HW activities. It is important that each of the disciplines involved in the drilling, completion, and operation of thermal HWs be aware of the specifications required for a successful outcome. For example, the steam assisted gravity drainage (SAGD) process holds tremendous potential for economically unlocking many large scale heavy oil deposits. Success requires stringent specifications:

1. Drilling—A form of SAGD requires 2 HWs to be drilled one on top of the other within as short a distance as 15 ft and that this distance be maintained over HW lengths greater than 1,500 ft. In many conventional HW recovery processes the vertical tolerance for HWs need not be as tight.
2. Completions—A form of SAGD requires uniform injection of steam along greater than 1,500 ft of HW section. Without uniform injection, sweep conformance can be diminished which influences the overall economic viability of the project.
3. Reservoir engineering—To be effective the SAGD process requires a constant balancing of injected volumes to produced fluids, otherwise process efficiency can suffer severely. Injection rates have to be speci-

fied to equal production rates. Conventional vertical well (VW) thermal recovery processes do not require such careful balancing of the fluids injected to fluids produced.

4. Operations—The SAGD process frequently requires that injection pressures be controlled constantly to within ± 5 psi, otherwise harmful steam breakthrough which can affect performance can occur at the producer HW. Similarly, the production of steam must be controlled such that a minimal amount of steam is produced from the HW at bottomhole conditions.

To achieve a successful project outcome requires coordination of these disciplines. Each must understand what the other disciplines are trying to achieve. The disciplines may have to employ different approaches to the norms for conventional HWs, if the project is to be viable.

Ultimately, for thermal recovery HW processes, patience is required. Conventional HWs invariably have their highest production rates near to the start of their productive life, then the rates decline over time. HW thermal recovery processes frequently require one to two years of reservoir heating before peak rates are achieved. Very low oil rates can be observed initially. Fortitude and a long-term commitment are required to recognize that productivity will eventually increase and that the economics associated with the fully realized project will be acceptable. A sound, fundamental understanding of the thermal recovery process is required to confidently predict that productivity will increase with time.

One area of activity associated with HW thermal recovery focuses on approaches to increase early productivity. Encouraging developments are being studied, primarily in laboratory environments (Joshi and Threlkeld, 1984; Toma et al., 1984; Anderson, 1988). The 1990s undoubtedly will see some of the more promising lab developments proceed to field scale evaluations.

OVERVIEW

Horizontal well recovery processes for heavy oils can be divided into two main categories:

1. Thermal recovery
2. Primary production

The first category involves the injection of heat into a reservoir to increase reservoir temperature. Usually steam at saturated conditions is used as the heating medium. Increasing reservoir temperature decreases the viscosity of the in situ oil phase. Reduced oil viscosity increases the rate of recovery of the oil phase. Figure 7-1 illustrates the relationship of saturated steam injec-

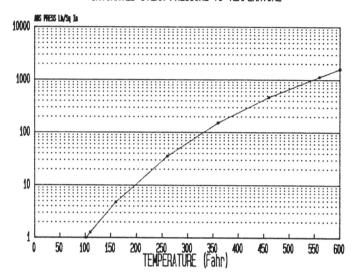

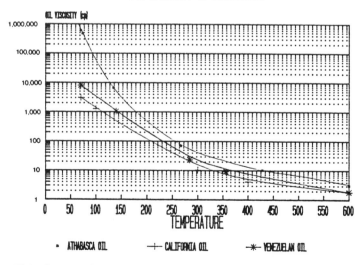

Figure 7-1. Saturated steam pressure vs temperature; oil viscosity vs temperature.

Thermal Recovery and Primary Production for Heavy Oils 213

tion pressure to that of steam temperature. In high viscosity oil reservoirs, without the injection of heat, oil productivity typically is very low.

The second category for heavy oil processes is similar to conventional recovery processes. In this group of reservoir environments the in situ oil is mobile enough to be produced by primary production means. A HW serves to substantially increase primary productivity via greater contact area with the reservoir, as is the case for conventional HW recovery. However, the productivity improvement for primary production HWs in homogeneous heavy oil reservoirs, over that of VWs, is theoretically much greater than the productivity improvement of HWs in conventional oil reservoirs relative to VWs. This greater improvement in recovery is discussed in the section entitled "Primary Recovery of Heavy Oils."

Primary production heavy oil HW processes can incorporate thermal recovery approaches to further improve oil productivity via the reduction in oil viscosity. As a consequence, there can be overlap between the two categories of heavy oil recovery processes.

In this chapter, the term heavy oil will define all oils that have an in situ gravity of less than 15° API, or a viscosity of greater than 500 cp at original reservoir conditions.

Example 7-1

Assess the difference in productivity for an Athabasca oil at original temperature of 74° F versus the pressuring up of the reservoir to 1,000 psia with steam. Assume all reservoir conditions except oil viscosity are the same for both cases.

Darcy's law for radial well inflow

$$q_o = \frac{7.08 \, Kh}{B_o \, \mu} \frac{(P_e - P_w)}{\ln(r_e/r_w)}$$

where $K = 2.0$ D
 $h = 50$ ft
 $P_e - P_w = 500$ psia
 $\ln(r_e/r_w) = \ln(500 \text{ ft}/0.3 \text{ ft})$
 $B_o = 1$ resbbl/stb

From Figure 7-1:

μ at virgin conditions = 1,000,000 cp

μ at 1,000 psia injection pressure corresponding to 540° F = 5 cp

214 Horizontal Wells

q_o at virgin conditions = 0.048 bbl/d
q_o at 1,000 psia steam injection pressure = 9,544 bbl/d

Productivity of oil increases by a factor of 200,000 by increasing the reservoir temperature from 70° F up to 540° F. It is preferable to heat the reservoir rather than accept low oil productivity at initial reservoir conditions.

BASIS FOR STUDYING HEAVY OIL RECOVERY PROCESSES

With the recent realization that the world's supply of conventional oil is limited, attention has begun to focus on the heavier, more viscous oils. The resource base for heavy oils is tremendous. It has been conservatively estimated that the volume of heavy oil in the world amounts to 4,070,000,000,000 (four trillion) barrels (Meyer and Fulton, 1982). This resource is widely dispersed throughout the world, making it attractive to many countries for reasons of internal security of oil supply. Deposits in Alberta, Canada and in the Orinoco Belt of Venezuela exceed 1 trillion barrels of oil in the ground. Table 7-1 provides a breakdown of the world's heavy oil resources by country.

Table 7-1
World Bitumen Resources

Country	Number of Deposits	OOIP (10^6 bbl)
Albania	1	371
Canada	7	2,432,600
Cuba	1	undetermined
Italy	4	14,155
Madagascar	2	24,800
Mexico	3	undetermined
Nigeria	1	undetermined
Peru	1	66
Romania	1	25
Trinidad and Tobago	1	60
USSR	10	557,024
USA	53	34,390
Venezuela	4	1,000,012
West Germany	4	undetermined
Zaire	1	1,925
Total	94	4,065,428

Source: Meyer and Fulton, 1982.

The challenge faced for exploitation of the heavy oil resource base has been to be economically cost competitive with conventional oil recovery operations. To date, only a select few of the heavy oil deposits have been exploited commercially (Carrigy, 1982). The low productivity derived from the highly viscous oil has been typically insufficient for economic profitability. HW heavy oil processes offer an opportunity for commercial exploitation of the large resource base.

Butler (1984) espoused the following advantages associated with HWs over conventional wells in heavy oil reservoirs:

Heavy Oil from Steamed Reservoirs

1. More production per well
2. Better recovery
3. Better oil steam ratio
4. Better economics

Primary Production

1. More contact with the reservoir
2. One well can replace several conventional ones

HW pilot projects have evolved as a consequence of the perceived benefits associated with HWs. The benefits of HWs were identified in:

1. The laboratory, in bench scale experiments;
2. The derivation of analytical expressions describing the dominant recovery mechanisms;
3. Numerical reservoir simulation studies.

This chapter discusses the advantages of HWs as identified in the lab, by analytical expressions, and from numerical simulation studies.

THERMAL RECOVERY PROCESSES

Thermal recovery processes are divided into two main categories. One group is known as steam assisted gravity drainage (SAGD) because the dominant reservoir mechanism is gravity drainage. The second group is referred to as displacement processes. There can be overlap between SAGD and displacement processes. The basis for employing SAGD HWs in reservoirs that contain essentially immobile oil at original reservoir conditions is to have a high level of control over the placement of heat in the reservoir. VWs in reservoirs with limited fluid mobility have had to typically fracture the reservoir to place acceptable quantities of heat into the reservoir. Fracturing of reservoirs frequently establishes recovery processes that make recovery prediction difficult and are difficult to control over the full life of the project. As a consequence, ultimate recovery of the original oil in place (OOIP) can be low. On the other hand, HWs enable the placement of heat in highly controlled directions that are often best suited to the particular geology of the reservoir.

Effective steam flooding of immobile oil reservoirs requires fluid communication between injection and production wells. With VW processes, the mobility contrast between the injected fluid and the initial immobile oil is typically so large that sweep conformance is very poor between injector and producer wells. The outcome is low recovery of the oil in place. Because HWs contact more reservoir compared to VWs, the potential exists to substantially improve the conformance and hence ultimate recovery from immobile oil reservoir environments. Example 7-2 illustrates the relationship of displacement recovery of OOIP relative to the viscosity differences between oil and the displacing water phase. For high oil to water viscosity ratios, recovery of the OOIP is low, based on fractional flow analysis.

VWs have been employed in reservoirs where higher water saturation strata can take injected fluids at acceptable rates. In this manner the reservoir need not be fractured. These natural communication paths can become plugged when heated: hot oil is cooled by being displaced into unheated reservoir. The plugging of strata by the viscous oil in VW processes diminishes the potential ultimate recovery of oil. Tam et al. (1984) describes the intrinsic difficulties associated with using natural communication paths for placement of heat from VWs. First, hot water, then steam needs to be injected into the mobile water strata to ensure that, between wells, heated oil does not cool in advance of the heat front and plug the communication path. HW processes enable communication paths to remain open because heat can be maintained along the length of HW throughout the entire life of the project.

In reservoirs that have negligible mobility, only SAGD, or a combination of SAGD and displacement processes merit investigation for HWs. In reservoirs with acceptable initial oil mobility either SAGD or displacement processes may be employed. The advantage of HWs over VWs is to reduce steam override in displacement processes by locating the HWs near to the base of the reservoir.

Thermal recovery processes for heavy oil reservoirs are evaluated principally by computer. Analytical expressions have been derived and computer programs written to enable ease of use for evaluating processes and different reservoir environments. In addition, numerical reservoir simulators have evaluated thermal processes. The numerical simulators are capable of evaluating a wide variety of reservoir environments, whether the environments are homogeneous or heterogeneous. The simulators have the flexibility of evaluating multiwell processes where the sequencing of injection fluids can have an influence on performance.

Thermal Recovery and Primary Production for Heavy Oils 217

In practice, output from analytical expressions and numerical simulators should be viewed as results that provide a directional indication of the performance to be expected, as opposed to an absolute indication of performance. That is, for each given reservoir environment, sensitivity studies on a variety of possible recovery processes will establish which process potentially will provide the superior performance. The actual productivity derived from the process is usually only attainable from field projects.

Examples 7-2 through 7-8 illustrate:

1. Recovery efficiency based on oil to water viscosity ratio
2. The need for fracturing VWs in highly immobile reservoir environments to place adequate quantities of steam in the reservoir
3. The volume of steam required to heat a given volume of reservoir to a temperature T
4. The volume of reservoir fluids required to be displaced by the injected steam
5. The time required to inject the steam to achieve a set reservoir temperature
6. The horizontal pressure drop per foot of reservoir relative to the vertical pressure drop per foot of reservoir

Example 7-2: Oil Recovery for Very Viscous Oils Based on Fractional Flow Analysis

Assume no forces due to gravity.
Assume no forces due to capillary pressure.

$$q_o = \left[\frac{k_o A}{\mu_o}\right]\left[\frac{\partial P_o}{\partial x}\right]$$ Darcy's Law for horizontal displacement in reservoir barrels of oil

and

$$q_w = \left[\frac{k_w A}{\mu_w}\right]\left[\frac{\partial P_{wa}}{\partial x}\right]$$ Darcy's Law for water

Because we assume no capillary force, $\partial P_{wa} = \partial P_o$.
Establish $q_t = q_o + q_w$

$$q_t = \left[\frac{k_o A}{\mu_o}\right]\left[\frac{\partial P_o}{\partial x}\right] + \left[\frac{k_w A}{\mu_w}\right]\left[\frac{\partial P_o}{\partial x}\right]$$

or $q_t = A \left[\dfrac{\partial P_o}{\partial x}\right] \left[\dfrac{k_o}{\mu_o} + \dfrac{k_w}{\mu_w}\right]$

Now $q_w = \left[\dfrac{k_w A}{\mu_w}\right] \left[\dfrac{\partial P_o}{\partial x}\right]$

Then $\dfrac{q_w}{q_t} = f_w = \dfrac{\left[\dfrac{k_w A}{\mu_w}\right] \left[\dfrac{\partial P_o}{\partial x}\right]}{A\left[\dfrac{\partial P_o}{\partial x}\right] \left[\dfrac{k_o}{\mu_o} + \dfrac{k_w}{\mu_w}\right]}$

Cancelling terms provides

$$\dfrac{q_w}{q_t} = f_w = \dfrac{\dfrac{k_w}{\mu_w}}{\dfrac{k_o}{\mu_o} + \dfrac{k_w}{\mu_w}}$$

or dividing the numerator and denominator by μ_w/k_w, the equation becomes

$$f_w = \dfrac{1}{\dfrac{\mu_w}{k_w}\dfrac{k_o}{\mu_o} + 1}$$

For simplicity, assume that k_o and k_w are straight line relationships (see Figure 7-2).

For the permeability relationship assumed, at the following oil-water viscosity ratios, the in situ water saturations at 99% fractional water flow are:

μ_o/μ_w	f_w	In situ Water Saturation (S_w) at 99% Water Cut
1,000	0.99	0.28
100	0.99	0.57
1	0.99	0.89

In other words

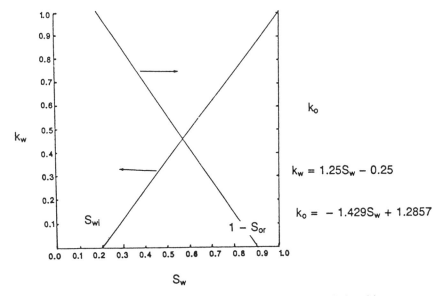

Figure 7-2. Oil and water idealized permeability relationship.

μ_o/μ_w	Pore Volume of Oil Recovered $[(S_w \text{ at } 99\%) - (S_w \text{ initial})]$	% Recovery of OOIP
1,000	0.08	10.00
100	0.37	46.25
1	0.69	86.25

Refer to Figure 7-3 for a depiction of the fractional water cut as a function of water saturation analysis.

For very viscous oils, recovery of OOIP is typically low when the displacing medium is water. It is only when the viscosity of oil approaches that of water that displacement efficiency, in terms of recovery of OOIP, is high.

Figure 7-3a depicts how the oil-water viscosity ratio changes as a function of temperature for 3 different gravity oils. For the higher API oils, the viscosity ratio is not excessive and thus fractional flow recovery should be acceptable if effective permeabilities of the oil and water phases are comparable. It is only by increasing temperature for the low API oils that the viscosity ratio begins to approach unity, which may provide an acceptable recovery efficiency.

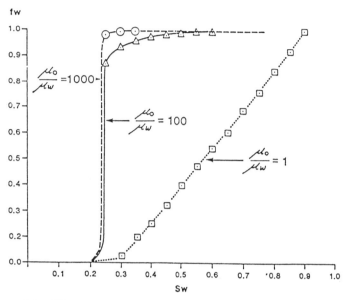

Figure 7-3. Oil water viscosity ratio as a function of fractional water flow.

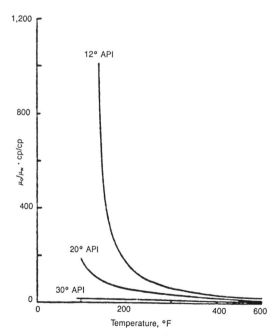

Figure 7-3a. The effect of temperature on oil-water viscosity ratio (Courtesy of Gates and Ramey and Dingley) (after Wu, 1977).

Thermal Recovery and Primary Production for Heavy Oils 221

Example 7-3: Steam Injection Rate Below Fracture Reservoir

Assess the maximum steam injection rate below fracture pressure for a VW, given the following:

K = 2.0 D fracture pressure = 1,000 psia
h = 50 ft original reservoir pressure = 300 psia
  oil viscosity = 100,000 cp
ln (re/rw) = ln (500 ft/0.3 ft) B_o = 1.0 resbbl/stb

Assume that the steam must displace oil in a pistonlike manner.

$$q \text{ steam} = \frac{7.08 \text{ Kh (Pe} - \text{Pw)}}{B_o \, \mu \, \ln (r_e/r_w)}$$

q steam maximum below fracture pressure = −0.67 bbl/d at reservoir conditions

The negative sign indicates fluid flowing into the reservoir. Typically, steam is injected at 500 bbl/d to 2,000 bbl/d (cold water equivalent on surface) into VWs, to heat the reservoir as quickly as possible.
For this example, to provide acceptable injectivity from the VWs the reservoir would have to be fractured.

Example 7-4: How Oil Saturation (S_o) and Water Saturation (S_w) Affect Heat Capacity of Oil Sands at Different Reservoir Temperatures

Given that the specific heat of oil C_o is:

$$C_o = (0.388 + 0.00045T)\Big/\sqrt{\rho_{0,T,1}}$$

C_o is in (BTU/lbm °F)
T is the desired temperature in (°F)
$\rho_{0,T,1}$ is the density of the oil at T in dimensionless units

$$\rho_{0,T,1} = \rho_{0,1} - \frac{0.3444T - 11.022}{1,000}$$

$\rho_{0,1}$ is at standard conditions of 60°F

$$\rho_{0,1} = \frac{141.5}{°API + 131.5}$$

Combining the above equations provides

$$C_o = \frac{(0.388 + 0.00045T)}{\sqrt{\dfrac{141.5}{°API + 131.5} - \dfrac{0.3444T - 11.022}{1{,}000}}}$$

Given that the specific heat of saturated water C_w is:

$$C_w = 1.0504 - 6.05 \times 10^{-4}T + 1.79 \times 10^{-6}T^2$$

C_w is in (BTU/lbm °F)

Given that the density of saturated water at T is:

$$\rho_{w,T} = \frac{1}{(0.01602 + 0.000023\,G)}$$

where $G = -6.6 + 0.0325T + 0.000657T^2$

$\rho_{w,T}$ is in lbm/ft^3

Given that the heat capacity of sand C_s is:

$$C_s = 0.00006T + 0.18$$

C_s is in (BTU/lbm °F)
Given that the density of sand (ρ_s) is:

sand (fine)	102 lbm/ft^3
sand (coarse)	109 lbm/ft^3

Assume that the density of sand is not a function of temperature.

The volumetric heat capacity of a fluid-saturated oil sand is established by the sum of the specific heat of its constituents.

$$M = \rho_{0,T} C_o S_o \phi + \rho_{w,T} C_w S_w \phi + \rho_s C_s (1 - \phi)$$

ϕ = porosity
$\rho_{0,T} = \rho_{0,T,1} \times 62.4$ lbm/ft^3

For T = 212°F and an 11°API oil:

$C_o = 0.501$ BTU/lbm °F $\rho_{o,T} = 58.09$ lbm/ft³
$C_w = 1.0026$ BTU/lbm °F $\rho_{w,T} = 59.86$ lbm/ft³
$C_s = 0.193$ BTU/lbm °F $\rho_s = 105$ lbm/ft³

For T = 500°F and an 11°API oil:

$C_o = 0.672$ BTU/lbm °F $\rho_{o,T} = 51.90$ lbm/ft³
$C_w = 1.1954$ BTU/lbm °F $\rho_{w,T} = 49.95$ lbm/ft³
$C_s = 0.210$ BTU/lbm °F $\rho_s = 105$ lbm/ft³

The following tables may be set up for different ϕ, S_o, and S_w:

For T = 212°F

$\phi = 0.15$			$\phi = 0.25$			$\phi = 0.35$		
S_o	S_w	M (BTU/ft³°F)	S_o	S_w	M (BTU/ft³°F)	S_o	S_w	M (BTU/ft³°F)
0.0	1.00	26.23	0.0	1.00	30.20	0.0	1.00	34.18
0.50	0.50	23.91	0.50	0.50	26.34	0.50	0.50	28.77
1.00	0.0	21.59	1.00	0.0	22.47	1.00	0.0	23.36

For T = 500°F

$\phi = 0.15$			$\phi = 0.25$			$\phi = 0.35$		
S_o	S_w	M (BTU/ft³°F)	S_o	S_w	M (BTU/ft³°F)	S_o	S_w	M (BTU/ft³°F)
0.0	1.00	27.70	0.0	1.00	31.47	0.0	1.00	35.23
0.50	0.50	25.84	0.50	0.50	28.36	0.50	0.50	30.89
1.00	0.0	23.97	1.00	0.0	25.26	1.00	0.0	26.54

The above "M" values are multiplied by either 212°F or 500°F to provide the heat content per ft³ of oil sand at the respective temperatures. Figure 7-3b depicts the heat capacities.

Note that at 212°F for 50% S_o and 50% S_w changing ϕ from 0.15 to 0.35 changes the value of M by 20.3%.

Note that at 500°F for 50% S_o and 50% S_w changing ϕ from 0.15 to 0.35 changes the value of M by 19.5%.

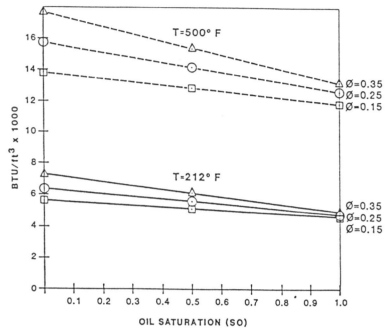

Figure 7-3b. Heat content of 1 ft³ of liquid saturated oil sand.

Changing the porosity ϕ by a factor of 2.33 times affects M by 20%. M does not change substantially for changes in ϕ.

Similarly, for a porosity of 25%, the value of M changes by 7.7% by evaluating M at 212°F relative to M at 500°F. M does not change substantially for changes in temperature.

Example 7-5: Volume of Steam Necessary to Heat 5 Acres of Reservoir to 212°F and 500°F

Given:

Original reservoir temperature	50°F
Porosity	25%
Oil API	11°
Oil saturation	70%
Water saturation	30%
Reservoir thickness	50 ft
Steam quality	70%

Thermal Recovery and Primary Production for Heavy Oils 225

- Assume that there is no heat loss to the overburden and underburden.
- Assume that steam is at saturated conditions and the gaseous steam condenses to 100% water at saturated conditions.
- Assume that the injected steam volume does not increase the water saturation over the 5 acres.
- Assume that the steam uniformly heats the 5 acres.

Using the information in Example 7-4:

M at 212°F is 24.41 BTU/ft^3 °F
M at 500°F is 26.81 BTU/ft^3 °F

There is heat capacity already within the reservoir at the original temperature of 50°F because the datum temperature is 32°F.
Following the approach of Example 7-4:

M at 50°F is 23.20 BTU/ft^3 °F

The bulk volume of reservoir is (5 acres)(43,560 ft^2/acre)(50 ft) = 10,890,000 ft^3
The number of BTUs originally in the reservoir at 50°F is

(10,890,000 ft^3)(23.30 BTU/ft^3 °F)(50°F − 32°F)

The datum for all heat capacity values is 32°F.
The number of BTUs in the reservoir at 212°F is

(10,890,000 ft^3)(24.41 BTU/ft^3 °F)(212°F − 32°F)

The number of BTUs in the reservoir at 500°F is

(10,890,000 ft^3)(26.81 BTU/ft^3 °F)(500°F − 32°F)

BTUs at 32°F over 5 acres	0
BTUs at 50°F over 5 acres	4,547,664,000
BTUs at 212°F over 5 acres	47,848,480,000
BTUs at 500°F over 5 acres	136,637,700,000

The number of BTUs from steam condensation to heat the reservoir to 212°F is

47,848,480,000 − 4,547,664,000 = 43,300,810,000

The number of BTUs from steam condensation to heat the reservoir to 500°F is

136,637,700,000 − 4,547,664,000 = 132,090,000,000

From Figure 7-1, saturated steam at 212°F is at a pressure of 14-7 psia. Also from Figure 7-1, saturated steam at 500°F is at a pressure of 680.9 psia. From steam tables at 212°F the heat content of the injected steam at 70% quality is

(0.7)(1,150.5 BTU/lbm gas) + (0.3)(180.16 BTU/lbm water)
= 859.4 BTU/lbm

After all of the steam condenses to saturated water at 212°F the heat content is

(1.0)(180.17 BTU/lbm water) = 180.17 BTU/lbm

The heat given up to the oil sand per lbm of steam is

859.4 − 180.17 = 679.23 BTU/lbm

Similarly, for the 500°F steam, from steam tables, the heat content of the injected steam at 70% quality is

(0.7)(1,202.2 BTU/lbm gas) + (0.3)(487.9 BTU/lbm water)
= 987.9 BTU/lbm

The heat given up to the oil per lbm of steam is

987.9 − 487.9 = 500.0 BTU/lbm

The lbm of steam needed to heat the reservoir to 212°F is

$$\frac{43,300,810,000 \text{ BTUs}}{679.23 \text{ BTUs/1bm}} = 63,749,861 \text{ lbm}$$

The cold water equivalent (CWE) volume of steam is

$$\frac{63,749,861 \text{ lbm}}{62.4 \text{ lbm/ft}^3} \times 0.178 \text{ bbl/ft}^3$$

The CWE needed to heat the reservoir to 212°F is 181,960 bbl.

Thermal Recovery and Primary Production for Heavy Oils 227

The lbm of steam needed to heat the reservoir to 500°F is

$$\frac{132{,}090{,}000{,}000 \text{ BTUs}}{500.0 \text{ BTU/lbm}} = 264{,}180{,}000 \text{ lbm}$$

The CWE volume of steam is

$$\frac{264{,}180{,}000 \text{ lbm}}{62.4 \text{ lbm/ft}^3} \times 0.178 \text{ bbl/ft}^3$$

The CWE needed to heat the reservoir to 500°F is 753,590 bbl.

Note that approximately 4 times the CWE of steam is needed to heat the reservoir to 500°F compared to heating the reservoir to 212°F. The benefits of using 4 times as much steam to heat the reservoir to 500°F over that of 212°F has to be weighed against the rate of production to be achieved for the oil at 500°F versus 212°F.

Following Example 7-1, and using Athabasca oil for the viscosity of the oil in Figure 7-1, the following are obtained:

μ_o at 212°F = 300 cp
μ_o at 500°F = 7.5 cp
q_o at 212°F = 159.1 bbl/d (assuming a 500 psia pressure drop)
q_o at 500°F = 6,362.4 bbl/d

By injecting 4 times the volume of steam to heat the reservoir to 500°F as compared to 212°F, the oil productivity will increase by a factor of 40.

In Example 7-5 it was assumed that the water saturation would remain constant at $S_w = 0.25$. Originally the volume of water within the 5 acres was

(5 acres)(43,560 ft²/acre)(50 ft)((0.25)(0.25)(0.178 bbl/ft³)
= 121,151.3 bbl of water

To heat the reservoir to 212°F or 500°F would involve increasing the water saturation substantially over the 5 acres because 181,960 bbl or 753,590 bbl of CWE steam would be necessary to heat the reservoir to 212°F or 500°F.

Either the oil would have to be displaced out of the 5 acres to place the CWE of steam into the reservoir or cycles of steam injected and CWE of steam produced would be necessary to heat the reservoir to either of the two temperatures.

Example 7-6: Volume of 70% Quality Steam at 212°F and 500°F Relative to the CWE Volume of Steam Required

At 212°F

The saturated liquid density = 59.81 lbm/ft^3
The saturated steam density = 0.037 lbm/ft^3

At 500°F

The saturated liquid density = 48.95 lbm/ft^3
The saturated steam density = 1.48 lbm/ft^3

1 bbl of CWE steam equates to (62.4 lbm/ft^3)(5.6146 ft^3/bbl) = 350.3 lbm/bbl
From Example 7-5, to heat 5 acres to 212°F requires 181,960 bbl of CWE steam.
Similarly, to heat 5 acres to 500°F requires 753,590 bbl of CWE steam.
For the 212°F case the CWE steam equates to 63,740,588 lbm.
For the 500°F case the CWE steam equates to 263,982,577 lbm.
The volume of steam occupied at 212°F is

$$(0.3 \text{ quality}) \left[\frac{1}{59.81} \frac{\text{ft}^3}{\text{lbm}} \right] + (0.7 \text{ quality}) \left[\frac{1}{0.037} \frac{\text{ft}^3}{\text{lbm}} \right]$$

= 18.92 ft^3/lbm

The volume of 70% quality steam at 212°F is

(63,740,588 lbm)(18.92 ft^3/lbm)
= (1,206,222,731 ft^3)(0.178 bbl/ft^3)
= 214,707,646 bbl at 212°F.

The CWE of steam has increased by 1,180 times by heating 70% quality steam to 212°F.
The volume of steam occupied at 500°F is

$$(0.3 \text{ quality}) \left[\frac{1}{48.95} \frac{\text{ft}^3}{\text{lbm}} \right] + (0.7 \text{ quality}) \left[\frac{1}{1.148} \frac{\text{ft}^3}{\text{lbm}} \right]$$

= 0.479 ft^3/lbm

The volume of 70% quality steam at 500°F is

(263,982,577 lbm)(0.479 ft^3/lbm)(0.178 bbl/ft^3) = 22,507,682 bbl

The CWE of steam has increased by 29.9 times by heating 70% quality steam to 500°F.

The volume of steam injected at in situ conditions is substantially higher than the CWE volume of steam reported on surface. As discussed in Example 7-5, the volume of oil needed to be displaced over 5 acres of reservoir so that the steam may be injected is greater than the CWE volume indicated, recognizing that the steam at the 212°F or 500°F conditions occupies a greater volume relative to the CWE value reported.

The factors for converting CWE to steam temperature volumes are the maximum increase in volumes for the particular temperature of 212°F and 500°F. As steam is injected into the reservoir it will condense, or partially condense. The volume of condensed steam to 100% water at saturated conditions would represent the lower limit of the expansion of steam at in situ conditions. For 212°F 100% water at saturated conditions the volume is 189,691 bbl which is 1.04 times the volume of CWE steam. For 500°F water at saturated conditions the volume is 959,937 bbl which is 1.27 times the volume of CWE steam. The increase in fluid volume at reservoir temperature of 212°F or 500°F will be somewhere between the upper limit which assumed no condensation of 70% quality steam and the lower limit which assumed full condensation of steam to 100% saturated water at 212°F or 500°F.

The volume of injected steam at reservoir conditions can be very large and can occupy a significant portion of the porous space available for flow. Occupying a significant portion of the porous space establishes the reservoir environment wherein fluid interference between wells can be observed. Fluid interference frequently is harmful to thermal recovery processes and attempts are often made to avoid interference, if at all possible.

Example 7-7: Length of Time Needed to Inject 70% Quality Steam into a 5-Acre Reservoir such that the Reservoir is Either 212°F or 500°F

Using the information in Examples 7-4 and 7-5, the volume of CWE steam needed to heat the reservoir is

Heat reservoir to 212°F   181,960 bbl CWE
Heat reservoir to 500°F   753,590 bbl CWE

230   *Horizontal Wells*

Assume that the injection rate is 1,000 bbl/d CWE of 70% quality steam. Assume that there are no heat losses to the overburden and underburden.
To heat the reservoir to 212°F would require 182 days (0.5 yr)
To heat the reservoir to 500°F would require 754 days (2.1 yr)
   A considerable length of time is required to heat the reservoir to 500°F and to have the reservoir oil viscosity sufficiently low such that high oil productivity is achieved.

Example 7-8: Magnitude of Pressure Drop per ft of Displacement in the Reservoir for Steam and Oil; Horizontal and Vertical Displacement

Horizontal Flow
Given:

100% steam quality flow rate	1,000 bbl/d CWE
Reservoir thickness	50 ft
Portion of reservoir that steam flows through	10 ft
Reservoir permeability to steam	2.0 D
Spacing per well	5 acres
(263.3 ft radius to center)	

From Figure 7-3c, steam viscosity = 0.0125 cp at 212°F
                                = 0.0185 cp at 500°F

Assume that all steam flows radially as depicted in Figure 7-3d. The pounds of steam injected per day are

   (1,000 bbl/d CWE)(5.618 ft^3/bbl)(62.4 lbm/ft^3) = 350,563 lbm/day

The density of steam at 212°F is 0.0373 lbm/ft^3
The density of steam at 500°F is 1.482 lbm/ft^3
The volume of steam flowing through the reservoir is
at 212°F:

   (350,563 lbm/day)(1/0.0373 ft^3/lbm)(0.178 bbl/ft^3) = 1,672,928 bbl/day

at 500°F:
   (350,563 lbm/day)(1/1.482 ft^3/lbm)(0.178 bbl/ft^3) = 42,105 bbl/day

Thermal Recovery and Primary Production for Heavy Oils 231

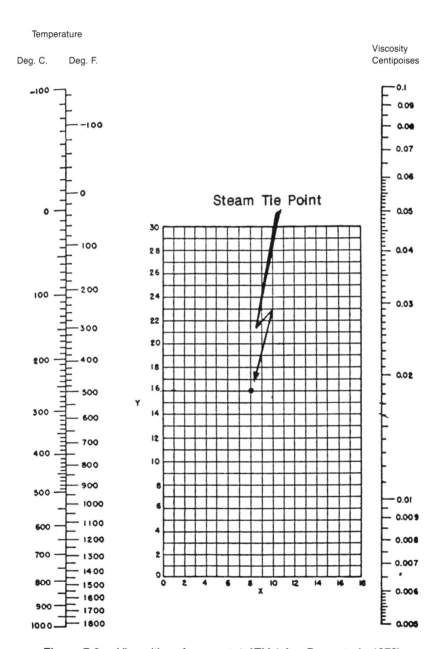

Figure 7-3c. Viscosities of gases at 1 ATM (after Perry et al., 1973).

Horizontal Wells

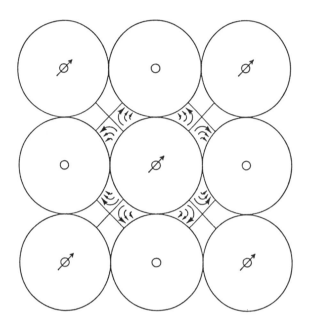

○ Producer
⌀ Injector
〈〈 Fluid is assumed to move instaneously from the injector radii to the produce radii at the points shown, for the purposes of the evaluation.

Figure 7-3d. Horizontal displacement assumed radial flow configuration.

The pressure drop *midway* between the injector and the producer is given by:
For 212°F

$$1{,}672{,}928 \text{ bbl/d} = \frac{(7.08)(2.0 \text{ D})(10 \text{ ft})(\Delta P \text{ per ft})}{(0.0125 \text{ cp}) \ln(263.3/262.3)}$$

$\Delta P = 0.562$ psi/ft for 1,000 bbl/d CWE steam at 212°F.

For 500°F

$$42{,}105 \text{ bbl/d} = \frac{(7.08)(2.0 \text{ D})(10 \text{ ft})(\Delta P \text{ per ft})}{(0.0185 \text{ cp})\ln(263.3/262.3)}$$

$\Delta P = 0.021$ psi/ft for 1,000 bbl/d CWE steam at 500°F.

Thermal Recovery and Primary Production for Heavy Oils 233

In radial terms the midpoint between the injector VW and producer VW is the location where the smallest pressure drop per ft is observed.

Injecting 1,000 bbl/d of CWE 100% quality steam at 212°F is not feasible because the pressure associated with 212°F is 14.7 psia. The pressure drop from the injector well to the midpoint between injector and producer is given by:
For 212°F

$$1{,}672{,}928 \text{ bbl/d} = \frac{(7.08)(2.0 \text{ D})(10 \text{ ft})(\Delta P)}{(0.0125 \text{ cp}) \ln(263.3 \text{ ft}/0.3 \text{ ft})}$$

$$P_{well} - P_{mid} = \Delta P = 1{,}000.9 \text{ psia}$$

The steam is at 14.7 psia for 212°F steam and would become undersaturated water at pressures between 1,000 and 2,000 psia at the injection well for a fluid temperature of 212°F.

Steam at 500°F equates to 680.9 psia pressure (from Figure 7-1).
For 500°F steam, the pressure drop between the injector well and midpoint between injector and producer is given by:

$$42{,}105 \text{ bbl/d} = \frac{(7.08)(2.0 \text{ D})(10 \text{ ft})(\Delta P)}{(0.0185 \text{ cp}) \ln(263.3 \text{ ft}/0.3 \text{ ft})}$$

$$P_{well} - P_{mid} = \Delta P = 37.28 \text{ psia.}$$

The reservoir bbl of steam at 500°F are capable of maintaining a practical pressure drop between the injector and producer wells because 100% quality steam at 500°F does not create an excessive pressure drop for the reservoir configuration assumed.

In reservoir terms it is more reasonable to observe displacement pressure drops for steam saturated flow paths on the order of 0.021 psi/ft as opposed to 0.562 psi/ft.

It is left to the reader to determine the flow rate for the remaining 40 ft of the 50 ft thick reservoir. It was assumed that the remaining 40 ft contained oil. Because oil has a substantially higher viscosity than steam, the flow rate from the 40 ft oil column, with such a small pressure gradient applied (as controlled by the steam zone pressure drop), would make oil productivity very low. Edmunds (1984) provides a detailed discussion on the steam drag effect for low mobility oils. Typically, very low oil to steam ratios are obtained because the pressure drop between injector and producer is low as a result of the minor pressure drop through the steam zone.

Vertical Flow

The maximum pressure drop that can be established vertically by gravity alone is given by:

$$\Delta P = \frac{(\rho_o - \rho_{steam}) \, gh}{g_c}$$

This expression assumes that the vertical permeability is large. If vertical permeability were small the pressure drop would be greater than indicated in the example.

Using the information from Examples 7-4 and 7-6:

At 212°F

$\rho_o = 58.09 \text{ lbm/ft}^3 \qquad \rho_{steam} = 0.037 \text{ lbm/ft}^3$

At 500°F

$\rho_o = 51.90 \text{ lbm/ft}^3 \qquad \rho_{steam} = 1.48 \text{ lbm/ft}^3$

At 212°F

$$\Delta P = [58.09 \text{ lbm/ft}^3 - 0.037 \text{ lbm/ft}^3] \left[\frac{\text{lbf} - \sec^2}{32.2 \text{ lbm-ft}} \right] [32.2 \text{ ft/sec}^2] \, [1 \text{ ft}]$$

$$\Delta P = (58.05 \text{ lbf/ft}^2)(\text{ft}^2/144 \text{ in}^2)$$

or

$$\Delta P = 0.403 \text{ psia per foot of vertical drop}$$

At 500°F

$$\Delta P = [51.90 \text{ lbm/ft}^3 - 1.48 \text{ lbm/ft}^3] \left[\frac{\text{lbf} - \sec^2}{32.2 \text{ lbm-ft}} \right] [32.2 \text{ ft/sec}^2] \, [1 \text{ ft}]$$

$$\Delta P = (50.42 \text{ lbf/ft}^2)(\text{ft}^2/144 \text{ in}^2)$$

or

$$\Delta P = 0.350 \text{ psia per foot of vertical drop}$$

Thermal Recovery and Primary Production for Heavy Oils 235

Vertical flow due to gravity is a maximum of 0.350 psia to 0.403 psia per foot of reservoir for the conditions assumed.

Horizontal flow due to a pressure gradient as a result of displacement is in the range of 0.562 psia to 0.021 psia per horizontal foot of displacement. As discussed, 0.562 psia per horizontal foot is not physically feasible for the conditions assumed.

For 500°F reservoir conditions the vertical gravity forces are 16.67 times greater than the horizontal displacing forces (0.35 psia/ft versus 0.021 psi/ft).

In oil sand reservoirs gravity may be a more effective means to recover oil relative to horizontal displacement of the oil in terms of the higher pressure drop per ft than can be obtained with gravity relative to horizontal displacement. Provided that means are available, it may be preferable to displace viscous oils down vertically rather than displace oil horizontally because of the higher pressure drops that can be achieved for vertical recovery versus horizontal recovery. Vertical permeability must be sufficient for the vertical movement of oil.

Examples 7-2 through 7-8 serve to illustrate the following:

1. Based on fractional flow analysis, displacement efficiency for very viscous oils can be low.
2. Typically, for highly viscous oil reservoirs, fracturing is necessary to inject reasonable volumes of steam—at least at the start of the injection process.
3. A large volume of steam is required to heat the reservoir to temperatures where the oil viscosity is substantially reduced.
4. The volume of steam injected to heat the reservoir to high temperatures is large. Consequently, a combination of the following occurs:

   - The reservoir is fractured to take the additional fluid volume.
   - Oil is displaced from the reservoir volume and/or
   - The reservoir porosity increases.

   Often, deleterious well fluid communication occurs because of the above 3 points, as a result of the viscosity difference between the oil and water phases. (That is, water flows through the reservoir far more easily than the oil phase and, typically, only water is observed at surrounding wells).

5. The pressure drop due to gravity head can be larger than the pressure drop due to horizontal displacement. A high pressure drop is required to displace oil horizontally. If steam breakthrough between wells occurs, there can be insufficient pressure drop for the oil phase to be mobilized between wells. This is due to the pressure drop in the steam phase between wells being lower than the pressure drop necessary to

236 Horizontal Wells

mobilize the oil. Displacement of oil by gravity offers a means to alleviate the difference in pressure drop between the oil and steam phases.

STEAM ASSISTED GRAVITY DRAINAGE (SAGD) PROCESS

SAGD Concept

SAGD is a reservoir recovery process whereby steam is introduced above a horizontal producer located near to the base of the reservoir. The manner in which steam can be injected is either via a HW or a VW well above the HW producer. Figure 7-4 depicts the process in cross section with the HW coming out of the page. Figure 7-4 illustrates an established SAGD reservoir environment. Steam is injected essentially continuously into a zone that has

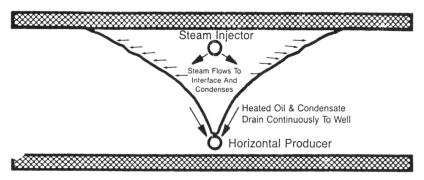

Figure 7-4. Steam-assisted gravity drainage with horizontal producer (adapted from Butler et al., 1979).

been depleted of oil via the SAGD process. The depleted oil region is saturated primarily with steam. Oil in the vicinity of the interface between the steam and oil zones is heated to steam temperature primarily via conduction heating. Through heating, the viscosity of the oil decreases and becomes substantially more mobile. As a consequence of the density difference between the steam and the heated mobile oil, the oil in the layer of reservoir that surrounds the steam zone drains downward by gravity to the producer HW and is removed from the reservoir along with the condensate from the steam that has condensed at the interface. Oil is removed from the reservoir at a rate such that steam is not produced by steam coning from above the steam injector location. That is, a liquid level is maintained to some degree above the HW producer such that steam is not produced. Steam replaces the

Thermal Recovery and Primary Production for Heavy Oils 237

oil in the region that has drained by gravity. Thus the process continues to perpetuate itself with the steam/water/oil interface growing outward into the oil zone. Figure 7-5 illustrates the mechanistic concept of SAGD. The outer extent that the steam zone may eventually reach can be limited by the boundary of the reservoir or by interference with drainage patterns of neighboring SAGD wells.

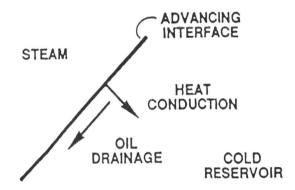

Figure 7-5. Gravity drainage mechanism (after Butler, 1988).

The concept of countercurrent gravity drainage flow within porous reservoir media is analogous to that of countercurrent flow within packed columns, which has been used for gas-liquid mass transfer contact. Countercurrent packed columns, as illustrated in Figure 7-5a, have been used extensively in the chemical and petroleum refining industries for decades. A packed countercurrent column consists of a liquid distribution system at the top of the column and a gas inlet at the bottom. Due to density differences between the gas and liquid, the gas rises up through the packed column and the liquid flows down and is collected at the base of the vessel. The column is essentially at constant pressure. Extensive correlations have been developed within the chemical engineering literature relating column packing configuration, packing material, gas flow rates up, and liquid flow rates down to the pressure drop characteristics of the column.

The concept of countercurrent flow within reservoir porous media has only recently been developed and, as a result, correlations of the physical process have not been as extensively established as correlations developed for packed columns. However, both concepts are similar, as is illustrated in Figure 7-5a. Both have gas flowing into the system from the bottom and liquid from the top draining down through packed porous material. Gas in the reservoir system is removed through steam condensation to water via heat transfer to reservoir material rather than explicitly being removed from the system, as is the case for surface packed vessels.

238   Horizontal Wells

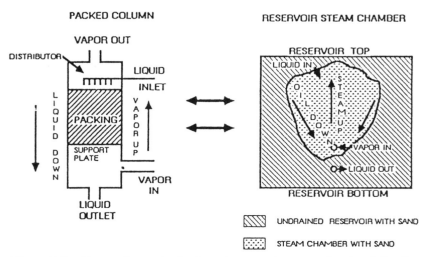

Figure 7-5a. Comparing packed column to reservoir steam chamber (after Suggett et al., 1985).

Countercurrent flow has been proven to be viable within porous media for petroleum refinery columns. Similarly, various oil sand physical model studies have been undertaken which have demonstrated that countercurrent flow is viable within oil sands material.

Benefits of SAGD

The perceived benefits of SAGD over VW thermal processes are:

1. Higher oil productivity relative to the number of wells employed
2. Higher volumes of oil produced to volumes of steam injected (oil to steam ratio, OSR)
3. Higher ultimate recovery of oil in place
4. Minimization of deleterious well interference or too early link up between wells
5. Reduced sand production

The section entitled "SAGD vs Vertical Well Cyclic Steam" discusses points 1–4 by using numerical simulation results.

Productivity. Because a HW has a substantially larger surface area open to flow than a VW, higher productivity is achieved. In addition, as a consequence of the oil being continuously produced at essentially steam temperature, high oil mobility is maintained, whereas in certain VW processes oil temperature, and hence mobility, decrease over time, often in a cyclical manner.

Thermal Recovery and Primary Production for Heavy Oils 239

OSR. The concentration of heat within the confines and vicinity of the steam zone minimizes heat losses to overlying and underlying strata. In addition, the concentration of heat applies to the oil zone. In VW processes there is a tendency for heat to be more widely diffused in the oil zone which does not provide for the highest possible oil production temperatures.

Ultimate Recovery. The SAGD process establishes high areal and vertical conformance via the high surface area of the HW in contact with the reservoir and via the recovery of oil by the gravity drainage mechanism.

Well Interference. In VW displacement processes or multiwell cyclic steam stimulation processes, because of the high mobility contrast between the oil and the steam, conformance can be poor. Undesirable steam production at production wells is often the result. (Refer to Example 7-2 to illustrate low conformance and hence low oil recovery.) The SAGD process intrinsically balances injection of steam to production of liquids over a localized region of the reservoir. As a consequence, pressure and temperature transients don't extend greatly beyond the region of the steam heated zone and there is minimal interference between wells until such time that the process has matured to a substantial extent and linkup between wells is desirable.

Sand Production. Due to the large surface area open to flow in the HW producer, relative to a VW producer, fluid velocities in the near wellbore region are reduced. HW thermal experiences have been that sand production is minimized when compared to adjacent VWs that produce copious volumes of sand.

Limitations of SAGD

The two principal shortcomings of SAGD are:

1. SAGD may not be applicable to reservoir environments that have a low absolute vertical permeability.
2. The maximum feasible length of the HW section is constrained by steam pressure drop considerations.

Vertical Permeability. The SAGD process relies on ease of movement of both steam upward and drainage of oil downward to the HW producer. If absolute vertical permeability is low, the rate of displacement by gravity forces will be deleteriously affected. Similarly, if vertical barriers exist in the reservoir the rate of gravity displacement will diminish and hence the rate of oil recovery will be reduced. Refer to the sections entitled "Growth of Steam

Zone for Different Vertical Permeabilities" and "Influence of k_v in Mobile Oil Reservoirs" later in this chapter for examples of the influence of vertical permeability on performance.

HW Length. For a pair of HWs—an injector and a producer—the maximum possible length of the HW injector is influenced by the pressure drop within the HW. Due to the close proximity of HW injector and HW producer, reservoir pressure drops between producer and injector are small. To ensure uniform steam conformance across the HW section, the pressure at the injection end of the HW injector must not be too high or steam could rapidly channel to the producer HW. Figure 7-6 illustrates the concept.

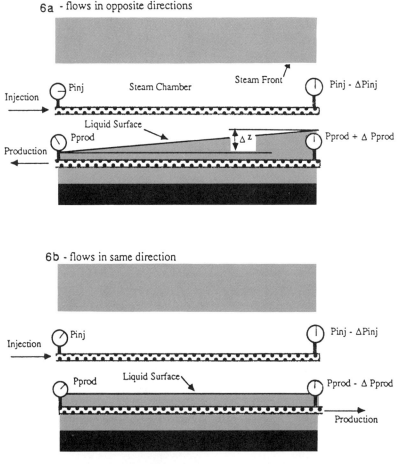

Figure 7-6. Effect of well pressure gradients on fluid distribution (after Edmunds et al., 1987).

Thermal Recovery and Primary Production for Heavy Oils 241

The net result of the pressure drop relationship is that for a given size of HW injector pipe the HW effective steam injector length is governed by the pressure drop across the HW injector. Otherwise, steam will channel between injector and producer at one end of the pair of wells. The diameter of the HW injector pipe determines the maximum length of the HW.

The use of VWs above and along the length of the HW producer could alleviate the HW length limitation. However, VW injectors do not place steam as uniformly above the HW producer as does a HW injector, unless the VW injectors are spaced very close to each other. The section called "SAGD Laboratory Experiments" describes the work by Joshi and Threlkeld (1984) that indicates HW injectors are superior to VW injectors.

Example 7-9

1. Calculate the pressure drop for a 6-inch diameter HW, 1,500 ft in length, with 300 bbl/d of 11° API oil and 700 bbl/d of water at saturated conditions flowing through it at a temperature of 500°F.
2. Calculate the pressure drop for a 6-inch-diameter HW, 1,500 ft in length, with 700 bbl/d of CWE steam flowing through it at a temperature of 500°F.

Solutions.

1. Assume that the viscosity of the oil and water mixture behaves as that of the oil phase.

    Assume that all oil and water flow through the entire length of pipe. The density of the mixture at 500°F is given by:

    oil = (300 bbl/d)(5.618 ft^3/bbl)(61.96 lbm/ft^3)

    water = (700 bbl/d)(5.618 ft^3/bbl)(52.4 lbm/ft^3)

    104,427 lbm/d oil at standard conditions

    245,394 lbm/d water at standard conditions

    From Example 7-4: the density of oil at 500°F is 51.90 lbm/ft^3
    the density of water at 500°F is 49.95 lbm/ft^3

    The volume of oil and water at 500°F is

    2,012.1 ft^3/d of oil   and   4,912.8 ft^3/d of water

Assume that the volumes of fluid are additive 6,924.9 ft³/d of total fluid flowing through the pipe.
The Reynolds number R_e is given by:

$$R_e = \frac{(\text{Diameter})(\text{Average velocity})(\text{Density})}{\text{Viscosity}}$$

From Figure 7-1 the assumed viscosity of the mixture is that of the oil viscosity, which is 7.5 cp.

$$Re = \frac{\left[\dfrac{6 \text{ in.}}{12 \text{ in./ft}}\right]\left[\dfrac{6{,}924.9 \text{ ft}^3/\text{day}}{((3/12)^2\pi)\text{ft}^2}\right]\left[\dfrac{\text{day}}{24 \text{ hr}}\right]\left[\dfrac{\text{hr}}{3{,}600 \text{ sec}}\right][50.54 \text{ lbm/ft}^3]}{(7.5 \text{ cp})[6.72 \times 10^{-4} \text{ lbm/ft-sec-cp}]}$$

Re = 2,046.8

The Reynolds number is approximately 2,000, which indicates that the fluid is in laminar flow.
For laminar flow the pressure drop (ΔP) across the HW is given by:

$$\Delta P = q \, \frac{8}{\pi} \, \frac{\mu}{(r_w)^4} \, \Delta x$$

where q = flow rate
r_w = radius of the pipe
Δx = length of the HW.

$$\Delta P = 6{,}924.9 \, \frac{\text{ft}^3}{\text{day}} \left[\frac{12 \text{ in.}}{\text{ft}}\right]^3 \left[\frac{\text{day}}{24 \text{ hr}}\right]\left[\frac{\text{hr}}{3{,}600 \text{ sec}}\right] \frac{8}{\pi} \left[\frac{7.5 \text{ cp}}{(3 \text{ in.})}\right]$$

$$1{,}500 \text{ ft} \left[\frac{\text{ft}}{12 \text{ in.}}\right] \left[2.09 \times 10^{-5} \, \frac{\text{lb ft-sec}}{\text{cp ft}^2}\right]$$

ΔP = 0.085 lbf/in² (psia) over the 1,500 ft length of the HW.

At temperatures of 500°F, in a 6-inch-diameter wellbore with 1,000 bbl/d of fluid flowing through 1,500 ft of HW, the pressure drop is negligible.

2. Pressure drop for 500°F steam over 1,500 ft of HW length.
   700 bbl/d of 100% quality steam CWE equates to 29,500 bbl/d at reservoir conditions

Thermal Recovery and Primary Production for Heavy Oils 243

From Example 7-8 the viscosity of steam is 0.0185 cp.
The Reynolds number is given by:

$$\frac{6 \text{ in.}}{12 \text{ in./ft}} \quad \frac{1.917 \text{ ft}^3}{\text{sec}} \quad \frac{\text{lbm}}{0.6749 \text{ ft}^3} \quad \frac{1}{\pi(0.25 \text{ ft})^2}$$
$$(0.0185\text{cp}) \ (6.72 \times 10^{-4} \text{ lbm/ft-sec cp})$$

$$R_e = 5.818 \times 10^5$$

The Reynolds number is greater than 4,000, which indicates that the steam is in turbulent flow.

$$\frac{\text{Pressure at inlet}}{\text{Density at inlet}} - \frac{\text{Pressure at outlet}}{\text{Density at outlet}} = -4f \frac{\Delta x}{D} \frac{(\text{Velocity})^2}{2}$$

Where f = the friction factor.

Assume that the surface roughness is 0.0009.
The relative roughness is

$$\frac{0.0009 \text{ ft}}{\text{Diameter}} = \frac{0.0009 \text{ ft}}{0.5 \text{ ft}} = 0.0018$$

From Figure 7-6a the friction factor f is 0.00575.
The pressure drop is given by:

$$(-4)(0.00575) \frac{1,500 \text{ ft}}{0.5 \text{ ft}} \left[1.917 \frac{\text{ft}^3}{\text{sec}} \frac{1}{\pi(0.25)^2 \text{ ft}} \right]^2 1/2$$

$$= 3,288.54 \text{ ft}^2/\text{sec}^2$$

Assume that the density remains constant for the pressure drop observed

$$\Delta P = \left[3,288.54 \frac{\text{ft}^2}{\text{sec}^2} \right] \left[1.4817 \frac{\text{lbm}}{\text{ft}^3} \right] \left[\frac{\text{lbf-sec}^2}{32.2 \text{ lbm-ft}} \right] \left[\frac{\text{ft}^2}{144 \text{ in.}^2} \right]$$

$$= 1.051 \text{ lbf/in}^2 \text{ over the 1,500 ft of HW length}$$

Provided that steam can be uniformly distributed over the 1,500 ft of HW length the pressure drop would not be considered large for a 6-inch-diameter pipe.

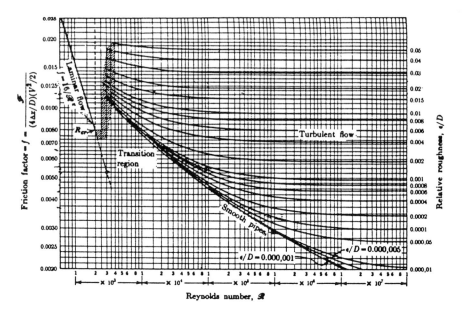

Figure 7-6a. Friction-factor plot for pipes (after Moody, 1944).

```
              681 psia                         680 psi
Steam in    ─────────────────────────────────────────
            |◄──────────── 1,500 ft ────────────►|

              679.9 psi                       680 psi
Oil out     ─────────────────────────────────────────
```

Note that there is a driving pressure gradient at one end of the HW. For 15 ft of separation between the HW injector and producer, the pressure head due to a steam column between the wells would amount to 0.154 psia. The pressure drop due to flow in the wells is in the order of 1.1 psia at the inlet end of the HWs. Steam would channel between the HWs because the driving force pressure could exceed the stable head of steam column between the wells.

To overcome the steam channel, the liquid level above the HW producer would have to be maintained higher than just at the the HW wellbore so that the 1.1 psia pressure drop would become small relative to the liquid column head. If a liquid level head were maintained to the base of the HW injector, the head would amount to 5.26 psia over the 15 ft of separation.

Thermal Recovery and Primary Production for Heavy Oils 245

The pressure drop would not be excessive relative to the head between the wells. Figure 7-6b is an illustration of the pressure drops.

If it is deemed important to maintain the liquid level at just above the HW producer to minimize the pressure drop along the injector HW, a larger HW diameter could be considered.

The pressure drops discussed above are considered the minimum pressure drops across the wells. Usually, tubular strings are run nearly to the end of the HW. Flow is either through the tubing or the annulus. In both instances, the surface area open to flow is smaller than the area considered for the above example. Thus, pressure drops could be higher than those already discussed. Balancing the potentially higher pressure drops is the fact that not

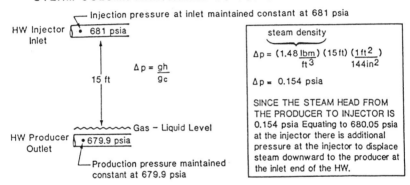

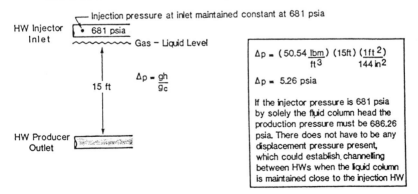

Figure 7-6b. Pressure distribution for different fluid columns above the HW.

246 *Horizontal Wells*

all fluid passes through the entire length of the HW. Rather, steam is usually injected into the reservoir along the entire length of HW and fluid produced from along the entire length of HW.

Field Applications of SAGD

One field project has applied the specific concept of SAGD. Other field projects have employed variations of the SAGD process. Here are four examples of field projects that have utilized the SAGD concept.

1. The Underground Test Facility (UTF) operated by the Alberta Oil Sands Technology and Research Authority (AOSTRA)
2. The Cold Lake HW projects operated by Esso Resources Canada Ltd.
3. The Tangleflags project operated by Sceptre Resources Ltd.
4. Yarega, USSR

AOSTRA UTF

In 1987, AOSTRA commenced operations of a HW pilot 40 mi west of Fort McMurray, Alberta. The pilot HWs access the reservoir from a tunnel placed beneath the target reservoir. Access to the tunnel is via two shafts 10 ft in diameter. The shaft depth from surface is 700 ft (Best et al., 1985).

The average HW length is 200 ft. The distance between the HW injector and HW producer averages 16 ft. The well spacing between the pairs of HWs is 80 ft. Figure 7-7 illustrates the well configuration. The reservoir characteristics of the target McMurray Formation are provided in Table 7-2.

Operations commenced with the preheating of the HW pairs. First, hot water, and then steam were circulated down the annulus and through the tubing until adequate temperature existed between the HW injectors and HW producers. At the juncture of adequate heating the SAGD process commenced with essentially continuous injection of steam into the upper HWs and continuous production from the lower HWs.

AOSTRA has been extremely encouraged by the performance to date: "The field results have been more than satisfactory and it appears that Phase A will be successful in all major test efforts" (Edmunds et al., 1988). AOSTRA plans to commence a 3-HW-pair pilot in 1991 with HWs that will average 1,500–3,000 ft in length. The forecast recovery from a commercial operation employing 1,500 ft long wells is 230,000 bbl per well pair at an oil to steam ratio of 0.27 bbl cold water equivalent steam per barrel of oil recovered. Ultimate recovery of the original oil in place is projected to be 54% (Best et al., 1985).

Thermal Recovery and Primary Production for Heavy Oils   247

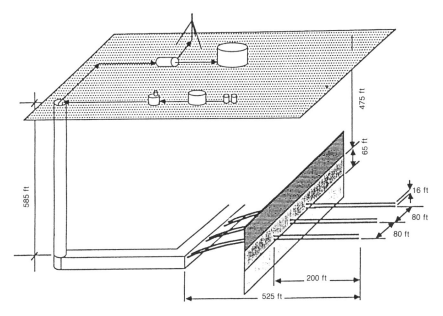

Figure 7-7. Schematic of UTF Phase A facilities (after Edmunds et al., 1987).

Table 7-2
Average Reservoir Characteristics of UTF Site

Depth	492 ft
Net pay	65 ft
Porosity	35 %
Oil saturation	81 %
Water saturation	19 %
Horizontal permeability	4 D
Formation pressure	65 psi
Formation temperature	52 °F
Oil density	6 °API

Source: Best et al., 1985.

Esso Cold Lake

Esso commenced an evaluation of the SAGD process at Cold Lake, Alberta in 1978. One 820 ft HW was drilled near to the base of the reservoir and 1 VW drilled above the HW. The true vertical depth of the HW is 1,560 ft. Steam has been injected into the VW in a cyclic manner. Through 1982, the HW had produced over 100,000 bbl of oil (Mainland and Lo, 1983). Operations are continuing at the well.

In 1985, Esso drilled a second HW at Cold Lake and placed 2 VWs above the HW producer. Approximately 3,430 ft of reservoir was intersected in the horizontal plane (MacDonald, 1985).

Sceptre Tangleflags

The Sceptre Tangleflags operation commenced in 1987. The project is located in Saskatchewan, Canada. The pilot consists of one 1,380 ft long HW producer and four VW injectors, spaced at different distances from the HW (Figure 7-7a). Reservoir characteristics at Tangleflags are depicted in Table 7-3. The oil is capable of flowing without the injection of heat into the reservoir. To increase oil productivity into the HW, steam is injected into a gas zone above the oil column. The steam blanket moves downward into the vicinity of the HW. A fluid level is maintained above the HW such that steam production is not severe. Beneath the HW exists a 3–15 ft water column. By maintaining higher pressure in the steam zone than in the underlying aquifer, water production from the aquifer is minimized. Figure 7-8 conceptually depicts the Tangleflags reservoir environment. Hence, SAGD is capable of working in reservoir environments that have a gas cap, an aquifer, or both present.

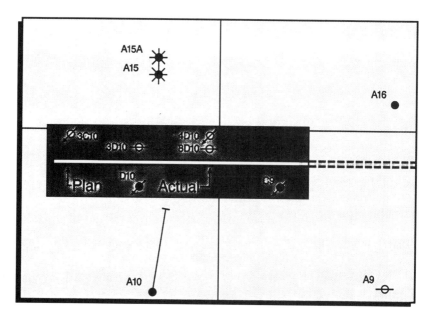

Figure 7-7a. Tangleflags North steamflood project area (after Jesperson, 1989).

Thermal Recovery and Primary Production for Heavy Oils

Table 7-3
Tangleflags North Steamflood Pilot

Reservoir & Fluid Properties	
Depth	1,475 ft
Oil zone thickness	89 ft
Porosity	33%
Oil saturation	80%
Permeability	5 Darcies
Initial temperature	66°F
Initial pressure	591 psig
Oil gravity	12–13°API
Oil viscosity	13,000 cp @66°F

After Jesperson, 1989.

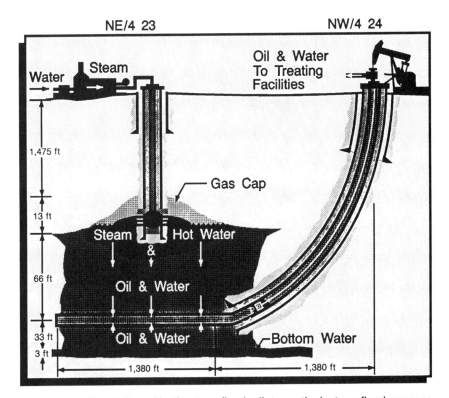

Figure 7-8. Tangleflags North steamflood pilot; vertical steamflood process schematic (after Jesperson, 1989).

250 Horizontal Wells

Performance from the pilot has been quite encouraging. After 12 months on production the 1,380 ft-long HW was producing at a rate of 285 bbl/d. Cumulative oil production for the first 13 months was 63,000 bbl oil and 460,000 bbl of water. The calendar-day oil rate from the HW equates to 160 bbl/d (Jesperson, 1989).

Yarega, USSR

At Yarega, in Russia, HWs have been drilled since 1972 from underground chambers and steam has been injected above the HWs from separate, nearly vertical wells (Butler, 1984). Oil is produced from the HWs until steam production from the HWs is observed. At the onset of steam production, the HWs are shut in to enable the liquid level to build up above the HWs. Experience dictates how long each HW should be shut in, until such time that production can commence again.

In the next section, Joshi and Threlkeld (1984) describe how, in the laboratory, steam production is controlled by monitoring production temperature. The lab production was stopped whenever the production temperature approached saturated steam temperature. When the production temperature had dropped, production resumed. The Yarega field approach to controlling steam production is comparable to that employed in the lab. Thus, in field operations the monitoring of bottomhole temperature may be a means to control steam production.

SAGD Laboratory Experiments

The basis for the understanding of the SAGD process evolved from lab work done mainly by Butler et al. (1979). Butler's bench scale lab observations led to the development of analytical expressions that describe the dominant SAGD mechanisms. The lab work by Joshi and Threlkeld (1984), Griffin and Trofimenkoff (1984), and Toma et al. (1984) added to the knowledge base on the variations of the SAGD process.

Butler (1980) described particular physical model studies in a subsequent paper:

> "The series of photographs shown in Figures 7-9 to 7-14 shows the sequential development of a steam chamber due to gravity drainage of oil during continuous steam flow into the reservoir. The steam chamber is the lighter region developing in the centre of the pictures. This is the portion of the reservoir from which the majority of the oil has drained. In this model the

text continued on page 254

Thermal Recovery and Primary Production for Heavy Oils

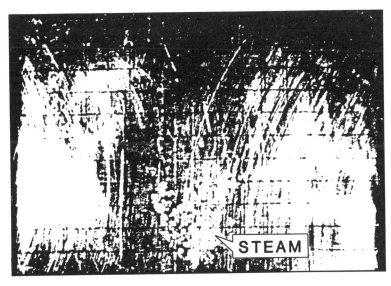

Figure 7-9. Development of the steam chamber during gravity drainage in the laboratory model (after Butler et al., 1980).

Figure 7-10. Development of the steam chamber during gravity drainage in the laboratory model (after Butler et al., 1980).

252 *Horizontal Wells*

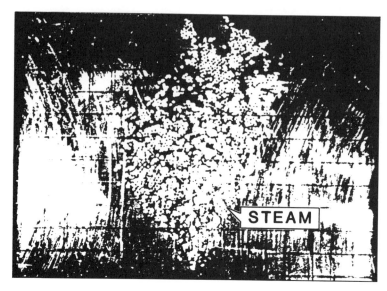

Figure 7-11. Development of the steam chamber during gravity drainage in the laboratory model (after Butler et al., 1980).

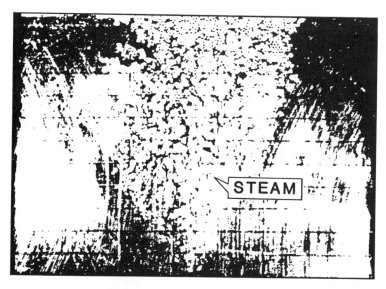

Figure 7-12. Development of the steam chamber during gravity drainage in the laboratory model (after Butler et al., 1980).

Thermal Recovery and Primary Production for Heavy Oils 253

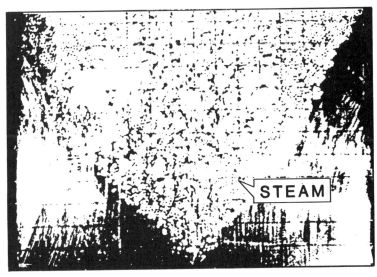

Figure 7-13. Development of the steam chamber during gravity drainage in the laboratory model (after Butler et al., 1980).

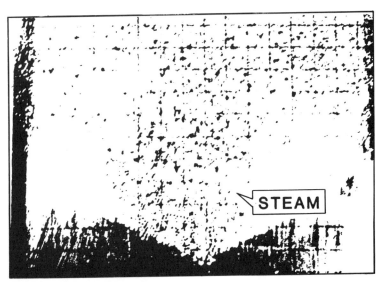

Figure 7-14. Development of the steam chamber during gravity drainage in the laboratory model (after Butler et al., 1980).

text continued from page 250

steam inlet and oil outlets are both positioned near the bottom of the reservoir with the inlet 1 in. above the outlet. Steam flows in at atmospheric pressure to replace the hot oil and condensate flowing out. The residual oil saturation in the steam chamber is typically about 5%."

In Figures 7-9 to 7-14, steam is observed to move upward *with no applied pressure gradient* other than the density difference between the steam and liquid phases. As oil is removed from the physical model, steam flows in to replace the volume of both the oil and steam condensate that are recovered. Butler's work indicates that physically countercurrent steam-oil gravity drainage processes can work. Conceptually, the process is simple in that through heat transfer to saturated oil sand a volume of steam condenses to water condensate. Additional steam flows into the steam zone to replace the collapsed steam volume. Furthermore, steam flows in to replace the oil and water that drain to the base of the chamber and are removed from the system. Butler has demonstrated the mechanisms impacting upon performance by developing appropriate analytical equations that have matched the performance from the physical model experiments. The analytical equations are described in the next section.

Since Butler's studies other researchers have confirmed the SAGD results. Joshi and Threlkeld (1984) investigated the gravity drainage process for three separate well configurations. They studied HW pairs, vertical injection wells above a horizontal production well, and a single vertical injection/production well which incorporated a packer between the injection and production points. Each configuration is illustrated in Figure 7-15. Joshi's physical model results indicated that the two-HW concept performed slightly better than the other two concepts, as is indicated in Figure 7-16. Joshi concluded that the two-HW pair recovered oil more effectively because "it heats the reservoir more uniformly." The similarity between Joshi's results for the three different well configurations may have been due to the actual physical size of the experiments. In a field application with wells spaced further apart than the lab wells there may be a larger difference in performance between each configuration.

Joshi's results further supported Butler's concept of the gravity drainage process in that he schematically depicted the mechanisms of gravity drainage in a similar manner to Butler (Figure 7-17). Furthermore, Joshi's physical model pictures of the growth of steam zone upward were similar to Butler's (Figure 7-18). Joshi summarized the manner in which steam production was minimized even though a negligible pressure drop existed between injection and production points.

text continued on page 258

Thermal Recovery and Primary Production for Heavy Oils 255

Scheme I: A Horizontal Well Pair

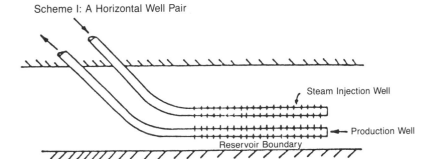

Scheme II: Vertical Steam Injection Wells and a Horizontal Production Well

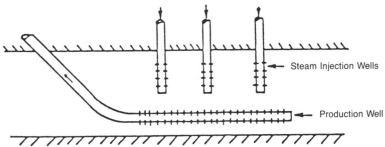

Scheme III: Single Vertical Injection and Production Well

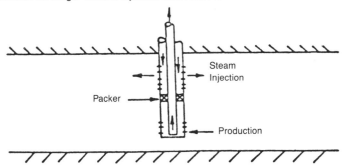

Figure 7-15. Well configuration schemes (after Joshi et al., 1984).

256 Horizontal Wells

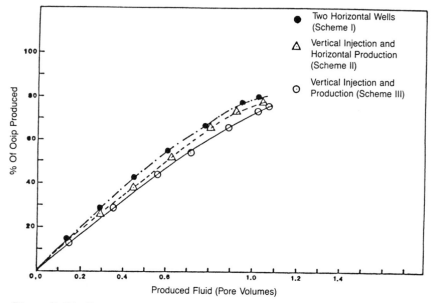

Figure 7-16. Percentage of OOIP recovered using different production schemes (after Joshi et al., 1984).

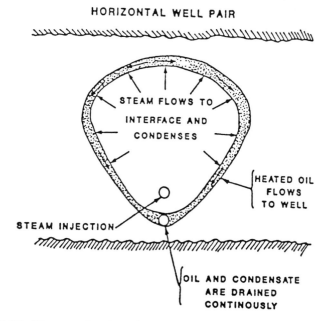

Figure 7-17. Steam assisted gravity drainage concept (after Joshi et al., 1984).

Thermal Recovery and Primary Production for Heavy Oils 257

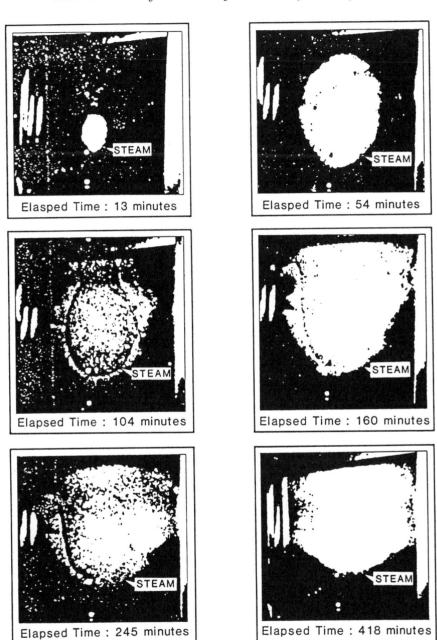

Figure 7-18. Typical steam bubble development in a uniform permeability model (after Joshi et al., 1984).

text continued from page 254

"In each experiment, 4-7 psig steam with 2-3°F of superheat was injected into the porous pack through the injection well. Although only a small pressure drop existed between the injection and production wells, there was a significant temperature difference between the two wells. At the beginning of steam injection, the production well temperature was about 71°F. As time progressed, the temperature of the produced fluid measured at the production well increased. This temperature was not allowed to exceed 180-190°F so that a liquid leg (sump) would always exist around it. As noted earlier, the throttle valve on the production well and the limited pressure drop between the injection and production wells prevented steam short-circuiting. This helped low density steam to rise upward and form a bubble above the injection well. The progress of the steam bubble front with time is shown in Figure 7-18. As seen in the figure, when most of the reservoir was swept, the steam broke through to the production well and the experiment was terminated."

Joshi also investigated the influence of placing vertical fractures between and above the HW pairs. Figure 7-18a illustrates the configurations studied, with the resultant recoveries compared to the situation where no fractures existed. With fractures, early time performance increases by a factor of two compared to the uniform reservoir case. In the event that uniform fractures can be placed along the entire length between HW pairs and vertically above the HW injector, early time productivity could be expected to increase over that of nonfractured reservoir environments. The fractures aid the movement of steam upward to establish a steam zone more quickly.

Griffin and Trofimenkoff (1984) supported Joshi's vertical injector/horizontal producer study and indicated that the steam zone could be established whether the injector were directly above the production drawoff point or at some distance along the HW (Figures 7-19a and 7-19b). A steam zone is established first around the vertical injector well section and grows out horizontally away from the injector occupying pore spaces vacated by oil that has gravity-drained along the length of the producer. The importance of gravity drainage for the process becomes apparent because if gravity were not to let the oil fall to the horizontal producing well then the process would not work because the pore space would not be available for steam. Griffin illustrated that placing the injection and production points at opposite ends of the HW increased production compared to the case where injection and production points were proximate.

For the two-HW steam process, Griffin illustrated the point of peak oil rate, which coincided with having the steam zone reach the top of the for-

Thermal Recovery and Primary Production for Heavy Oils 259

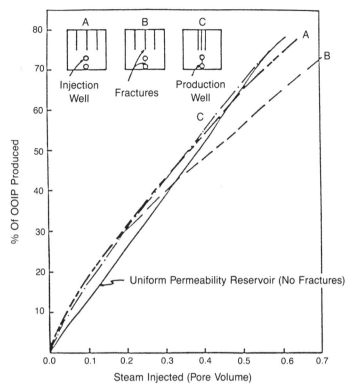

Figure 7-18a. Influence of fracture distribution on percentage of OOIP recovered (after Joshi et al., 1984).

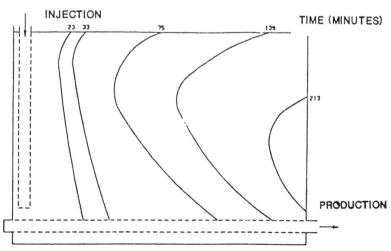

Figure 7-19a. Visual model hot horizontal well steam chamber growth (after Griffin et al., 1984).

260   *Horizontal Wells*

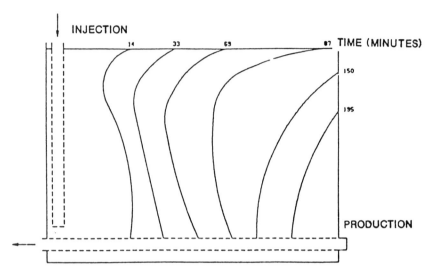

Figure 7-19b. Visual model cold horizontal well steam chamber growth (after Griffin et al., 1984).

mation (Figure 7-20). In both the physical model and the analytical model the peak oil rate is observed at the point when the steam zone approaches the top of the formation. Field operations should be able to determine when the steam zone is at the top of the reservoir by observing the point of peak oil rate. This will be important for determination of the rate of steam growth upward for extrapolation to other HWs in expanded operations. Numerical simulation results discussed later in this chapter produced the same peak oil rate at the point when the steam zone reached the top of the reservoir, for wells on close spacing.

Toma et al. (1984) reported that oil recovery could occur from one single HW through the use of short duration pressure cycles. A three-dimensional model with Athabasca oil sands was used for the studies. It was speculated that a blowdown phase was required to enable the water surrounding the sand grains at a specific distance from the HW to flash to steam upon depressurization and hence, through steam expansion, displace the oil within the porous space towards the producing end of the HW. Toma concluded:

1. A period of *preheat continuous* steam injection was required before any oil would be produced. This normally represented one-fourth of the total time to reach 40% recovery of the oil in place. Extrapolation of the preheat phase to field conditions may diminish substantially the length of preheat relative to the duration of the project.
2. The rate of heat transfer to the oil sand material was 10–100 times higher than could be transferred by conduction alone. This suggests that convection heat transfer at the steam-oil interface dominates the

Thermal Recovery and Primary Production for Heavy Oils 261

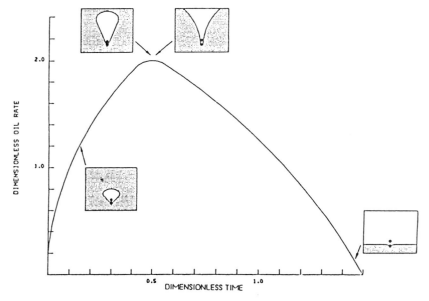

Figure 7-20. Theoretical visual model oil rate (after Griffin et al., 1984).

heat transfer mechanism and ensuing oil production rate. Oil recovery will be higher than just assuming that conduction heating occurs, provided that steam pressure cycling is employed.
3. The addition of naphtha to the steam decreased the time to establish oil production by a factor greater than two. The use of additives may spur the initial startup preheat period.

Toma's results provide the direction which may be required to successfully operate HW processes in the field. Pressure cycling may be required to ensure acceptable early time productivity. In addition, the use of an additive such as naphtha may establish initial acceptable early time oil productivity.

In summary, physical model bench scale studies have been conducted for different HW and VW configurations and all indicate that mechanistically and physically the countercurrent gravity drainage mechanism can be operated on a continuous or cyclic basis.

Analytical Derivations Describing SAGD Mechanisms

The work of Butler et al. (1979) established the initial analytical expressions that described the SAGD mechanisms. The laboratory observations of the SAGD process, which are described in the previous section, served as the basis for developing the expressions.

Darcy's Law Applied to SAGD

Consider a small section of a mature SAGD operation as depicted in Figure 7-20a. At the steam-oil interface, steam condenses and heat is liberated. A thermal gradient is established via conduction between the steam temperature at the interface and the original reservoir temperature. The steam temperature within the steam zone is denoted as Ts and the original reservoir temperature is denoted as Tr. Oil flows approximately parallel to the steam-oil interface down towards the HW producer (somewhere below and to the left in Figure 7-20a). The interface is inclined at an angle Θ to the horizontal. At a distance ξ from the steam oil interface, where the viscosity of the oil is μ, and the kinematic viscosity of the oil is ν, Darcy's law may be written for a steam zone section of unit thickness, as:

$$dq = \frac{K(d\xi)\,(\rho_o - \rho_g)\,g\sin\Theta}{\mu} \qquad (7\text{-}1)$$

The potential gradient is $(\rho_o - \rho_g)g \sin\Theta$. Because ρ_g is smaller than ρ_o, ρ_g is neglected. (Usually ρ_o is 25–1,500 times greater than ρ_g depending upon

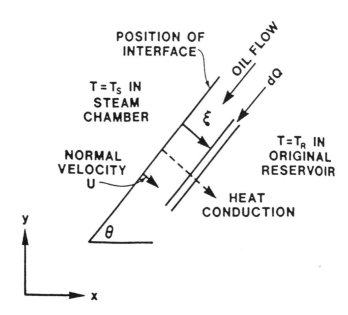

Figure 7-20a. Small vertical section of interface (after Butler, 1985).

Thermal Recovery and Primary Production for Heavy Oils

the steam zone pressure.) The term μ/ρ_o is set equal to ν, the kinematic viscosity of the liquid. It is assumed that there is no pressure gradient affecting flow. Rather, gravity is the driving mechanism for flow. Equation 7-1 becomes:

$$dq = \frac{Kg \sin\Theta}{\nu} d\xi \qquad (7\text{-}2)$$

Equation 7-2 gives the rate of drainage of liquid, dq, within the element $d\xi$.

As liquid drains via gravity out of the element $d\xi$, steam moves in to replace the liquid. Consequently the steam-oil interface moves at a velocity U perpendicular to the steam-oil interface. Transfer of heat from the steam-oil interface is via conduction to the oil. The temperature profile ahead of the interface for steady state conduction is given by:

$$\frac{T - T_r}{T_s - T_r} = e^{-U\xi/\alpha} \qquad (7\text{-}3)$$

High values of the velocity U, measured normal to the interface, result in the temperature falling rapidly with distance from the interface. Low values of U establish a slowly falling temperature profile from the interface, because the conduction mechanism has a longer period of time to heat the reservoir.

For an unheated reservoir the differential flow for oil would be given by:

$$dq_r = \frac{Kg \sin\Theta}{\nu_r} d\xi \qquad (7\text{-}4)$$

To account for flow due solely to the conduction heating mechanism, Equation 7-4 is subtracted from Equation 7-2.

$$dq - dq_r = Kg\sin\Theta \, (1/\nu - 1/\nu_r) \, d\xi \qquad (7\text{-}5)$$

The inclusion of nonheated liquid flow prevents the total flow determined by integration in Equation 7-5 from being infinite, because ν_r must be finite, even though ν_r itself can be very large.

Redefining dq as $dq - dq_r$, Equation 7-5 becomes:

$$dq = Kg\sin\Theta \, (1/\nu - 1/\nu_r) \, d\xi \qquad (7\text{-}6)$$

Integration of Equation 7-6 results in:

$$q = Kg\sin\Theta \int_0^\infty \left[\frac{1}{\nu} - \frac{1}{\nu_r}\right] d\xi \tag{7-7}$$

To determine q in Equation 7-7 requires an understanding of how the oil viscosity μ changes as a function of temperature and hence distance from the steam-oil interface. Viscosity as a function of temperature can be written in the form:

$$T - T_r = (T_s - T_r) e^{-U\xi/\alpha} \tag{7-3}$$

Differentiating Equation 7-3 gives:

$$dT = (T_s - T_r)(-U/\alpha) e^{-U\xi/\alpha} d\xi \tag{7-8}$$

Placing Equation 7-3 into Equation 7-8 gives:

$$dT = -\frac{U}{\alpha}(T - T_r) d\xi \tag{7-9}$$

Note that the temperature profile is a function of the interface velocity. Placing Equation 7-9 into Equation 7-7 gives:

$$q = \frac{Kg\alpha \sin\Theta}{U} \int_{T_r}^{T_s} \left[\frac{1}{\nu} - \frac{1}{\nu_r}\right] \frac{dT}{(T - T_r)} \tag{7-10}$$

By definition, let:

$$\frac{1}{m\nu_s} = \int_{T_r}^{T_s} \left[\frac{1}{\nu} - \frac{1}{\nu_r}\right] \frac{dT}{(T - T_r)} \tag{7-11}$$

Therefore, Equation 7-10 becomes:

$$q = \frac{Kg\alpha\sin\Theta}{m\nu_s U} \tag{7-12}$$

The integrand of Equation 7-11 needs to be evaluated. Figure 7-21 is an example of an evaluated integrand for an oil that has a kinematic viscosity of 129 cs at 212°F and 6.8 cs at 424°F for reservoir temperatures varying from 32°F to 104°F. In the figure, the kinematic viscosity of the oil phase at each steam temperature has been factored into the represented curves to give the

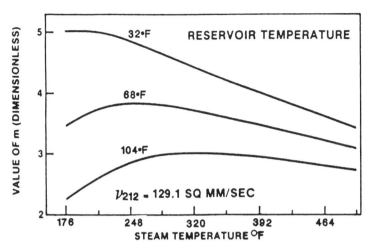

Figure 7-21. Effect of steam temperature on parameter m (after Butler, 1985).

values of m. Note that m is a function of reservoir temperature and steam zone temperature and should be evaluated for each reservoir oil that is considered.

Equation 7-12 is an intermediate equation because $Sin\Theta$ and U are unknown variables. Equation 7-12 indicates that increasing the permeability K, thermal diffusivity α, or the angle to the horizontal Θ, increases the flow rate q. Conversely, the higher the kinematic viscosity at steam temperature ν_s, or the higher the rate of advance of the steam-oil interface U, the lower the flow rate.

Material Balance

A material balance for the region depicted in Figure 7-20a is:

Flow into − Flow out = Accumulation
Region   of Region   within Region   (7-13)

or

$$(t_1 - t_2)q_{in} - (t_1 - t_2)q_{out} = (\phi S_o xyz)_{t1} - (\phi S_o xyz)_{t2} \quad (7\text{-}14)$$

z is assumed to be unity as a unit length along the HW is being considered. t_1 and t_2 are two separate points in time. Let Δ represent the changes in the parameters

Horizontal Wells

$$\Delta q = \frac{\Delta(\phi S_o xy)}{\Delta t} \tag{7-15}$$

Rearranging Equation 7-15 and taking partial differentials of both sides of the equation yields:

$$\left[\frac{\partial q}{\partial x}\right]_t = \phi \Delta S_o \, (\partial y/\partial t)_x \tag{7-16}$$

This expression accounts for the changing dimension of the steam zone as it expands at different rates vertically downward and horizontally across. In addition, the change in oil saturation ΔS_o is taken to *not* be a function of the x and y locations within the steam zone. Rather, it is assumed that there is a step change in oil saturation, from initial oil saturation to residual oil saturation right at the steam-oil interface.

Interface Velocity

The velocity of the interface can be defined by:

$$U = - \text{Cos}\Theta \, (\partial y/\partial t)_x \tag{7-17}$$

Figure 7-22 shows Equation 7-17 pictorially. Equation 7-17 indicates that the advance of the steam-oil interface is coupled to the rate of drainage downward in the y direction with time.
Place Equation 7-17 into Equation 7-12

$$q = \frac{- Kg\alpha \text{Sin}\Theta}{m\nu_s \text{Cos}\Theta(\partial y/\partial t)_x} \tag{7-18}$$

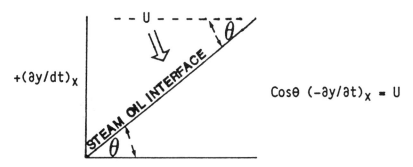

Figure 7-22. Relationship of U with Y as a function of t.

Thermal Recovery and Primary Production for Heavy Oils 267

$$\text{Sin}\Theta/\text{Cos}\Theta = (\partial y/\partial x) \qquad (7\text{-}19)$$

Placing Equation 7-19 into Equation 7-18 gives

$$q = \frac{-Kg\alpha \partial y}{m\nu_s \partial x (\partial y/\partial t)_x} \qquad (7\text{-}20)$$

Rearranging the material balance equation (Equation 7-16) provides

$$(\partial y/\partial t)_x = \frac{(\partial q/\partial x)_t}{\phi \Delta S_o} \qquad (7\text{-}21)$$

Placing Equation 7-21 into Equation 7-20 yields:

$$q = \frac{Kg\alpha \partial y (\phi \Delta S_o)}{m\nu_s \partial x (\partial q/\partial x)_t} \qquad (7\text{-}22)$$

The negative sign has been removed to account for a different direction in coordinate axis. Removing the ∂x terms and rearranging Equation 7-22 results in:

$$q = \frac{Kg\alpha \phi \Delta S_o}{m\nu_s} (\partial y/\partial q)_t \qquad (7\text{-}23)$$

Rearranging the variables in Equation 7-23 and integrating yields the following:

$$\int_o^q q\,dq = \int_o^{h=y} \frac{\phi \Delta S_o Kg\alpha}{m\nu_s}\,dy \qquad (7\text{-}24)$$

Integration gives:

$$\frac{q^2}{2} = \frac{\phi \Delta S_o Kg\alpha (h-y)}{m\nu_s} \qquad (7\text{-}25)$$

or

$$q = \sqrt{\frac{2\phi \Delta S_o Kg\alpha (h-y)}{m\nu_s}} \qquad (7\text{-}26)$$

Equation 7-26 provides the flow rate q at any height y in the reservoir. The equation is for one-half of a steam zone. To account for both sides of the

268   Horizontal Wells

steam zone, for the full flow rate from the steam zone, Equation 7-26 must be multiplied by 2.

$$q = 2\sqrt{\frac{2\phi \Delta S_o K g \alpha (h-y)}{m \nu_s}} \quad (7\text{-}27)$$

Assuming that the HW producer is at the bottom of the steam zone (where y = 0 in Equation 7-27) results in the equation for the total flow rate from a HW producer in the SAGD process:

$$q = 2\sqrt{\frac{2\phi \Delta S_o K g \alpha h}{m \nu_s}} \quad \text{SAGD predicted production rate} \quad (7\text{-}28)$$

Note in Equation 7-28 that all variables have equal weighting. Increasing or decreasing any variable by a factor of two changes the HW flow rate by the square root of two. Furthermore, the production rate q is only a function of the height of the slope and not the angle of recline Θ. High angle Θ slopes have large gravity components parallel to the steam-oil interface slope but have short slope lengths exposed to steam. Similarly, shallow angle slopes have lower gravity components parallel to the slope but have longer slopes exposed to steam. Thus, the two factors Θ and slope length cancel each other out and the oil rate is only a function of the height of the slope and not of the slope recline or slope length.

An important observation related to Equation 7-28 is that basically only one variable in the equation can be operationally varied. That variable is ν_s, the kinematic viscosity of the oil phase at steam temperature. The pressure in the steam zone can be operationally controlled and hence the steam zone temperature at the saturated steam pressure can be controlled. In this manner ν_s can be changed. To obtain as large an oil rate q as possible, ν_s should be made as small as feasible. In other words, the steam zone pressure and temperature should be made as large as possible. Figure 7-23 depicts how steam temperature influences drainage rate for three oils with different viscosities. For example, a 100 cs oil at 424°F has an oil recovery rate of 0.38 bbl/d per foot of well. The same oil at a steam zone temperature of 636°F has a recovery rate of 0.75 bbl/d per foot of well. Increasing the steam zone temperature by 212°F increases the rate of recovery by a factor of 2.

All of the remaining parameters within Equation 7-28 are established by the characteristics of the reservoir. Operationally, little can be done with just straight steam injection into the steam zone to influence the flow rate q for the SAGD process. Rather, the unalterable reservoir characteristics dictate how the SAGD process is to perform in terms of flow rate. Operationally controllable parameters such as distance between HW injector and producer, well spacing distance between pairs of HWs, and location of HW in-

Thermal Recovery and Primary Production for Heavy Oils 269

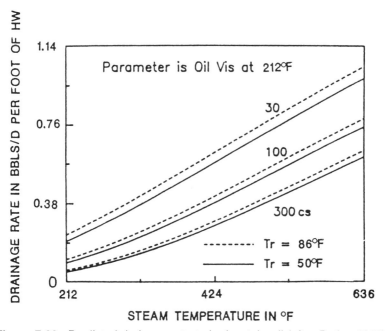

Figure 7-23. Predicted drainage rate to horizontal well (after Butler, 1988).

jector and HW producer within the reservoir all have a bearing on productivity performance and are discussed in under the heading "SAGD Processes in Reservoirs with Mobile Oil." The basic SAGD equation describes the ideal SAGD environment where heat losses and well spacing are not considered. In addition, Equation 7-28 assumes that there is no maximum limit for HW length. Provided that fluid can flow into the HW producer at all points along the HW, the HW length could approach infinity.

An alternate way to write Equation 7-28, to take into account the density of the gas phase, is:

$$q = 2\sqrt{\frac{2\phi\Delta S_o Kg\alpha h \rho_{os}}{m\mu_{os}}} \qquad (7\text{-}28A)$$

where ρ_{os} = density difference between the oil and the steam
μ_{os} = viscosity difference between the oil and steam

For SAGD environments that use high steam pressures, the density of steam begins to approach that of the oil phase. The density of steam cannot be neglected in these situations and so the difference between oil and steam densities must be included in Equation 7-28.

Example 7-10

Establish the rate of oil recovery for a 1,500 ft SAGD well operating at a steam zone temperature of 500°F.

Given: $\phi = 0.35$
$\Delta S_o = 0.60$
$K = 2.0$ D
$h = 50$ ft
$\rho_o = 51.90$ lbm/ft^3
$\rho_{steam} = 1.48$ lbm/ft^3
$\mu_o = 7.5$ cp
Reservoir Temp = 68°F
$m = 3.2$, from Figure 7-21
$g = 32.2$ ft/sec^2
$\alpha = 0.9418$ ft^2/day (follow the method given in Example 7-12)
$M = 26.81$ BTU/ft^3 °F, from Example 7-4
$S_w = 0.30$

$$q = 2\sqrt{\frac{(2)(0.35)(0.60)(2.0\text{ D})\left(32.2\frac{\text{ft}}{\text{sec}^2}\right)0.9418\frac{\text{ft}^2}{\text{day}}(50\text{ ft})50.42\frac{\text{lbm}}{\text{ft}^3}\left(1.0764\times10^{-11}\frac{\text{ft}^2}{\text{D}}\right)}{(3.2)(7.5\text{ cp})(6.72\times10^{-4}\text{lbm/ft-sec./cp})(24\text{hr/day})(3{,}600\text{ sec/hr})}}$$

$q = (4.4545 \times 10^{-5} \text{ft}^2/\text{sec})$ (1 ft/1 ft) (0.178 bbl/ft^3) (3,600 sec/hr) (24 hr/d)

$q = 0.6851$ bbl/d – ft

For a 1,500-ft-long HW, the recovery would be 1,027.6 bbl/d.

Position of Steam Oil Interface

The horizontal velocity at the interface is as follows:

$$(\partial x/\partial t)_y = \frac{(\partial y/\partial t)_x}{(\partial y/\partial x)_t} \qquad (7\text{-}29)$$

Combining Equation 7-29 with Equation 7-16 yields:

$$(\partial x/\partial t)_y = \frac{(\partial q/\partial x)_t}{\phi \Delta S_o \, (\partial y/\partial x)_t} \qquad (7\text{-}30)$$

Cancelling terms provides:

$$(\partial x/\partial t)_y = \frac{1}{\phi \Delta S_o} (\partial q/\partial y)_t \tag{7-31}$$

Taking the partial derivative of q with respect to y in Equation 7-26 results in:

$$(\partial q/\partial y)_t = 1/2 \left[\frac{(2\phi \Delta S_o K g \alpha)}{m \nu_s}\right]^{1/2} (h-y)^{-1/2} \tag{7-32}$$

Combining Equation 7-32 with Equation 7-31 results in:

$$\left[\frac{\partial x}{\partial t}\right]_y = \frac{1}{\phi \Delta S_o} \left[1/2 \sqrt{\frac{2\phi \Delta S_o K g \alpha}{m \nu_s}}\right] \frac{1}{\sqrt{h-y}} \tag{7-33}$$

$$(\partial x/\partial t)_y = \sqrt{\frac{K g \alpha}{2\phi \Delta S_o m \nu_s (h-y)}} \tag{7-34}$$

If it is assumed that y is a constant by assuming that the steam zone is a vertical plane above the production well, then Equation 7-34 can be integrated to yield:

$$x = t \sqrt{\frac{K g \alpha}{2\phi \Delta S_o m \nu_s (h-y)}} \tag{7-35}$$

Earlier in the chapter, Joshi describes a vertical fracture placed above the HWs, which is equivalent to the assumption of an initial steam zone above the HW being a thin vertical plane.

Equation 7-35 provides that at every height y in the reservoir the position of the interface in the x horizontal direction is determined for all values of time t. Note that as y approaches the full height of the reservoir h that x becomes very large because $(h - y)$ approaches zero in the denominator of Equation 7-35. The equation indicates that the bulk of the steam-oil interface will be near the top of the reservoir. In other words, steam has a propensity to rise in the steam zone relative to the denser oil phase. As a consequence the steam interface is primarily at the top of the reservoir. Equation 7-35 may be solved for y, with the result:

$$y = h - \frac{K g \alpha}{2\phi \Delta S_o m \nu_s} (t/x)^2 \tag{7-36}$$

Equation 7-36 may also be represented in dimensionless form:

272 Horizontal Wells

$$Y = 1 - 1/2(T^*/X)^2 \quad (7\text{-}37)$$

where $Y = y/h$
$X = x/h$

$$T^* = t/h\sqrt{\frac{Kg\alpha}{\phi\Delta S_o \nu_s mh}}$$

Values of Y calculated from Equation 7-37 have been plotted against X in Figure 7-24. Note in Figure 7-24 that as time increases, the steam-oil interface moves away from the (X,Y) = (0, 0) point, where the horizontal producer well is. The steam zone in the figure becomes larger as oil drains by gravity out of the system. Eventually, after a long period of time, the reservoir has been depleted of oil by gravity drainage and only a steam zone exists.

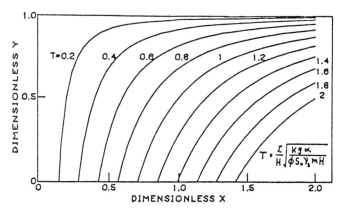

Figure 7-24. Interface curves—original theory (after Butler and Stephens, 1980).

TANDRAIN

The basic SAGD analytical expression does not take into account how the heated oil flows horizontally to the point (X,Y) = (0,0) as the steam-oil interface moves away horizontally from (X,Y) = (0,0). In reality the steam-oil interface will frequently stay at the horizontal production well as the steam zone grows larger above the well, rather than moving horizontally away from the HW. A simple way to keep the interface at the HW is to maintain it attached to the HW at all times as the steam zone expands the steam-oil

Thermal Recovery and Primary Production for Heavy Oils 273

interface above the well. Figures 7-24 and 7-25 illustrate the differences between the original theory and the modification of keeping the interface at the horizontal well. The modification is referred to as TANDRAIN (Butler and Stephens, 1980). Basically, what TANDRAIN does is draw a tangent from the HW producer location to the steam-oil interface curves for particular points in time.

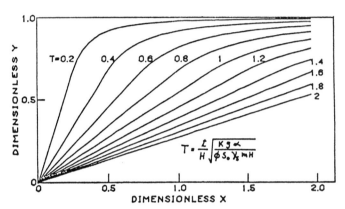

Figure 7-25. Interface curves with Tandrain assumption (after Butler and Stephens, 1980).

The TANDRAIN assumption decreases the theoretical rate of oil which is recovered at any particular time relative to the original theory. Figure 7-26 illustrates the difference via the shaded area between the two curves. The difference between the two curves is about 13%. Using TANDRAIN, the rate is 87% of the original theory. Or, in equation form, roughly:

$$0.87 \, q_{old} = \sqrt{\frac{1.5}{2}} \, q_{old} = q_{Tan} \qquad (7\text{-}38)$$

Multiplying both sides of Equation 7-28 by $\sqrt{\frac{1.5}{2}}$ yields:

$$q_{Tan} = 2 \sqrt{\frac{1.5 \phi \Delta S_o K g \alpha h}{m \nu_s}} \quad \text{TANDRAIN Equation} \qquad (7\text{-}39)$$

The TANDRAIN Equation is slightly more representative of the laboratory bench scale results described earlier. Figure 7-27 illustrates the similarity between lab performance and TANDRAIN results.

274 Horizontal Wells

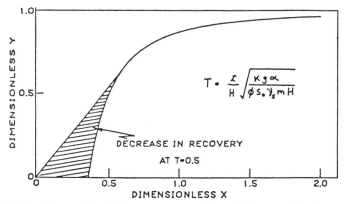

Figure 7-26. Effect of Tandrain on recovery (after Butler and Stephens, 1980).

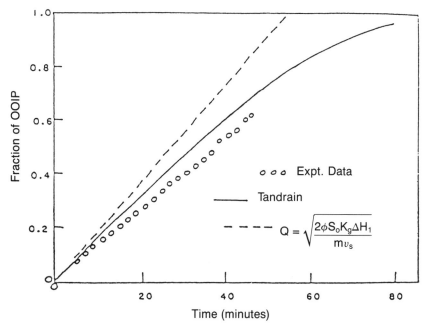

Figure 7-27. Predicted and actual cumulative oil production vs time (after Butler and Stephens, 1980).

Thermal Recovery and Primary Production for Heavy Oils 275

No-flow Boundary Considerations

These SAGD equations have assumed a single steam zone process for the entire reservoir. In reality, to exploit a heavy oil reservoir with the SAGD process, more than one steam zone will typically be required. For multiple steam zones, at some point in time the pressure and temperature transients from the steam zones will meet. At this juncture, mirror image representations of the steam zones would exist, as depicted in Figure 7-28.

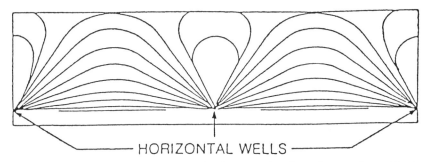

Figure 7-28. Diagram showing growth of steam chamber above adjacent horizontal wells (after Butler et al., 1979).

Butler and Stephens (1980) developed an analytical approach to depict the effect of no-flow boundaries. For a no-flow boundary at a time t the steam-oil interface shall be as depicted in Figure 7-29. At the no-flow boundary (x = w) the interface will be at y_w. As the oil drains from the confined well, y_w decreases and moves in the downward direction.

Consider the point P on the interface curve. P is a very small distance δy below y_w and is a distance δx from the no-flow boundary. For a constant y the point P moves to the right at the velocity given by Equation 7-34.

$$(\partial x/\partial t)_y = \sqrt{\frac{Kg\alpha}{2m\nu_s(y_w - y)\phi \Delta S_o}} \tag{7-40}$$

Note that y_w has replaced h in the equation and that y_w changes downward with time because the steam zone at the boundary depletes the oil and as a consequence moves downward. Assume that y_w is changing at a rate of $\dot{y}_w$ over time δt. Thus

$$y_w - y = \delta y + \dot{y}_w \, \delta t \tag{7-41}$$

Horizontal Wells

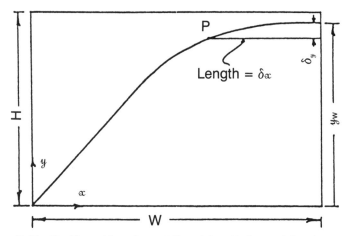

Figure 7-29. Position of interface at time (after Butler and Stephens, 1980).

It is important to note that initially $y_w - y$ equates to δy, and then decreases with time because of the decrease of y_w.

Placing Equation 7-41 into Equation 7-40 gives

$$(\partial x/\partial t)_y = \sqrt{\frac{Kg\alpha}{2m\nu_s\,(\delta y + \dot{y}_w \delta t)\phi\Delta S_o}} \qquad (7\text{-}42)$$

The time δt required for P to move horizontally a distance δx may be found by integrating Equation 7-42.

$$\int_0^{dx} d(x) = \int_0^{\delta t} \sqrt{\frac{Kg\alpha}{2m\nu_s\,(\delta y + \dot{y}_w \,\delta t)\phi\Delta S_o}}\, d(t) \qquad (7\text{-}43)$$

Integration of Equation 7-43 results in:

$$\delta x = \sqrt{\frac{Kg\alpha}{2m\nu_s\phi\Delta S_o}} \left| \frac{2\sqrt{\delta y + \dot{y}_w\,\delta t}}{\dot{y}_w} \right|_{t=0}^{t=\delta t} \qquad (7\text{-}44)$$

Since $\dot{y}_w$ is defined as $(-\delta y/\delta t)$, Equation 7-44 may be rearranged:

$$\delta x = \sqrt{\frac{Kg\alpha}{2m\nu_s\phi\Delta S_o}} \left| \frac{2\sqrt{\delta y - \delta y}}{-(\delta y/\delta t)} - \frac{2\sqrt{\delta y}}{(\delta y/\delta t)} \right|$$

Thermal Recovery and Primary Production for Heavy Oils 277

or

$$\delta t = \delta x \sqrt{\frac{\phi \Delta S_o m \nu_s \delta y}{2K g \alpha}} \quad (7\text{-}45)$$

Equation 7-45 gives the time δt required for the point P to move horizontally a distance δx to the no-flow boundary. Define $x_0, x_1, x_2, x_3, \ldots x_n$ to be equally spaced values of the ordinate $y_0, y_0 + \xi, y_0 + 2\xi, \ldots y_0 + n\xi$. Define $x_n = w$ and $y_n = y_0 + n\xi = y_w$. In this manner at time $y_n + \delta t$, x_{n-1} becomes equal to w. Equation 7-45 becomes:

$$\delta t = (w - x_{n-1}) \sqrt{\frac{\phi \Delta S_o m \nu_s \xi}{2K g \alpha}} \quad (7\text{-}46)$$

In dimensionless variables Equation (7-46) becomes:

$$\delta T^* = (1 - X_{n-1}) \sqrt{\frac{E}{2}} \quad (7\text{-}47)$$

where $T^* = t/w \sqrt{\frac{Kg\alpha}{\phi \Delta S_o \nu_s m h}}$

$Y = y/h$
$E = \xi/h$
$X = x/w$

In dimensionless form Equation 7-40 becomes:

$$(\delta X_i / \delta T^*)_{Yi} = \frac{1}{\sqrt{2(Yw - Yi)}} \quad (7\text{-}48)$$

Equation 7-48 may be integrated over the interval δT^* by substituting:

$$Y_w - Y_i = E(n - i - \tau/\delta T^*) \quad (7\text{-}49)$$

τ is the dimensionless time starting at the beginning of the time period and ending with the value δT^* at the end. The integral to Equation 7-48 becomes:

$$\delta X_i = \int_0^{T^*} \frac{d\tau}{\sqrt{2E(n - i - \tau/\delta T^*)}} \quad (7\text{-}50)$$

Integration yields =

$$\delta X_i = \delta T^* \sqrt{2/E} \, (\sqrt{n-i} - \sqrt{n-i-1}) \qquad (7\text{-}51)$$

Equation 7-47 is substituted with Equation 7-51 for δT^* to give:

$$\delta X_i = (1 - X_{n-1}) (\sqrt{n-i} - \sqrt{n-i-1}) \qquad (7\text{-}52)$$

Equations 7-47 and 7-52 are used repetitively to calculate positions of the interface. The time element to drop the peak of the interface one increment is calculated from Equation 7-47 and the new position of each interface point is calculated from Equation 7-52. At every intermediate stage, the time is calculated from the cumulative time of the δT^* and the position of each point on the interface from the cumulative sum of δX_i. At each stage of the iteration, the peak of the interface drops one increment and n is decreased by 1.

The calculation is made starting at X_{n-1} and moving down the interface. At each step, it is determined whether the slope is less than that of a straight line drawn to the production well. At the point where they become equal the rest of the interface is drawn as a straight line.

The total recovery is calculated by integrating the interface curve over distance and establishing the drained area above it. Figure 7-30 illustrates the shape of the steam zone in a bounded steam zone pattern as dimension-

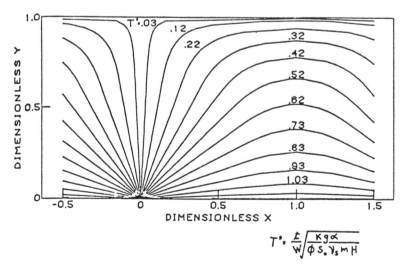

Figure 7-30. Confined horizontal well interfaces (after Butler and Stephens, 1980).

Thermal Recovery and Primary Production for Heavy Oils 279

less time T* varies from 0 to 1.0. Note that, in the figure, essentially the entire bounded reservoir volume is depleted of oil over total time.

Figure 7-31 illustrates one example of how a bounded well pattern influences recovery. The interference of bounded patterns at mid dimensionless time T* of 0.5 diminishes the rate of oil recovery thereafter, relative to a reservoir environment that has only one pair of HWs situated in it. Thus, in field pilots, it is worthwhile to consider bounded HW patterns such that the bounded patterns would be representative of large-scale commercial operations.

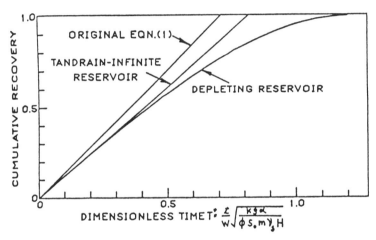

Figure 7-31. Cumulative drainage to horizontal well (after Butler and Stephens, 1980).

SAGD Analytical Expression Refinements

Refinements have been developed to the basic SAGD analytical expressions. One refinement has been to account for the vertical growth of the steam zone (Butler et al., 1980). The previous theory assumed the steam zone would be initiated from a hot vertical plane. Incorporating the growth of the steam zone upward decreases the early time oil rate because time is required to build the steam zone to the top of the reservoir. Figure 7-31a depicts the difference in rate as a function of recovery fraction between accounting for the growth of the steam zone upward and the original TANDRAIN analytical expression. The two rising steam zone curves in Figure 7-31a are for confined SAGD wells with the width of the well spacing between SAGD pairs equal to the height of the reservoir or equal to twice the height of the reservoir.

280   *Horizontal Wells*

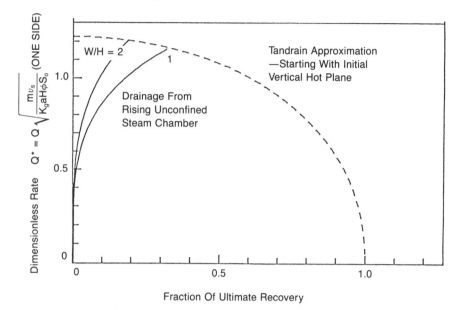

Figure 7-31a. Calculated drainage rates to a horizontal well (after Butler et al., 1984).

Example 7-11

Evaluate the sensitivity of reservoir and operational parameters for the SAGD process for the early time of recovery associated with SAGD.

Butler's analytical expressions have been formulated into a model that is run on personal computers. The Drilling Engineering Association Project Number 44 (DEA-44) has access to Butler's PC model. A sensitization of the SAGD process follows.

Table 7-3a provides the reservoir parameters that are maintained constant for all of the sensitivity runs. Table 7-3b provides the reservoir parameters that were varied from the base case run and also provides the incremental change in cumulative oil recovered and OSR from the base case.

The model runs all assume a HW injector and HW producer. All runs assume a rising steam zone from the HW injector. The HWs are assumed finite in length for each case. Heat losses to the overburden are accounted for. The analytical run sensitizations indicate:

1. Where feasible, it is preferable to drill long HWs
2. The vertical permeability to horizontal permeability ratio (k_v/k_h) has a significant bearing on productivity and oil steam ratio (OSR)

Thermal Recovery and Primary Production for Heavy Oils   281

Table 7-3a
Reservoir Characteristics Maintained Constant
for All SAGD Analytical Cases

Reservoir temperature	50°F
Steam quality fraction	0.70
Reservoir porosity	0.35
Residual oil saturation	0.15
Reservoir thermal conductivity	18.24 BTU/ft-d-°F
Reservoir rock density	162.2 lbm/ft³
Reservoir rock heat capacity	0.23 BTU/lbm-°F
Oil gravity	11.5°API
Overburden heat capacity	0.20 BTU/lbm-°F

Table 7-3b
Sensitization Runs of SAGD Analytical Expressions

Base Case:		
	Steam temperature	= 500°F
	Reservoir thickness	= 50 ft
	Well spacing	= 5 acres
	Oil saturation	= 0.7
	Horizontal permeability	= 1,000 md
	Oil viscosity at 212°F	= 100 cs
	k_v/k_h	= 0.5
	HW length	1,500 ft
Base Case Results:	Cumulative oil at 1 year	= 187 mbbls
	Cumulative oil at 2 years	= 285 mbbls
	Cumulative OSR at 1 year	0.28 bbls/bbl
	Cumulative OSR at 2 years	0.27 bbls/bbl

Parameter Varied to Base Case	Change in Cumulative Oil Recovery to Base Case		Change in Cumulative OSR to Base Case	
	1 year (mbbl)	2 years (mbbl)	1 year (bbl/bbl)	2 years (bbl/bbl)
HW length = 500 ft	− 132	− 130	+ 0.03	− 0.02
HW length = 3,000 ft	+ 73	+ 83	+ 0.03	0.00
Steam temperature = 400°F	− 81	− 61	+ 0.08	+ 0.05
Steam temperature = 600°F	+ 37	+ 49	− 0.06	− 0.06
Reservoir thickness = 25 ft	− 57	− 109	− 0.08	− 0.08
Reservoir thickness = 75 ft	0	+ 77	+ 0.04	+ 0.04
Well spacing = 2.5 acres	− 57	− 100	+ 0.03	0.00
Well spacing = 7.5 acres	+ 10	+ 60	0.00	− 0.02
Oil saturation = 0.5	− 57	− 85	− 0.10	− 0.10
Oil saturation = 0.8	+ 7	+ 45	+ 0.06	+ 0.05
Horizontal permeability = 500 md	− 69	− 60	− 0.01	− 0.02
Horizontal permeability = 2,000 md	+ 38	+ 50	+ 0.02	+ 0.02
Oil viscosity at 212°F = 10 cs	+ 93	+ 85	+ 0.06	+ 0.02
Oil viscosity at 212°F = 30 cs	+ 38	+ 50	+ 0.03	+ 0.03
k_v/k_h ratio = 0.1	− 114	− 98	− 0.06	− 0.04
k_v/k_h ratio = 1.0	+ 33	+ 30	+ 0.01	0.00

282 *Horizontal Wells*

Numerical Simulation of SAGD Process

A thermal numerical reservoir simulator offers an alternate means of evaluating the SAGD process. Simulators provide the versatility of investigating particular process parameter sensitivities on performance which analytical expressions to date are incapable of doing. As an example, a thermal simulator is capable of investigating the impact on performance of:

1. Injector HW to producer HW distance
2. Reservoir heterogeneities
3. Steam blow down phase
4. Multiple steam chamber interference if one steam chamber is shut in and others continue operation

In addition, simulators are capable of comparing performance between HW processes, VW processes, and combinations of HW and VW processes.

A drawback of thermal numerical simulation is that building HW reservoir models can be manpower intensive compared to the relative ease of running analytical models on computers. However, once the HW reservoir model is built, sensitivities comparing recovery processes generally can be run quickly.

Approach to Conduct SAGD Simulations

An effective means to evaluate the SAGD process for a particular reservoir environment is given below. The approach is by no means the only method to carry out simulations.

The first step is to amass the rock and fluid properties associated with the reservoir. If certain properties are unavailable for the simulation, then sensitivities on the parameters can be undertaken to determine the influence of the parameter on performance. If a small change in the parameter has a significant influence on the outcome, consideration should be given to obtaining an accurate value for the parameter. If, on the other hand, the parameter has little influence on the result it may not be necessary to obtain an accurate value for the parameter. Table 7-4 depicts the influence of particular parameters for the AOSTRA UTF reservoir. In the table, note that a doubling of the water relative permeability curve has a negligible influence on both cumulative oil produced and cumulative steam oil ratio, at the UTF site. Similarly, decreasing the water relative permeability values by a factor of 225 has a minor influence on production and oil steam ratio. Thus, for the UTF site, the sensitization of water relative permeability through the use of numerical simulation suggests that an accurate water relative permeability curve may not be critical to obtain a reasonable projection of SAGD per-

Thermal Recovery and Primary Production for Heavy Oils 283

Table 7-4
Two-dimensional Steam Chamber Sensitivities
for Reservoir Characteristics at 600 Operating Days

Reservoir Sensitivity	Change from Base Case	Change in Cumulative Produced Oil from Base Case	Change in Cumulative Oil Steam Ratio from Base Case
Decrease residual oil saturation from 0.25 to 0.10	1.27	1.58	1.37
Double water relative permeability curve values	2.0	1.01	0.96
Decrease water relative permeability curve values by a factor of 225	0.0044	0.72	0.94
Reduce vertical permeability from 4 Darcy to 1 Darcy	0.25	0.28	0.56
Decrease oil relative permeability curve by a factor of 2	0.5	0.46	0.60
33-ft wide impermeable shale barrier 16 ft above steam injector	—	0.94	0.93

Source: Best et al., 1985.

formance. On the other hand, decreasing the oil relative permeability curve by a factor of two affects performance by a similar factor of two. Thus, oil relative permeability is an important parameter and an accurate knowledge of oil relative permeability is required to both forecast and understand performance.

These examples of the influence of relative permeability on SAGD performance apply to the UTF reservoir. In other reservoirs, the influence of each reservoir parameter may differ from that which is noted for the UTF reservoir. For each reservoir environment it is worthwhile to undertake sensitivity studies on the reservoir parameters to assess which parameters are required accurately for a good representation of field performance. In the case of the UTF project, a field pilot was required to obtain certain key petrophysical parameters related to the gravity drainage recovery of oil.

If field performance is available from HWs and/or VWs, the actual production rates are history matched in an attempt to determine reservoir and fluid parameters. It is important that a history match phase of field performance be undertaken to calibrate the perceived reservoir mechanisms that are incorporated in the simulator to actual field productivity.

284 Horizontal Wells

For the purpose of carrying out an evaluation of SAGD performance relative to sensitizing reservoir parameters, only a small element of symmetry of the SAGD reservoir need be used. Typically, a unit slice of the reservoir is used with the top and base of the reservoir included. If the SAGD process to be investigated is for a series of steam zones, then an element of symmetry is employed for a mirror image representation between parallel pairs of SAGD wells (Figure 7-32).

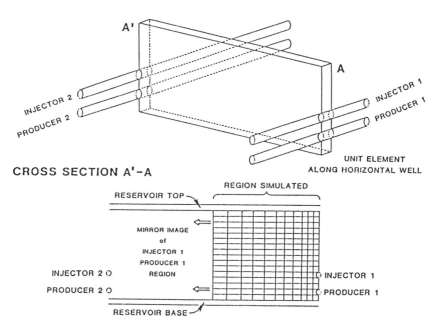

Figure 7-32. Element of symmetry for multiple pairs of horizontal wells.

In the parametric sensitivity studies of an element of symmetry it is important to ensure that the simulator grid block sizes do not influence the performance results. Typically, a series of simulator runs are made with different grid block sizes and an appropriate grid block size is selected at the point where performance results converge as grid block size becomes smaller (Figure 7-33). Experience has indicated that grid blocks 3 ft by 3 ft are more than acceptable for representing an element of symmetry of the SAGD process (Figures 7-33 a and b). In Figure 7-33b the fluctuation in oil rate for the large 12 ft square grid blocks is due to the temperature dispersion over the grid block volume. Time is required to heat a grid block to oil mobilization temperature. Because the grid block is too large, the time to heat the block to mobilization temperature is indicated by a fluctuation in oil rate.

Thermal Recovery and Primary Production for Heavy Oils 285

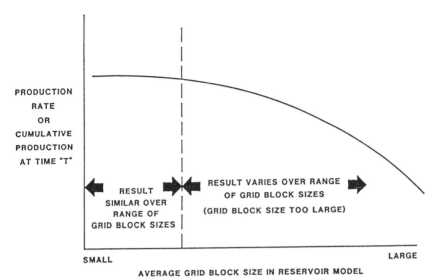

Figure 7-33. Conceptual grid block size sensitivity evaluation for SAGD process.

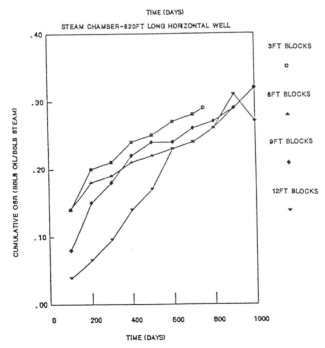

Figure 7-33a. OSR for grid block study—UTF Project (after Suggett et al., 1985).

286 Horizontal Wells

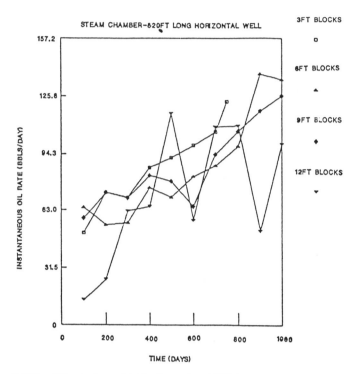

Figure 7-33b. Oil rate for grid block study—UTF Project (after Suggett et al., 1985).

For reservoirs where the in situ fluid mobility is negligible, analytical calculations are required to determine the length of time necessary for heat to be conducted from both the HW injector and the HW producer via circulation of steam in each of the wells. The section entitled "Placement of Steam Along HW Injectors" describes the manner in which steam can be effectively circulated through the HWs. Without the steam circulation phase for conduction heating, it may not be possible to initiate the SAGD process in a controlled manner.

The SAGD process requires fluid communication between the HW injector and the HW producer. Otherwise, fluids are unable to drain from the steam zone to the production draw off points and as a consequence the process is unable to initiate itself. Fluid mobility is required between the wells because otherwise, upon injection of steam, the reservoir pressure would build to the extent that the reservoir would fracture. Example 7-3 illustrates why fracturing would occur in reservoirs with low initial fluid mobility. Conduction heating enables the reservoir temperature to increase around the HWs, which decreases the viscosity of the oil phase around the HWs. At the point in time when conduction heating has progressed to where the average temperature between the HW injector and HW producer is sufficiently high that the oil phase is mobile, then the SAGD process can be initiated by the displacement of oil with steam from the injector HW down to the producer HW. Figure 7-34 illustrates the length of time for conduction heating to occur for different temperature isotherms. The necessary temperature between HW injector and producer will be dependent upon the viscosity temperature relationship of the oil phase, governed by the point in the viscosity curve where the oil phase is adequately mobile for fluid displacement to commence. In Figure 7-34 it is evident that there is a trade-off between injector-producer well distance and the time necessary for interwell reservoir heating to occur. It is beneficial to space the HW injector and producer as far apart as feasible to establish as large a pressure drop as possible between the wells. However, this means a trade-off on the well spacing distance relative to the length of time necessary to heat the reservoir between the wells.

Steam circulation need not be the means to establish fluid mobility between wells. The use of electrical resistance heating in either or both HWs has been suggested as a means to heat the reservoir between the wells. Solvent diffusive processes have been suggested also as means to establish fluid communication between wells.

For immobile fluid reservoirs, fracturing could be considered as a means of establishing communication between steam injector(s) and the HW producer. Laboratory experimentation by Joshi, described earlier, suggests that if controlled fracturing between injector and producer can be established, the rate of recovery of oil can be significantly increased. Furthermore, where VWs are employed as the injector points above the HW producer, the concept is to fracture down to the producer in a plane parallel to the HW producer. Figure 7-44 illustrates the concept. Due to the uncertainty in fracture configuration between injector and producer, it is preferable to employ conduction heating methods over an attempt to uniformly fracture along the entire length of the HW.

The temperature distribution between injector and producer HWs, determined from analytical calculations, is placed in the reservoir model in the initialization of the reservoir model stage. Thus, the reservoir model is initialized with sufficient temperature between the wells that the displacement of the oil between the wells can start. The time at which the numerical reservoir model starts is the time that the conduction heating phase ends.

Another way to establish communication in the reservoir simulator between the injector and producer, in a reservoir with negligible fluid mobility, is to alter the fluid saturation distribution in a localized region between the wells by increasing the water saturation. The higher water saturation region will have mobility to enable the drainage of the oil in the SAGD pro-

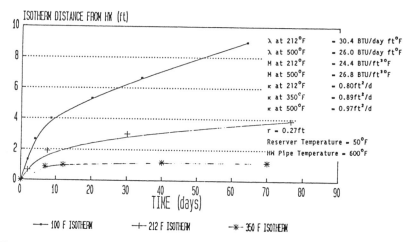

Figure 7-34. Conduction heating temperature as a function of distance from HW (following the approach of example 7-12).

cess to occur downwards to the HW producer. Increasing the water saturation between the wells influences the overall results to a slight degree. The increasing water saturation approach is deemed to be not as accurate as undertaking conduction heating calculations, but analysis has indicated that the overall results between the two approaches are similar.

For reservoirs in which fluid mobility originally exists, it is not necessary to alter water saturation or incorporate a temperature gradient in the reservoir grid blocks between the wells. Rather, with virgin fluid mobility, fluid displacement is capable of initiating the SAGD process.

Once steam communication is established between HW injector and HW producer the SAGD process may be operated continuously. At the onset of steam production the numerical simulator is placed on a steam production constraint. That is, the simulator continuously adjusts production drawdown such that neglible steam is produced at downhole conditions. If the simulator does not have a steam production constraint, then a maximum produced fluid temperature constraint may be employed. The maximum temperature would be established by the average pressure in the steam zone which would equate to an average temperature. To ensure that steam is not produced, the maximum production temperature would be set slightly below the steam zone temperature. Joshi's work describes how produced fluid temperature was used to control steam production and fluid rate in lab experiments. Setting a negligible steam production constraint ensures a high thermal efficiency for the process relative to the situation where steam is produced.

For the HW steam injector the constraint is pressure. The injection pressure is preferably below formation fracture pressure. Unless desired otherwise, the injector pressure is maintained constant throughout most of the life of the SAGD process, until very close to when the process has recovered the bulk of the oil in place. The injector maintains an essentially constant pressure in the steam zone. For every unit volume of fluid withdrawn from the reservoir, an equivalent volume of steam is injected into the steam zone. Thus, the process can balance the production rate to the injection rate and as a consequence there is negligible leak-off of fluid from the steam zone environs. The negligible leak-off equates to minimal well interface between steam zones until the point in time at which the thermal fronts of individual steam zones overlap.

The attractiveness of the SAGD process is that it can be operated continuously at essentially constant pressure and temperature. Operationally, there are no changes in pressure or temperature, and this can minimize:

1. Thermal stresses
2. Sand influx
3. Chemical deposit buildups at the wells (as opposed to situations where fluid rates, pressures, and/or temperature frequently change with time which is a conducive environment for chemical constituents to deposit out of solution)

The conceptual sequence of production and injection in the reservoir simulator is as follows:

1. Liquid is produced at the HW producer until steam is produced.
2. At the onset of steam production the production rate is reduced until steam is no longer produced.
3. A liquid level is established to a slight degree above the HW producer.
4. As oil and steam condensate drain down from the steam zone, the liquid level rises.
5. The simulator tests if steam is coming from the producer.
6. The simulator produces as much fluid as feasible until the liquid level decreases to the top of the wellbore and steam begins to be produced. The simulator slightly throttles back the production rate such that no steam is produced. (As discussed in Example 7-8 and described in Figure 7-6b, because of pressure drop considerations it may be preferable to maintain the liquid level higher than just right at the HW producer.)
7. Because fluid is withdrawn from the system, the pressure in the steam zone drops slightly.
8. There is a pressure gradient between the constant pressure steam injector and the steam zone. As a consequence, steam flows into the

steam zone at the same volumetric rate as the rate at which the fluid is withdrawn.

9. The rate at which oil can be produced from the zone is dependent upon the rate of heat transfer at the steam-condensate-oil interface and the viscosity-temperature relationship of the oil phase. (In addition, the permeability characteristics of the reservoir influence the rate of production.)

10. The process approaches that of a steady state condition and is controlled by steam rate at the producer and pressure at the injector.

It is important to decrease the viscosity of the oil phase to its lowest possible point while maintaining the highest steam zone pressure possible, in reservoir environments where there is negligible fluid mobility, without exceeding formation fracture pressure. However, in reservoirs where fluid mobility exists, deleterious fluid leak-off with concomitant heat loss could merit injection at pressures somewhat below formation fracture pressure so that an improved thermal efficiency can be obtained. Heat will have less of a tendency to leak away if reservoir injection pressure is lower. SAGD in reservoirs with mobile oil is discussed later in this chapter.

The SAGD process ends at the point where the bulk of the oil phase has been recovered from the steam zone. Frequently, simulations assess the sensitivity of the point at which steam injection can be stopped and the steam zone blown down. Usually there is a period of time before the SAGD process ends when there is sufficient heat in the reservoir to maintain further conduction heating of the oil phase such that steam injection is no longer required. Figure 7-35 conceptually depicts the blowdown phase.

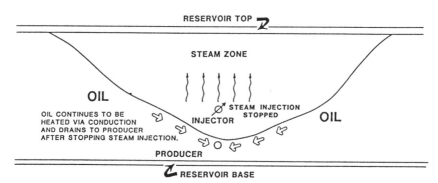

Figure 7-35. Cross section of SAGD process at possible blowdown stage.

Example 7-12

Establish the time to heat by conduction a 2.5 ft radius around a HW to a temperature of 250°F.
Refer to Figure 7-35a (Carslaw and Jaeger, 1959).
Given: Original reservoir temperature = 50°F

V = 500°F (Temperature at HW)
v at r = 250°F − 50°F
a = 0.25 ft (Radius of HW)
r = 2.5 ft

Therefore $\dfrac{v}{V} = 0.4$ $\log_{10}(r/a) = 1$

$$\varkappa = \dfrac{\lambda}{c}$$

$$\lambda \text{ at } 125°F = 0.5778 \left[1.272 - 2.250 \, \phi + \dfrac{0.390 \, \lambda_s \, \sqrt{S_w}}{0.5778} \right]$$

(From Somerton et al., 1974)

Given: $\lambda_s = 4.275 \dfrac{BTU}{hr \, ft \, °F}$

$S_w = 0.35$

$\phi = 0.35$

$\lambda \text{ at } 125°F = 1.2663 \dfrac{BTU}{hr \, ft \, °F}$

$$\lambda \text{ at } 250°F = 0.5778 \left[\dfrac{\lambda 125°F}{0.5778} - 0.002304 \left[\dfrac{(T_{250°F} - 32)}{1.8} - 51.7 \right] \right.$$

$$\left. \times \left[\dfrac{\lambda 125°F}{0.5778} - 1.419 \right] \right]$$

$\lambda \text{ at } 250°F = 1.195 \dfrac{BTU}{hr \, ft \, °F}$

Horizontal Wells

λ at 250°F in days = $28.678 \dfrac{\text{BTU}}{\text{day ft °F}}$

Given that the specific heat of the material is $35 \dfrac{\text{BTU}}{\text{ft}^3 \text{ °F}}$

then $\varkappa = \dfrac{28.678 \text{ BTU/day ft °F}}{35 \text{ BTU/ft}^3 \text{ °F}} = 0.819 \dfrac{\text{ft}^2}{\text{day}}$

From Figure 7-35a, for $\dfrac{v}{V} = 0.4$ and $\log_{10}(r/a) = 1$

$\dfrac{\varkappa t}{a^2} = 1{,}000$

$\dfrac{0.819}{(0.25)^2} (t) = 1{,}000 \qquad t = 76.3 \text{ days}$

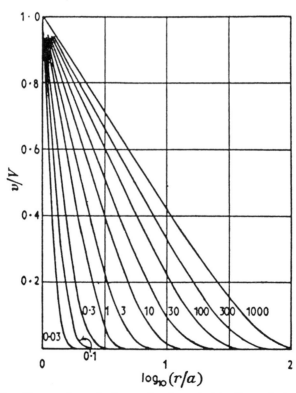

Figure 7-35a. Temperatures in the region bounded by the cylinder r = a, with zero initial temperature and constant surface temperature V. The numbers on the curves are the values of Kt/a^2 (after Carslaw and Jaeger, 1959).

Thermal Recovery and Primary Production for Heavy Oils 293

Therefore, it would require 76.3 days to heat oil sand 2.5 ft in radius away from the HW to 250°F with the HW temperature set at 500°F.
If the radius of the HW is doubled to 6 in. (a 12-in.-diameter HW) then, log (r/a) becomes 0.7
Assuming the above,

$$\frac{\varkappa t}{a^2} = 100 \quad t = 30.5 \text{ days}$$

It would require 30.5 days to heat oil sand 2.5 ft in radius away from the HW to 250°F with a HW temperature of 500°F and a wellbore diameter of 1 ft. The time is reduced 2.5 times by doubling the diameter of the HW.

Example 7-13

Generally, in thermal numerical reservoir simulators Cartesian coordinate grid blocks are employed as opposed to carrying out the simulations in radial coordinates. Cartesian coordinate models enable the evaluation of multiwell processes whereas radial coordinates are generally employed when only one well is studied. In Cartesian coordinates it is necessary to move fluids from the rectangular gridblocks into a radial wellbore. The term productivity index (PI) is used to carry out the transformation from a rectangular grid block to a radial wellbore. For a cross section of a HW to be simulated, the PI can be written:

$$PI = \frac{0.007082 \ (k_h \ k_v)^{1/2} \ L}{\ln \ [GF \ (r_e/r_w) + \text{Skin}]}$$

where k_h = horizontal permeability
 k_v = vertical permeability
 L = length of horizontal well section simulated
 GF = geometric shape factor dependent upon the shape of the rectangular blocks and the location of the HW within one of the blocks
 r_e = $(\Delta x \Delta z)^{1/2}$ See Figure 7-35b for an explanation of Δx and Δz
 r_w = HW wellbore radius
 Skin = skin factor

PI is in units of resbbl cp/day psi

294 *Horizontal Wells*

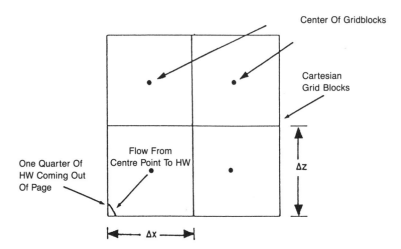

Figure 7-35b. Cartesian coordinate grid block designation for depicting how PI is calculated.

Given $k_h = k_v = 1.0$ D (1,000 md)
$L = 1$ ft
$GF = 0.721$ (a corner quarter symmetry well)
$r_e = [(10)(10)]^{1/2}$
$r_w = 0.3$ ft
Skin = 0.0

Determine the PI for the 10 ft square grid block:

PI = 2.227 resbbl cp/day psi for 10 ft square grid block.

Similarly,

PI = 8.076 resbbl cp/day psi for 1 ft square grid block
PI = 38.55 resbbl cp/day psi for 0.5 ft square grid block

Note that as Δx and Δz approach r_w in size, PI becomes very large. With PI very large, the PI tends to not control performance. Rather, the reservoir grid blocks surrounding the well grid block tend to control influx to the HW. Thus, if it is desired that PI not influence productivity, the grid blocks surrounding the HW should approach the size of the HW and hence, because PI is so large, it does not control productivity.

Reservoir Simulator Results Compared to Analytical SAGD Model

Both the reservoir simulator and the analytical model provide a means to predict SAGD performance in homogeneous reservoirs. Using the identical reservoir environment described in Table 7-4a for the simulator and analytical model, the simulator performance tended to be 20% lower in terms of cumulative oil production after 2 years of operation (Figure 7-36). The reservoir simulator grid configuration is described in Figure 7-37. In terms of steam injection, both the numerical simulator and analytical expression techniques provide similar profiles with time (Figure 7-38).

Table 7-4a
Numerical Simulator and Analytical Expression
Reservoir and Operating Conditions—UTF Reservoir

	Numerical Simulator	Analytical Model
Steam injection pressure (psi)	72.50	72.50
Distance from lower well to top of reservoir (ft)	62.30	62.30
Well spacing (ft)	85.30	85.30
Initial oil saturation	0.81	0.81
Reservoir porosity	0.35	0.35

The difference in results may be due to the use of a constant effective oil permeability for the analytical approach vs varying the oil permeability as a function of fluid saturations for the numerical simulator. The analytical solution assumes that there are no relative permeability effects between the steam condensate and the oil, because each phase basically flows in separate paths to the production well. On the other hand, the simulator averages fluid saturations in each grid block. Consequently, with both water and oil flowing to the producer, the saturation in the grid blocks that have mobile fluids contain both water and oil. The result is that the effective permeability of both water and oil is reduced and as a consequence the flow rate of both phases is diminished. Thus, the simulator rate is lower than the analytical solution. As discussed in the previous section, the size of the grid blocks in the simulator can affect the result. To approach Butler's analytical results, infinitesimally small grid blocks may be required. Practically, there is a trade-off between running time for the simulator and desired accuracy from the model.

In the TANDRAIN section (Figure 7-27) the analytical solution results were higher than the physical laboratory results. The lab investigators attributed the differences to heat losses to the surroundings in the lab being much higher than what would occur in the reservoir. This may be the case,

296 *Horizontal Wells*

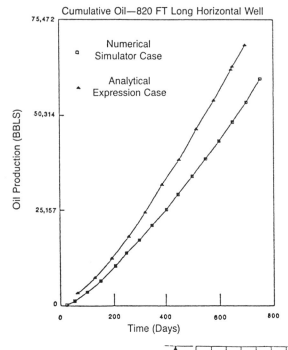

Figure 7-36. Simulator comparison to Butler's analytical solution (after Suggett et al., 1985).

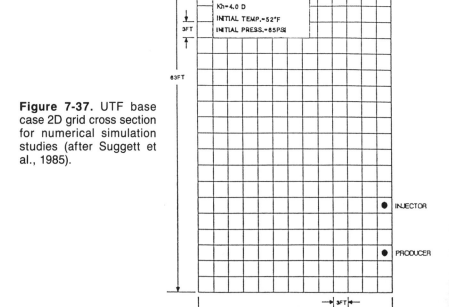

Figure 7-37. UTF base case 2D grid cross section for numerical simulation studies (after Suggett et al., 1985).

Thermal Recovery and Primary Production for Heavy Oils 297

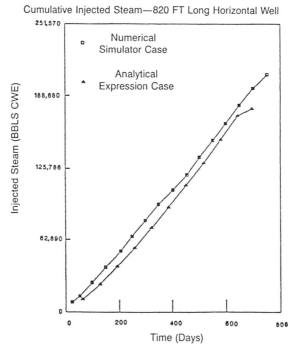

Figure 7-38. Simulator comparison to Butler's analytical solution-UTF project (after Suggett et al., 1985).

but other considerations could account for the differences, such as lab scale heterogeneities. The simulator results tend to be more in line with the physical model observations and may account for the nonideality associated with actual reservoir environments.

Growth of Steam Zone for Different Vertical Permeabilities

Edmunds et al. (1988) assessed the impact of absolute vertical permeability on SAGD performance using a simulator for an Athabasca deposit oil. Figures 7-39 and 7-40 illustrate the rate of growth of the steam zone with time for two different absolute vertical permeabilities. In the isotropic reservoir, the steam zone grows as observed in the physical model experiments and described in the analytical expressions. For the situation where the vertical permeability is reduced by a factor of 50, the steam zone tends to grow outward more than upward. In the low permeability case, the steam zone links up with the adjacent mirror image steam zone first, before reaching the top of the reservoir. The study by Edmunds illustrates that an understanding of reservoir characteristics is required before defining well spacing. For reservoirs with lower absolute vertical permeability it may be preferable to

298 Horizontal Wells

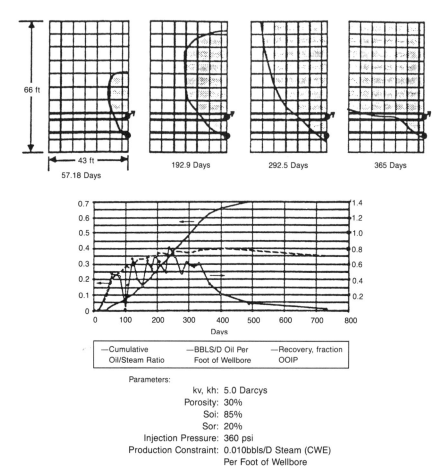

Figure 7-39. Steam chamber growth in homogeneous and isotropic sand (after Edmunds et al., 1988).

space wells a greater distance apart than for reservoirs which have higher absolute vertical permeability.

Edmunds' work indicates that by reducing the absolute vertical permeability k_v by a factor of 50, the recovery efficiency after 500 days of operation is reduced from 0.7 recovery of OOIP down to 0.125 recovery of OOIP. Thus performance rate is severely hindered with low k_v. Similarly, the cumulative oil to steam ratio at 500 days for the 0.1 D k_v case is half that of the $k_v = 5$ D case. The simulations indicate that it is preferable to operate the SAGD process in reservoir environments that have permeabilities greater than 0.1 D and that the k_v/k_h ratio should preferably approach that of unity.

Thermal Recovery and Primary Production for Heavy Oils   299

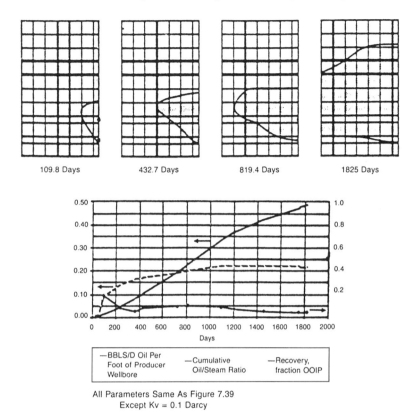

Figure 7-40. Steam chamber growth in homogeneous, anisotropic sand (after Edmunds et al., 1988).

SAGD vs Vertical Well Cyclic Steam

Khosla and Cordell (1984) assessed the SAGD process compared to that of cyclic steam stimulation for the Athabasca deposit. Two reservoir environments were considered and are described in Figures 7-41 and 7-42. One environment had a high mobile water saturation at the base while the second environment had essentially immobile fluid saturations throughout. Figures 7-43 and 7-44 depict the VW and HW well spacings and operating strategies employed. One-fourth of a repeatable pattern was simulated for the

text continued on page 302

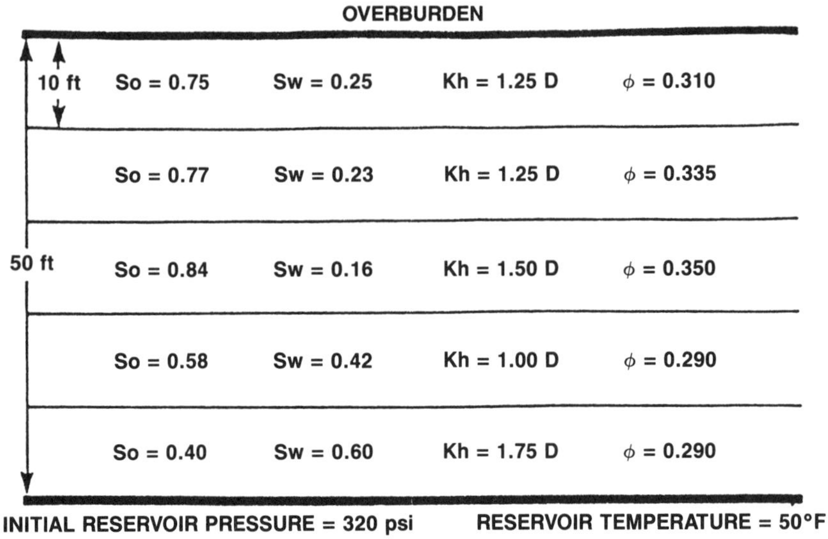

300 *Horizontal Wells*

		OVERBURDEN		
10 ft	So = 0.75	Sw = 0.25	Kh = 1.25 D	ϕ = 0.310
	So = 0.77	Sw = 0.23	Kh = 1.25 D	ϕ = 0.335
50 ft	So = 0.84	Sw = 0.16	Kh = 1.50 D	ϕ = 0.350
	So = 0.58	Sw = 0.42	Kh = 1.00 D	ϕ = 0.290
	So = 0.40	Sw = 0.60	Kh = 1.75 D	ϕ = 0.290

INITIAL RESERVOIR PRESSURE = 320 psi   RESERVOIR TEMPERATURE = 50°F

Figure 7-41. Reservoir characterization—bottom water case (after Khosla and Cordell, 1984).

		OVERBURDEN		
10 ft	So = 0.75	Sw = 0.25	Kh = 1.25 D	ϕ = 0.310
	So = 0.77	Sw = 0.23	Kh = 1.25 D	ϕ = 0.335
50 ft	So = 0.84	Sw = 0.16	Kh = 1.50 D	ϕ = 0.350
	So = 0.82	Sw = 0.18	Kh = 1.00 D	ϕ = 0.290
	So = 0.82	Sw = 0.18	Kh = 1.75 D	ϕ = 0.290

INITIAL RESERVOIR PRESSURE = 320 psi   RESERVOIR TEMPERATURE = 50°F

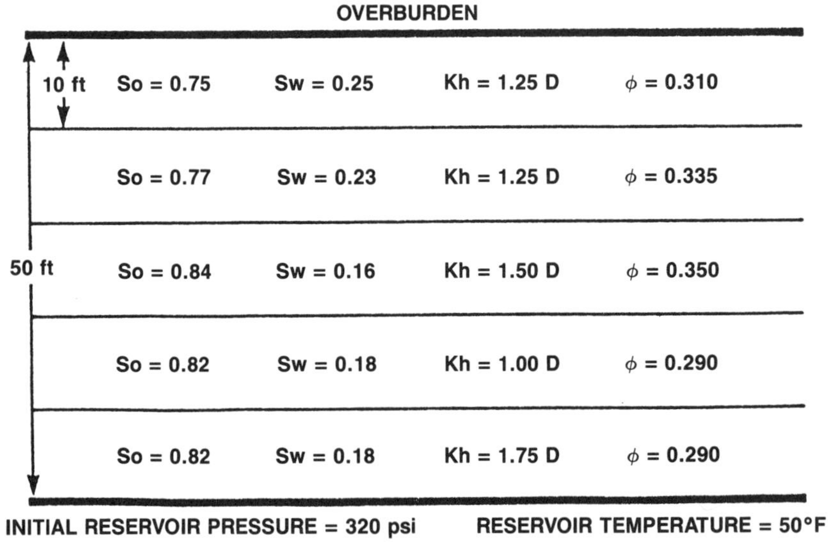

Figure 7-42. Reservoir characterization—oil case (after Khosla and Cordell, 1984).

Thermal Recovery and Primary Production for Heavy Oils 301

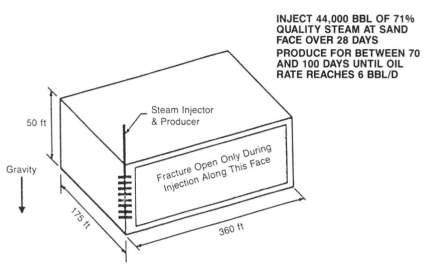

Figure 7-43. Cyclic steam reservoir depiction and operating strategy (1/4 of an isolated vertical well simulated—not to scale) (after Khosla and Cordell, 1984).

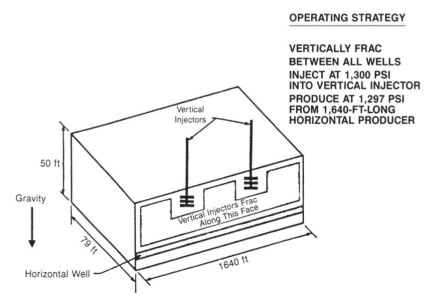

Figure 7-44. Gravity assisted steam drainage into horizontal wells (1/2 of repeatable pattern simulated—not to scale) (after Khosla and Cordell, 1984).

302 *Horizontal Wells*

text continued from page 299

VW while half of a repeatable pattern was simulated for the HW. Both VW and HW were on the same spacing. The steam zone for the HW was maintained by the injection of steam from 2 VWs above the HW producer. For reasons discussed in the "SAGD Laboratory Experiments" section, the practicality of uniformly fracturing between VW injectors and HW producers is suspect. It may be best to employ a HW injector to assure uniform placement of steam above the entire length of the HW producer. Nevertheless the comparison between VW cyclic steam and the SAGD process is valid as the VW injectors for the SAGD case, with the idealized fracture, accurately represent the SAGD process.

Table 7-5 compares the performance for the two processes. The HW SAGD processes are described as "Gravity Drainage" in the table. In the case where water mobility exists the SAGD process recovers 7.5 times that of the VW cyclic steam process. In addition, the OSR is 45% higher for the SAGD process relative to the VW process.

For the reservoir environment that has limited fluid mobility (high oil saturation case), the performance difference between HW and VW cyclic steam results in a 7.1-fold increase in oil production and an 11% increase in

Table 7-5
Comparison Between Horizontal and Vertical Well Performance per Repeatable Pattern

Process	Operating Life (yrs)	Cumulative Steam Injected (bbl)	Cumulative Oil Produced (bbl)	Cumulative Water Produced (bbl)	Cumulative Oil-Steam Ratio (bbl/bbl)	Calendar Day Oil Rate (bbl/d)
Vertical						
Cyclic steam high water sat.	3	440,000	47,170	490,500	0.11	42.8
Cyclic steam high oil sat.	3	352,000	61,650	357,230	0.18	56.6
Horizontal						
Steam drive no variations	3	1,679,000	386,150	1,937,000	0.23	352.8
Steam drive permeable path	3	748,000	75,470	993,700	0.10	69.2
Steam drive double spacing	3	1,730,000	242,150	2,389,900	0.14	221.4
Gravity drainage high water sat.	4	2,176,000	357,230	2,161,500	0.16	244.7
Gravity drainage high oil sat.	4	2,150,000	437,750	2,056,600	0.20	300.0

Source: Khosla and Cordell, 1984.

Thermal Recovery and Primary Production for Heavy Oils   303

OSR. The improvement in performance for the SAGD process may warrant the additional costs of drilling 1 HW and 2 VWs or 2 HWs compared to drilling 1 VW on the same spacing.

There are three major explanations for the improvement in performance with vertical injectors and horizontal producers over that of straight vertical cyclic steam wells. The first is that the reservoir operating temperature is always maintained close to the steam injection temperature for the HW case because steam is continuously injected above the horizontal producer and correspondingly, the oil viscosity is maintained low. On the other hand, for the cyclic steam VWs, by drawing the pressure of the reservoir down to close to 30 psi during production, the corresponding saturated steam temperature is much lower, which in turn results in a high oil viscosity. Oil does not flow as readily to the VWs because of its lower temperature and higher viscosity in comparison to maintaining a high steam zone temperature through continuous injection of steam, for the SAGD recovery strategy.

The second reason for better performance is that steam is never produced from the horizontal production well whereas during the early phase of production from the VW, for each cycle of operation, steam is produced which decreases the oil-steam ratio because of the inefficiency associated with producing steam. The vertical injector-horizontal producer concept, because of gravity segregation, maintains a liquid level of steam condensate and oil above the HW with the steam zone above it, and hence the injected steam is efficiently used.

The third reason is that a constant, high oil saturation is maintained in the vicinity of the HW producer for the SAGD process. In the VW cyclic steam process on the injection cycle phase, during the later cycles the oil is pushed away from the injector. During the production phase, time is required to resaturate the steam zone with oil. Furthermore, a certain residual volume of the oil which resaturates the steam zone may become residual immobile oil. Also, the oil saturations in the resaturated region around the well are frequently low and so productivity to the oil phase is low because of relative permeability considerations. The SAGD process need not have to contend with varying oil productivity over time.

Table 7-6 depicts the energy efficiency between SAGD and the VW cyclic steam process. The horizontal-vertical well gravity drainage process produces 20% less energy than the cyclic steam VWs and hence uses the 20% additional energy, in part, to heat more oil. Table 7-6 illustrates that heat losses from all processes are comparable, within the 20–30% range of the total energy injected at the sand face, for each case study.

Note, in Table 7-5, the cumulative balancing of the volumes of CWE steam injected to oil and water produced for the "Gravity Drainage" cases. For the high oil saturation reservoir, 2,150,000 bbl of CWE steam were injected and 2,494,350 bbl of oil and water were produced. The higher pro-

Table 7-6
Process Comparison for the Recovery of Reserves in Place

Process	% of Oil in Place Recovered	% of Energy Injected Lost to Over & Under Burden	% of Energy Injected that was Produced
Vertical			
Cyclic steam high water at bottom	9.7	22	63
Cyclic steam	10.7	22	58
Horizontal			
Steam drive no areal variations	85.0	29	42
Steam drive permeable path at bottom	15.8	27	46
Steam drive double spacing—perm. path	25.4	28	43
Gravity drainage—high water at bottom	73.8	29	41
Gravity drainage—high oil at bottom	77.0	28	42

After Khosla and Cordell, 1984.

duced fluids volume relative to the steam injected volume is related to the thermal expansion characteristics of each of the phases at steam zone temperature conditions. A part of the CWE steam is gaseous, in situ, which has a greater volume associated with it relative to the cold water equivalent. Thus, in reservoir condition terms, the volume injected to volume produced is similar. The balancing of injection to production fluids minimizes the prospects of deleterious interwell interference. The SAGD process can achieve high oil recovery efficiencies because of the limited prospect of steam bypassing oil between pairs of SAGD wells.

SAGD Processes in Reservoirs with Mobile Oil

Camilleri and Mobarak (1987) compared SAGD performance to that of VW steam drive and HW steam drive. The section entitled "Displacement Thermal Processes" discusses the HW steam drive results. Reservoir characteristics for the comparison were provided by Aziz et al. (1985) and are depicted in Table 7-7. The oil viscosity at original reservoir temperature is 500 cp, which is sufficient for oil displacement enabling flood processes to be considered.

Thermal Recovery and Primary Production for Heavy Oils 305

Table 7-7
Reservoir Characteristics of Mobile Oil Reservoir

Depth from surface to top of reservoir	1,500 ft
Reservoir thickness	80 ft
Absolute permeability	2.0D (Unless varied)
Porosity fraction	0.3
Thermal conductivity	24 BTU/ft-day°F
Heat conductivity	35 BTU/ft-day°F
Oil density	60–68 lbs/ft^3
Oil saturation	55%
Water saturation	45%
Reservoir temperture	125 °F
Pressure at top of reservoir	75 psia

Temperature (°F)	Viscosity (cp)
75	5,780
100	1,380
150	187
200	47
250	17.4
300	8.5
350	5.2
500	2.4

After Aziz et al., 1985.

A pattern area of 2.5 acres was used for all of the simulation studies. Figure 7-45 depicts the 3 well configurations studied with the element of symmetry that was studied for each case. For the steam chamber process, the injector HW was located 20 ft above the producer HW, unless otherwise noted. The reservoir thickness was 80 ft with four equally dimensioned gridblocks of 20 ft in the vertical direction. On repeating well patterns the number of full wellbores associated with each process is:

1. VW Nine Spot = 1 injector and 3 producers
2. Steam Chamber = 2 HW injectors and 2 HW producers
3. HW Line Drive = 1 HW injector and 1 HW producer

For the configurations studied, the steam chamber might have to produce twice as much oil as the line drive displacement process to be cost competitive, recognizing that twice as many wells are required. Similarly, the steam chamber process would have to produce more oil, or more oil in a timely manner, to be cost competitive with the VW nine spot configuration, assuming that the cost to drill 4 HWs is greater than the cost to drill 4 VWs.

306 Horizontal Wells

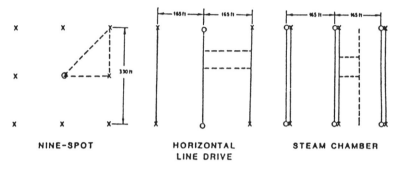

Figure 7-45. Areal diagrams of different development options and the elements of symmetry used for simulation (after Camilleri and Mobarak, 1987).

The operating conditions for the HW processes are provided in Table 7-8. The maximum bottomhole injection pressure is 1,000 psia, presumably below formation fracture pressure. The maximum steam production rate at the producer is controlled at 30 stb/day.

Figure 7-46 depicts the cumulative recovery fraction (crf) for the 3 processes in an isotropic reservoir with 2,000 md absolute permeability throughout. At 0.2 pore volume (PV), cold water equivalent steam injected, the steam chamber process has recovered approximately twice as much oil as

Table 7-8
Operating Conditions for Steamflood

The operating conditions given here apply to horizontal wells and are designed to give the same injection and withdrawals as the vertical wells on a per pattern basis.

Injector: Steam with a heat content of 972.35 BTU/lb subject to the following constraints:
  1. Maximum bottomhole pressure of 1,000 psia
  2. Maximum rate of 300 STB (CWE)/day

Producer: Produce well subject to the following constraints:
  1. Minimum bottomhole pressure of 17 psia
  2. Maximum rate of 3,000 STB/day of total liquids
  3. Maximum steam rate of 30 STB/day

After Camilleri and Mobarak, 1987.

Thermal Recovery and Primary Production for Heavy Oils 307

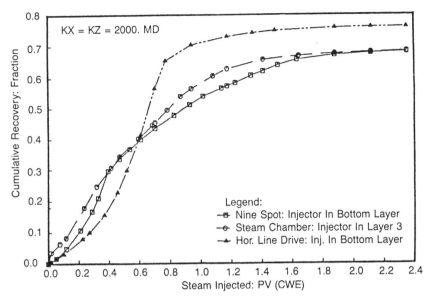

Figure 7-46. Cumulative recovery for different well patterns in an isotropic reservoir (after Camilleri and Mobarak, 1987).

the VW flood and HW drive processes (0.2 cumulative recovery fraction (crf) vs 0.1 crf). During the early life the SAGD process produces oil at a higher rate compared to the other processes. By the time that 0.4 CWE of steam is injected, the SAGD and nine spot VW processes have recovered an identical volume of oil, whereas the HW drive has recovered one-half the oil of the other processes. After the injection of 0.4 CWE PV of steam, the SAGD and VW processes are similar in terms of crf. Whether the SAGD process is viable in this particular reservoir environment compared to a VW nine spot will be dependent upon:

1. The length of time to inject 0.4 CWE PV of steam and recover 0.3 pore volume of the oil in place for each process
2. The benefit of a higher recovery rate for SAGD relative to the VW process during the early life of the process, up to a point of 0.4 PV of CWE steam injected

Huang and Hight (1986) indicated that recovery time is shortened when employing HWs versus VW processes. This is due to injection and production rates associated with HWs being potentially higher than VWs. Thus in terms of the time value of money the crf may be reached sooner with HW processes and the return on capital may occur more rapidly as a result.

The early time enhanced performance for SAGD over the VW process in a mobile oil reservoir is due in part to the fact that a mini steam drive is established between HW injector and HW producer at the onset of operations, with the HW injector and producer so close. In the example case, 20 ft of separation exists between the injector and producer. The displacement between wells occurs with relative ease and as a consequence the time required to sweep the oil between the two wells is very short. Thus, in mobile oil reservoirs, the SAGD process has a higher initial oil recovery efficiency and maintains that edge until steam breakthrough occurs at the producers in the VW process.

Figure 7-47 compares the injection well pressure for the 3 processes. For SAGD, steam breakthrough occurs almost instantaneously whereas for the VW process steam breakthrough occurs at 0.4 PV CWE steam injected. It is at the 0.4 value that cumulative recovery performance is similar for the two processes.

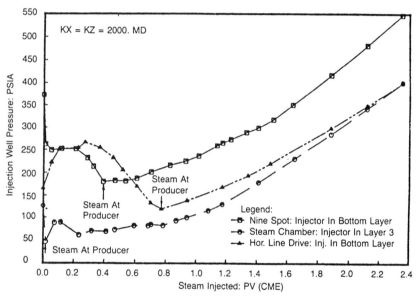

Figure 7-47. Injection pressure for different well patterns in an isotropic reservoir (after Camilleri and Mobarak, 1987).

Of note in Figure 7-47 is the low injection pressure for the SAGD process (less than 100 psi) relative to the higher pressures for the other two processes (250 psi). For the operating constraints employed and described in Table 7-8, the injection pressure is constrained because of the 300 STB maximum injection rate constraint. It is necessary to maintain the injection rate constraint. If the rate were allowed to go higher the process would become inefficient. *Any additional injected fluid above 300 bbl/d CWE steam would*

Thermal Recovery and Primary Production for Heavy Oils 309

displace oil away from the HW producer because the oil in the example reservoir has mobility. The producer could not take the additional steam because it is constrained by the 30 bbl/d of steam that it can produce. The steam chamber could not take more steam because its rate of growth is established by the rate of conduction heating away from the steam-oil interface. If additional steam were employed, the steam would *push the oil away from the steam zone*, thereby establishing a larger steam zone but at the expense of decreasing the oil saturation around the producer. The steam would be displacing heated oil away from the drawoff point at the HW well beneath the injector. Heat would be inefficiently diffused over a larger volume and not concentrated within the environs of the steam zone.

In Figure 7-47, the steam injector pressure does gradually build as a small volume of the injected fluid is allowed to build reservoir pressure within the confines of the repeatable element of symmetry pattern. That is, there are other neighboring SAGD patterns that are mirror representations of the single pattern studied which confine fluids within the single pattern. The gradually increasing pressure for SAGD is indicative of the approach of attempting to balance injection of fluids to withdrawal of fluids such that leak-off of fluid and heat from the steam zone is minimized. Correspondingly, by minimizing fluid and heat leak-off the possibility of linking up adjacent well pairs is reduced and thus the impact of perhaps deleterious well interference is diminished.

To employ SAGD in mobile reservoirs it is important to balance the injection of fluids to withdrawal of fluids. Otherwise, productivity and thermal efficiency will be impaired. The balancing of injection to production will result in lower steam zone pressures which equates to a higher viscosity for the oil phase relative to immobile reservoirs, where, because of limited leak-off, the steam zone pressure can approach that of formation fracture pressure. The lower pressure and temperature in the steam zone for mobile oils negate some of the benefits of SAGD in mobile oil reservoirs relative to immobile oil reservoirs. Often, the viscosity of the oil in the vicinity of the steam zone is similar for an immobile and mobile reservoir. Thus productivity can be similar, provided that the rate of conduction heating is similar.

Influence of k_v in Mobile Oil Reservoirs

As with immobile oil reservoirs, absolute vertical permeability k_v affects SAGD performance. Figure 7-48 compares the two cases where absolute vertical permeability is set at 2,000 md and 200 md. At 0.4 PV of CWE steam injected, reducing the vertical permeability by a factor of 10 reduces the crf by roughly a factor of 2 (from 0.30 down to 0.15). This same rough factor of 2 remains up to 1.0 PV of CWE steam injected. Thus, in mobile oil reservoirs k_v has a bearing on the cumulative recovery fraction relative to PV

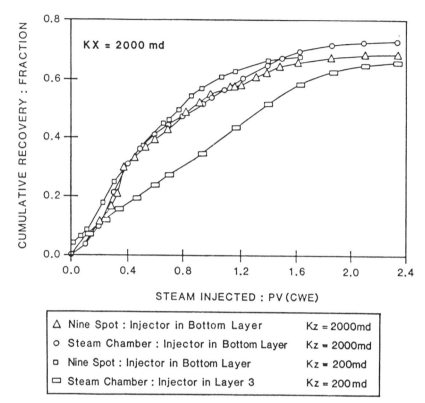

Figure 7-48. Cumulative recovery for different well patterns in anisotropic and isotropic reservoirs (adapted from Camilleri and Mobarak, 1987).

of CWE steam injected. The influence of k_v on performance is similar for both mobile and immobile reservoirs.

Figure 7-48 also illustrates that VW steam floods are not influenced to the same degree as SAGD when k_v is reduced. An understanding of k_v is required for any reservoir environment where SAGD is being considered. Vertical permeability will influence project performance and it may be that in low k_v environments alternate processes to SAGD should be employed.

Influence of Injection Location on Performance

Figure 7-49 illustrates that by moving the HW injector from 20 ft above the producer up to 60 ft above the producer, crf performance improves. When the injector is located higher in the reservoir, the downward steam flood to the producer displaces more oil. This greater "nonthermal" displacement of oil is maintained over that of the lower available "nonthermal" displacement when the injector and producer wells are located 20 ft apart.

Thermal Recovery and Primary Production for Heavy Oils   311

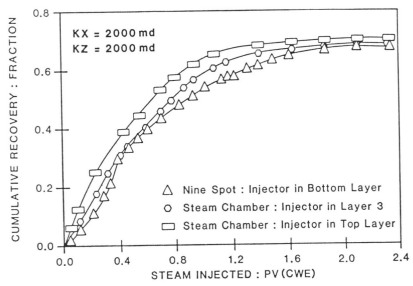

Figure 7-49. Cumulative recovery for different well patterns in an isotropic reservoir (adapted from Camilleri and Mobarak, 1987).

For the case where k_v is reduced from 2,000 md down to 200 md, locating the injector 60 ft from the producer substantially improves performance over locating the injector 20 ft from the producer (Figure 7-50). SAGD performance is on a par with the VW performance. Locating the injector high when k_v is lower enables higher pressures in the steam zone to be established before steam breakthrough at the producer occurs (Figure 7-50a). Higher steam pressures mean higher steam temperatures which equates to a lower oil viscosity. At the onset of steam at the producer (0.2 PV of CWE steam injected), the recovery of oil thereafter is better with the injector near the top of the reservoir because the oil viscosity in the vicinity of the steam zone is lower. It is important when comparing SAGD with VW processes in anisotropic reservoirs to sensitize the location of injector because this can influence performance.

COMBINATION DISPLACEMENT AND GRAVITY DRAINAGE PROCESSES

SAGD provides a means for high recovery efficiency of low mobility oils. A drawback of SAGD is the diminished recovery efficiency for those reservoirs where absolute vertical permeability reduces the rate of recovery of oil. A combination of SAGD and displacement processes has been devised whereby the effects of low vertical permeability are reduced.

312   *Horizontal Wells*

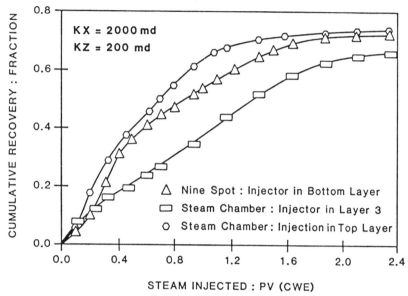

Figure 7-50. Cumulative recovery for different well patterns in an anisotropic reservoir (adapted from Camilleri and Mobarak, 1987).

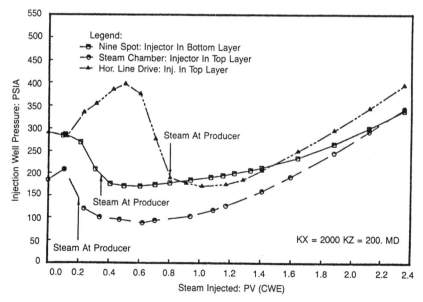

Figure 7-50a. Injection pressure for different well patterns in an anisotropic reservoir (after Camilleri and Mobarak, 1987).

Thermal Recovery and Primary Production for Heavy Oils 313

Heated Annulus Steam (HAS) Drive Process

The heated annulus steam (HAS) drive process consists of utilizing a horizontal, cased, unperforated pipe running between a vertical steam injection well and a vertical production well (Figure 7-51). High-pressure, high-tem-

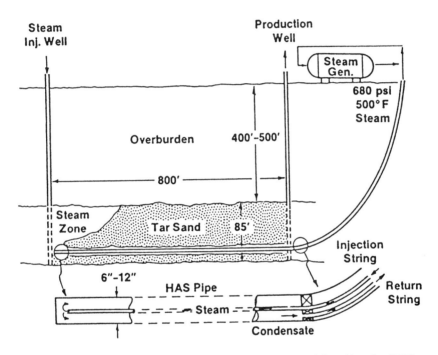

Figure 7-51. HAS drive (heated annulus steam drive) (after Hsueh, 1986).

perature steam is first circulated through the HAS HW which heats the surrounding reservoir by conduction. Over a period of time, as a result of ongoing steam circulation in the HAS pipe, a heated region is formed around the pipe. When the oil around the HAS pipe has been heated enough to become adequately mobile, steam injection below formation fracture pressure commences at the vertical injection well while production is initiated at the producer VW. The heated region around the HAS pipe serves as a means to horizontally displace oil between the injection and production wells.

The steam that is injected via the VW contacts the oil and begins to form a rising SAGD zone. As with the SAGD process, in the HAS process the oil and steam condensate drain to the HAS pipe. The difference between SAGD

314 Horizontal Wells

and HAS is that, with HAS, the fluid that drains to the HAS pipe is swept horizontally to the vertical production well by the pressure gradient maintained between the production and injection wells, whereas with SAGD, the fluid drains into a horizontal producer at the base of the reservoir. The incorporation of horizontal displacement suggests that absolute vertical permeability can be low, as separate SAGD steam zones can be formed with the oil being horizontally displaced to the vertical producer well (Figure 7-52).

While horizontal displacement is occurring, steam circulation in the HAS pipe is maintained to keep the communication path between the vertical production and injection wells open throughout the life of the project. Without steam circulation in the HAS pipe, there is the possibility of having less mobile fluids drain to the HAS pipe which could establish steam fingering to the producer well because of the mobility contrast between the less mobile oil and the steam and condensate phases. (Refer to Example 7-2 to illustrate the influence of the mobility ratio on recovery efficiency.)

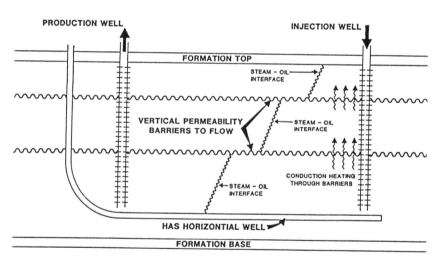

Figure 7-52. HAS process with vertical permeability barriers.

Benefits of HAS Horizontal Pipe

An unperforated HAS pipe provides the means to establish a controlled thermal process. First, steam can be circulated through the unperforated HW pipe at pressures above formation fracture pressure. This enables the highest possible temperature at the HAS pipe wall to establish high rates of

Thermal Recovery and Primary Production for Heavy Oils

conduction heating. Second, the unperforated HAS pipe does not have to deal with the possibility of sand influx which could cause plugging of the HW. And the third benefit of the unperforated HAS pipe is that the distribution of steam along the length of the HW is not a concern because steam is just circulated through its entire length.

HAS Drive Simulation Results

Numerical reservoir simulations have been conducted to assess the benefits of the HAS process (Hsueh, 1986). An example of performance to be achieved via the HAS process is provided below. Reservoir properties for the example are highlighted in Table 7-9. The simulations assumed a multiwell repeated pattern of HAS pipe wells spaced 136 ft apart (Figure 7-53). Each HAS pipe is 800 ft in length. One HAS pipe can drain a maximum area of 2.5 acres for the symmetry assumed (800 ft by 136 ft).

Table 7-9
Reservoir Data Used in the HAS Simulation

Pattern area, acres	5.0
Distance between injector and producer, ft	800
Sand thickness, ft	85
Porosity, % bulk volume	35
Permeability	
Horizontal, Darcy	3.0
Vertical, Darcy	1.5
Initial reservoir pressure, psia	350
Initial reservoir temperature, °F	60
Initial water saturation, % pore volume	35
Initial oil saturation, % pore volume	65
Injected steam temperature, °F	445
Pressure, psia	400
Max. rate BPD/Ac-ft (CWE)	1.5
Compressibility of oil, psi^{-1}	1.03×10^{-4}
Water, psi^{-1}	3.23×10^{-6}
Formation, psi^{-1}	5.0×10^{-5}
Thermal conductivity, BTU/(ft-day-°F)	
Rock	31.5
Water	8.6
Oil	1.85
Gas	0.70
Volumetric heat capacity, BTU (ft^3 – °F)	35.0
Thermal expansion coefficient	
Oil, 1/°F	0.33×10^{-3}
Water, 1/°F	0.44×10^{-3}

After Hsueh, 1986.

316   *Horizontal Wells*

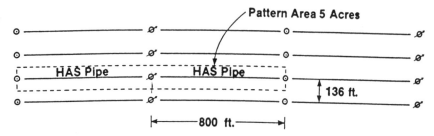

Figure 7-53. Base case well pattern (after Hsueh, 1986).

Prior to the horizontal steam displacement process between the VWs, steam is circulated within the HAS pipe to heat and maintain the pipe surface at 500°F. Heat is transferred by conduction from the pipe surface to the surrounding reservoir, at original reservoir temperature. Based on an analytical solution given in Carslaw and Jaeger (1959), the time required to heat a 5-ft-diameter volume of reservoir with a line heat source in the center at 500°F (the HAS pipe) to a temperature of at least 260°F is approximately 40 days. At 260°F, the oil viscosity is sufficiently low and oil mobility is high enough to allow reasonable horizontal displacement of the heated oil between injector and producer VWs. Continuous steam injection into the vertical injection well starts after 40 days of HAS pipe steam circulation. Heated oil is produced at the vertical production well.

Cumulative oil production as a function of the percent of the original oil in place (OOIP) is depicted in Figure 7-54. The rate of recovery is low for the first 3 years of the HAS process because the volume of steam injected (and hence heat added to the reservoir) is initially low and gradually increases with time. The steam injection rate into the vertical injector gradually increases as the heated volume around the HAS pipe increases in diameter. The oil within the heated volume is sufficiently mobile to be displaced. As the heated surface area normal to the HAS pipe increases, steam can be injected at higher rates because the SAGD process rate of drainage increases. A greater volume of steam can be injected because the surface area for heat transfer in the steam zone is larger, which enables a higher rate of steam condensation. The higher rate of steam condensation allows higher steam injection rates without the breakthrough of steam at the producer VW.

The cumulative oil production is 35% of the OOIP at the onset of steam breakthrough at the production VW. At the end of the tenth year of steam injection, recovery is more than 70% of the OOIP. The high recovery efficiency is due to:

1. The high density of wells, both horizontal and vertical, in confined patterns. (Essentially, a hot plate is located at the base of the reservoir.)

2. The efficient displacement of oil through a combination of the highly efficient SAGD process and horizontal displacement of the oil phase

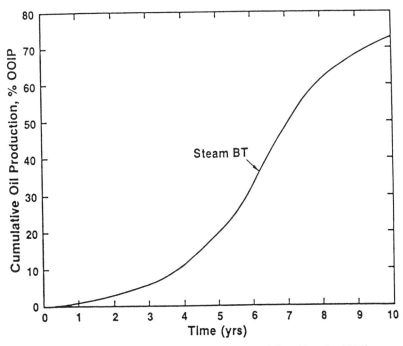

Figure 7-54. HAS drive projected recovery (after Hsueh, 1986).

HAS Drive Steam Oil Ratio (SOR)

The annual steam injected is the sum of the steam injected into the reservoir via the vertical injection well and the steam circulated through the HAS HW to maintain an open heated path between injector and producer VWs. Figure 7-55 depicts the annual volume of steam consumed for the HAS pipe and vertical injection well. The annual SOR decreases from 14 bbl/bbl in the first year to a low of 2.55 bbl/bbl at the time of steam breakthrough. The overall SOR for the pattern configuration for 10 years of operation is 4.55 bbl/bbl.

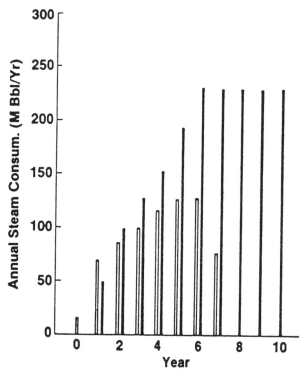

Figure 7-55. Annual steam consumption in pattern HAS pipes (open bars) and in injection well (solid bars) (after Hsueh, 1986).

Increasing Early HAS Time Recovery

As depicted in Figure 7-54, the rate of oil production is low for the first two years of operation because the communication path and steam zone require time to develop. Recognizing that thermal processes are capital intensive projects, methods have been suggested to increase the rate of oil recovery to improve project economics.

Drill a Larger HAS Pipe Hole. Prior to the steam drive, the communication path between injector and producer VWs is established by thermal conduction from the HAS pipe. The rate of heat conduction decreases rapidly as the diameter of the communication channel grows. As discussed previously, 40 days are required to heat a 5-ft-diameter channel around the HAS pipe to 260°F or higher, by using a 6-in.-diameter unperforated HAS pipe and 500°F steam. In employing a 6-in. pipe and 500°F steam, a full year would be required to increase the 260°F channel diameter to 10 ft.

Thermal Recovery and Primary Production for Heavy Oils 319

Using a larger diameter HAS pipe and higher temperature steam is a means to increase the rate of heat growth of the communication channel. For example, using a 12-in.-diameter HAS pipe and 600°F steam would create a 10-ft-diameter 260°F channel in 77 days. Oil production is significantly higher in the early years by using the larger diameter pipe and higher temperature steam, as depicted in Figure 7-56. By the end of the second year of operation, the volume of oil recovered is approximately double the recovery relative to that where the HAS pipe diameter is one half the 12-in.-diameter HAS pipe case.

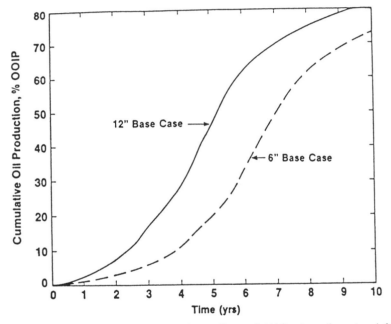

Figure 7-56. Cumulative oil production, effect of HAS pipe diameter (after Hsueh, 1986).

Offset HAS Pipe with a Series of Vertical Producer Wells. For reservoirs that exist at shallow depths, drilling a series of VW producers along the HAS HW increases the rate of oil recovery. In addition, well spacing can be enlarged by drilling a series of producer wells. Figure 7-57 depicts an example of this approach to increasing the rate of recovery. Oil is first produced from the vertical production well located 190 ft away from the vertical injection well. The producer is located 7 ft offset from the HAS pipe. After steam breakthrough occurs (in two years), the producer is shut in. A second vertical production well is then drilled 17.5 ft from the HAS pipe and 550 ft away from the injection well. This producer is also shut in at the onset of

320 *Horizontal Wells*

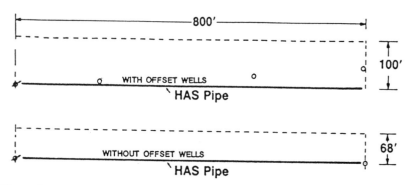

Figure 7-57. One quarter HAS drive pattern, aerial views, with and without offset wells (after Hsueh, 1986).

steam breakthrough. A third and fourth production well are then drilled 40 ft on each side of the HAS pipe and 800 ft from the injection well. The third and fourth wells are produced until the end of the project's life. Figure 7-58 shows the increase in cumulative oil recovery for the offset producer wells vs the base case with only one producer. Over ten years, the offset producer well case produces a total of 397,000 bbl oil vs 266,000 bbl oil for the base case. (Figure 7-58 represents one half of the production from a full pattern

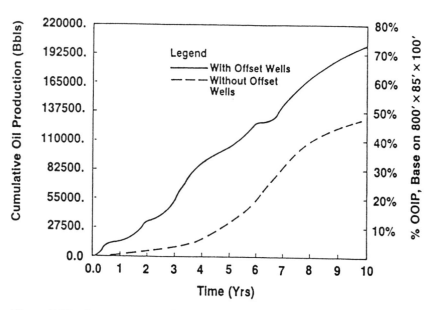

Figure 7-58. Cumulative oil production for multiple offset production wells in the HAS drive process (after Hsueh, 1986).

Thermal Recovery and Primary Production for Heavy Oils

HAS configuration. One-half element of symmetry was simulated.) It is necessary to drill three more production wells to recover an additional 131,000 bbl oil. The 50% increase in oil production between the two cases is a result of a 50% increase in well spacing for the offset well case. (Offset well case well spacing is 3.67 acres vs 2.5 acres for the base case). A trade-off may exist between the higher initial oil recovery for the offset well case and having to drill more wells to obtain a lower ultimate oil recovery over the base case. In certain reservoir environments the base case single producer well may be the preferred avenue for project development because fewer wells are required.

HAS Process Sensitizations

Various alternate HAS well configurations have been evaluated. All provided indications of poorer overall performance. These included:

1. Combining the steam injection well and the HAS pipe into one HW
2. Using a horizontal injection well
3. Reversing the injection and production wells at every other location

As an example, Figure 7-59 depicts the poorer performance obtained when using the HAS pipe as the injection well. The lower the recovery compared to that of the base case is a result of short-circuiting of steam between the injection and production wells. Because of the short distance from the end of the horizontal injection well to the vertical production well, oil near the production well is first mobilized, establishing a short flow path. Water and steam are short-circuited through this path, making oil recovery low.

HAS Drive Bench Scale Laboratory Observation

A series of HAS Drive lab experiments have been reported (Anderson, 1988). The laboratory findings are consistent with the numerical simulation results, in terms of the basic recovery mechanisms. Recovery efficiencies for various laboratory configurations ranged from 30–60% of the OOIP. Oil and sand from the Athabasca deposit were used in the experiments. Two key observations from the experiments were:

1. At the onset of water or steam breakthrough, oil production diminished substantially and the experiment ended (Figure 7-60). The lab work suggests that at steam breakthrough the process ends. The numerical simulation work indicated that the process would continue at economic oil rates after steam breakthrough. The difference between

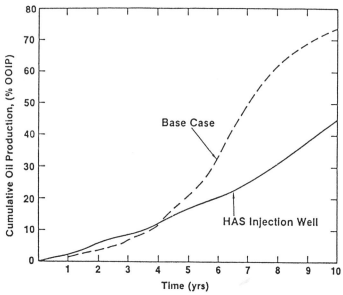

Figure 7-59. Cumulative oil production for HAS injection well (after Hsueh, 1986).

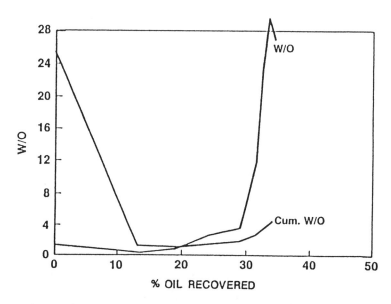

Figure 7-60. HAS drive run #3:30 ft × 6 in. vessel, water-oil ratio vs recovery (after Anderson, 1988).

lab and simulation is attributed to the lab configuration being unable to incorporate the SAGD mechanisms of gravity drainage of oil into the heated path around the HAS pipe. Via the SAGD mechanisms, the oil continually blocks or plugs the heated path and prevents water and steam from channeling to the vertical producer well. Without the drainage of oil from above to plug the heated path, water and steam would channel to the vertical producer. The lab experiments used a 1/4-in. OD HAS pipe in a 6.25-in.-diameter reservoir. The lab HAS pipe was placed 1 in. from the base of the 6.25-in.-diameter reservoir. Scaling up from a 1/4-in. HAS pipe to a 6-in. HAS pipe for the field equates the height of the reservoir to 10.5 ft in the lab experiment. Thus, in the lab, gravity drainage from a thick oil accumulation from above was not incorporated. The lab results indicate that without continuous feeding of oil from above the HAS pipe, at the onset of water breakthrough oil production will diminish substantially. Thus, it may be very important in field environments that absolute vertical permeability be reasonable, or else at the onset of water breakthrough the process may end. The contention that absolute vertical permeability need not be high for the HAS process to work effectively will have to be fully assessed in field tests.

Recognizing that the HAS process rate of recovery may be strongly influenced by the SAGD mechanisms, it is critical that the rate of injection of steam be carefully monitored and controlled to balance with the production of oil. If steam is injected at too high a rate to that necessary for steam condensation at the steam-oil interface of the SAGD steam zone, the additional steam will just channel to the vertical production well.

2. A second observation of the lab experiments was that at the onset of water breakthrough, with concomitant drop in oil rate, the recovery efficiency could be increased by employing a surfactant flood. Figure 7-61 shows the increase in recovery by using a surfactant containing polyoxethylated isoctyl phenol (Triton X-100). The results in the figure were obtained using a one-half pore volume of surfactant with the following aqueous composition:

Triton X-100	0.5%
Sodium hydroxide	2.0%
Sodium chloride	5.0%

Thus, in HAS processes, at the onset of water breakthrough surfactants may be employed to further enhance recovery.

In addition, surfactants may further increase ultimate recovery in the SAGD processes. The use of surfactants need not be limited to HAS processes.

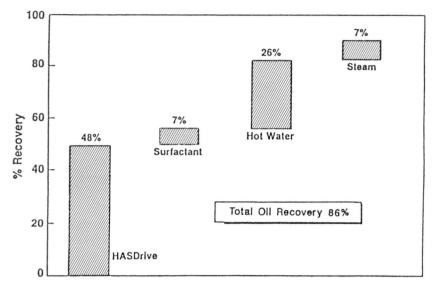

Figure 7-61. Post HAS drive surfactant flood (after Anderson, 1988).

HAS Drive Field Evaluation

To assess the merits of the HAS Drive process a single HAS well field test has been ongoing at AOSTRA's Underground Test Facility since November, 1987 (Duerksen, 1988). Figures 7-62 and 7-63 depict the pilot layout. The HAS pipe test well is 200 ft in length. Performance from the test well is intended to be extrapolated to commercial HAS pipe lengths. For the 200-ft-long HAS well, reservoir simulations have projected average oil rates (for the first two years) of 20 bbl/d; increasing to 40 bbl/d by the fourth year of operation. Figures 7-64 and 7-65 depict the forecast performance for the HAS pilot based on numerical reservoir simulations. The cumulative SOR of 3.5 bbl/bbl for the HAS process, depicted in Figure 7-65, suggests a highly efficient thermal process.

Actual performance from the pilot has not been released as of this writing.

DISPLACEMENT THERMAL PROCESSES

A variety of thermal displacement processes have been studied via the use of numerical reservoir simulation and reported in the literature. A caveat is placed on displacement processes in highly immobile oils to the extent that

Thermal Recovery and Primary Production for Heavy Oils 325

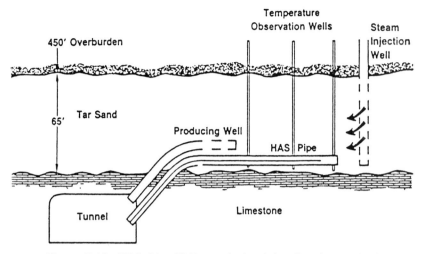

Figure 7-62. HAS drive UTF test design (after Duerksen, 1988).

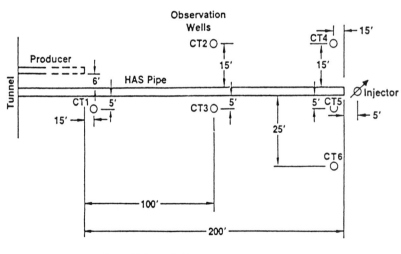

Notes: (1) Not to Scale
(2) Observation Wells Drilled and Completed
10 ft Deeper Than Bottom of Tar Sand
(3) Injector Completed and Perforated
Through Entire Tar Sand Interval

Figure 7-63. Location of HAS drive UTF observation wells (plan view) (after Duerksen, 1988).

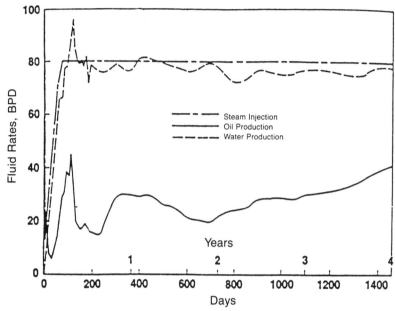

Figure 7-64. HAS drive UTF pilot—fluid rates for base case (HASB6).

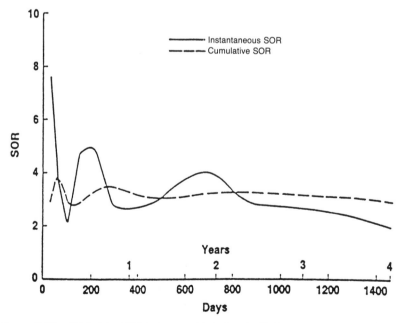

Figure 7-65. HAS drive UTF pilot—steam/oil ratio (SOR) for base case (HASB6) based on drive steam only (after Duerksen, 1988).

Thermal Recovery and Primary Production for Heavy Oils

early breakthrough of steam at any point along the producer HW could short-circuit the further recovery of oil at acceptable rates. That is, the steam viscosity is usually substantially lower than the viscosity of the in situ oil phase. Consequently, at the onset of breakthrough the mobility contrast between the injected fluid and the in situ oil phase can be so large that, essentially, it is primarily the lower viscosity injected steam that is produced at the point of breakthrough. As discussed earlier, to reduce the possibility of deleterious breakthrough the rate of displacement of the injected phase should balance the rate of growth of the steam zone such that oil mobilized at the steam zone interface is swept continually in advance of the injected fluid. Displacing at a rate higher than the rate of steam growth will establish the possibility of early breakthrough of the injected fluid. Most of the reservoir simulation studies reported do not take into account the possible early breakthrough due to nonideality of reservoir permeability in actual reservoir environments.

HW Displacement with Mobile Water Zones

For immobile oils at original conditions, it is necessary that a mobile water zone exist for HW displacement to be viable. Without the mobile water the wells would have to be spaced very, very closely together, or formation fracture pressure would have to be exceeded, to enable acceptable fluid injection rates.

Examples of horizontal well displacements follow:

1. For the reservoir characteristics described in Figures 7-41 and 7-42, 1,640-ft-long HWs spaced 148 ft apart (Figure 7-66) were placed at the base of the reservoir in a zone that had water mobility. Steam is injected at 1,730 bbl/d CWE steam into the HW injector and fluid produced at 2,200 bbl/d at the producer. The third line in Table 7-5 provides the projected performance for the given reservoir environment in which the repeatable pattern spacing gives a well spacing of 5.5 acres per pattern. For the short (3 yr) life of the project, 386,000 bbl of oil are projected to be produced at an oil to steam ratio of 0.23 bbl/bbl. The recovery of OOIP is 85%.

2. In the above reservoir description it was assumed that the reservoir characteristics were areally uniform. Thus, the displacement between injector and producer HW would be expected to be essentially uniform along the entire length of the HWs. In reservoir environments where areal heterogeneities exist, displacement may not be uniform which could influence performance. Figure 7-67 depicts a reservoir environment where, simplistically, the absolute horizontal permeabil-

328 Horizontal Wells

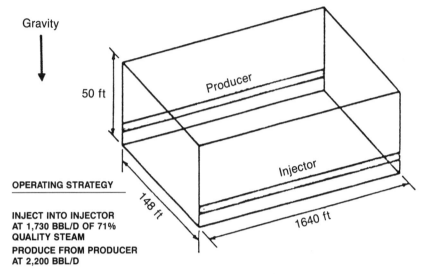

Figure 7-66. Horizontal well interwell steam drive for bottom water reservoir (½ of horizontal wells simulated—not to scale) (after Khosla and Cordell, 1984).

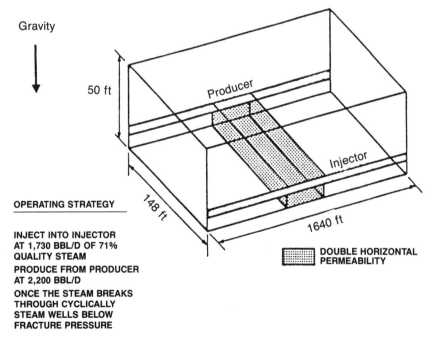

Figure 7-67. High permeability channel between injector and producer (½ of horizontal wells simulated—not to scale) (after Kholsa and Cordell, 1984).

ity along one fifth of the HW is double the permeability of the rest of the reservoir at the given reservoir depth. Except for the doubling of permeability, all reservoir characteristics are the same as in the first example. The study is intended to assess how performance is affected by steam breakthrough along just a part of the well. With breakthrough along just a length of the well, performance is severly impaired (Table 7-5, 4th row). Over a 3-year life, cumulative oil recovered decreases by a factor of 5 and oil to steam ratio drops by a factor of 2. Unless the locale where steam has broken through prematurely can be effectively plugged over the duration of the project, performance deteriorates substantially. (Plugging the breakthrough zone at the producer may not be sufficient. The possibility exists for steam to channel parallel to the HW producer and short-circuit the displacement process a short distance from where the HW producer well was mechanically plugged.)

3. If areal variations in reservoir characteristics are thought to exist for a particular reservoir environment, it is important to assess the magnitude of the variations before locating a series of injector and producer HWs. (Perhaps drill one HW and inject a radioactive tracer to assess the uniformity of injecting fluids along the HW.) If it is determined that displacement will not be uniform from the HW, it is suggested that studies be undertaken to determine how best to establish uniform displacement between injector and producer. It may be that the perforation density along the HWs (both injector and producer) may be varied to account for the variation in reservoir quality.

Alternately, with variations in reservoir quality, well spacing might be increased to lessen the influence of the areal variations in reservoir quality. By doubling the well spacing in the previous example from 5.5 acres per pattern up to 11 acres per pattern, the recovery of oil improves by a factor greater than 3 and the OSR improves by 40%. The reason for the improvement in recovery is that the steam breakthrough time increases by a factor of 3. This greater length of time to breakthrough allows a larger steam zone to be established above the higher permeability water zone at the base. The larger steam zone displaces more oil into the high permeability path and hence slows up the rate of advance of the steam in the high permeability path. The steam rises more above the high permeability path into the steam zone and mostly steam condensate and oil are displaced through the high permeability path. Hence, cumulative recovery is improved over that of the smaller well spacing. With areal variations in reservoir quality, for HW displacement processes, performance may be anticipated to be affected relative to displacement in homogeneous environments. Note, in Table 7-5, that recovery of OOIP is reduced by at least a factor of 3 between the homogeneous reservoirs and reservoirs where there are variations in horizontal absolute permeability along the length of the HW. Caution is recommended when considering HW injectors and producers for displacement processes in highly immobile oils.

4. Performance from a vertical well that is fractured and cyclically steam stimulated on 5.5 acres per well spacing is provided in Table 7-5 as a comparison to the HW performance. (See Figure 7-43 for a description of the VW fracture configuration.) In terms of recovery of OOIP, the HW displacment process is indicated to be superior. The incremental recovery would have to be weighed against the incremental increased cost of drilling and operating HWs.

Performance from a SAGD process is included in Table 7-5 also. Figure 7-44 depicts the SAGD process used in the comparison. Based on the simulation results for a homogeneous reservoir which has a mobile water zone, it may be preferable to consider a displacement process rather than SAGD. The reason is that SAGD works effectively in reservoir environments where there is negligible fluid mobility. In environments where mobility exists, the injected heat is not concentrated as effectively within the steam zone. Rather, upon pressurization of the reservoir, fluid and hence heat are displaced away from the SAGD wells. The SAGD process is designed such that the fluid is best displaced to the producer HW at the base of the reservoir. The displacement of heat away from the SAGD well tends to create a more diffused heated zone, thereby reducing the energy efficiency of the process.

Table 7-5 illustrates the improvement in performance for SAGD when no mobile water zone exists in situ. When there is no mobile water, cumulative oil increases by 23% and OSR improves by 25%. Thus for reservoir environments that contain initial fluid mobility it is worthwhile to consider both displacement and SAGD processes as a potential HW recovery process.

5. One means to possibly alleviate short-circuiting of steam is to selectively complete the HW injectors and producers. Jain and Khosla (1985) used slightly different reservoir characteristics than the preceding four examples to investigate selective completion. The reservoir characteristics are depicted in Tables 7-10 and 7-11. Five-hundred-ft-long HWs were spaced 200 ft apart (Figure 7-68). The wells were located approximately 11 ft from the base of the reservoir in a zone that had initial water mobility. The study investigated one HW injector and producer and assumed repeated HW patterns. Steam was initially injected at below formation fracture pressure 12 ft from one end of one producer HW for 100 days, while producing the injector well at a bottomhole pressure of 145 psi. Between 100 and 730 days, steam was injected into the injector well while producing fluids from the first well.

Thermal Recovery and Primary Production for Heavy Oils 331

Table 7-10
Reservoir Description

Layer #	Thickness ft	S_o	S_w	S_g	K_h md	K_v md	ϕ
1	9	0.43	0.55	0.02	1,000	400	0.29
2	8	0.73	0.25	0.02	1,500	600	0.33
3	5	0.83	0.15	0.02	1,500	600	0.33
4	3	0.68	0.30	0.02	1,000	400	0.34
5	9	0.81	0.17	0.02	1,500	600	0.35
6	12	0.83	0.15	0.02	1,500	600	0.35
7	9	0.68	0.30	0.02	1,500	600	0.33
8	2.5	0.43	0.55	0.02	500	200	0.25
9	7	0.78	0.20	0.02	1,500	600	0.35
10	3	0.0	1.0	0.0	2,000	800	0.25

Reservoir rock heat capacity = 245.8 BTU/bbl°F
Reservoir thermal conductivity = 24 BTU/d-ft°F
Overburden heat capacity = 224 BTU/bbl°F
Overburden thermal conductivity = 30 BTU/d-ft°F
Underburden heat capacity = 168.4 BTU/bbl°F
Overburden thermal conductivity = 24 BTU/d-ft°F

After Jain and Khosla, 1985.

Table 7-11
Viscosity-Temperature Relationship

Temperature °F	Oil Viscosity (cp)
50	2,730,000
150	2,030
250	75.2
350	13.3
450	4.69
550	2.32
650	1.38
750	0.916
850	0.659
950	0.502

After Jain and Khosla, 1985.

From 730 to 900 days, the reservoir was blown down to facilitate the movement of the tubing string in the injector from the 12 ft location to 295 ft along the length of the well. Packers were used around the tubing string at 98 ft so that steam channelling at the 12 ft location of each well would not occur further. Steam was injected for the period from 900 to 2,400 days. At 1,200 days, the tubing in the producer was moved to the 295 ft location and a packer placed between tubing and

332 Horizontal Wells

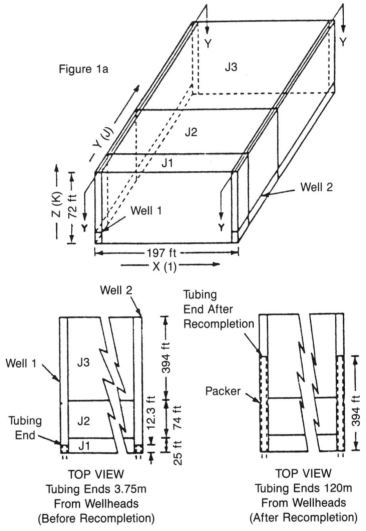

Figure 7-68. Placement of tubing in reservoir model (after Jain and Khosla, 1985).

HW annulus at the 98 ft location. At 2,400 days (6.6 years), the bulk of the oil was recovered from the flood.

The basis for recompleting the injector and producer wells is as a result of steam breakthrough at the producer HW. Figure 7-69 depicts that by 730 days oil production dropped substantially as a result of the injection of steam through tubing at one location along the well. The injector tubing was moved at 900 days. Similarly, at 1,200 days oil

Thermal Recovery and Primary Production for Heavy Oils   333

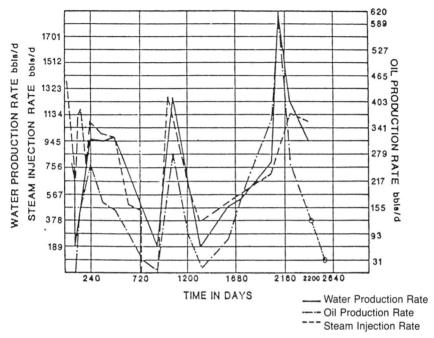

Figure 7-69. Oil and water production and steam injection rates vs time in days (after Jain and Khosla, 1985).

production once again dropped. The producer tubing was moved at 1,200 days. In both instances the steam broke through over 98 ft of the total 500 ft of HW length—at the vicinity of the producer draw-off location. To handle the steam breakthrough, first the injector tubing was moved; then after a further steam breakthrough, the producer tubing was moved further into the well.

This operation produced 333,000 bbl oil at a cumulative SOR of 5.5 bbl/bbl, over a 6.5-year operating life. Approximately 58% of the OOIP was recovered.

Two reasons can exist for moving the tubing strings at different points in time. First, if it is determined through injectivity tests that there are areal variations in injectivity, it may be necessary to isolate (with packers) sections of the injection HW and inject into somewhat poorer injectivity intervals such that displacement efficiency will be more uniform. After an established period of injection into the poorer injectivity intervals, the higher injectivity intervals may be opened to steam.

Second, if concern exists about sand influx into the HWs it is advisable to keep the tubulars out of most of the length of the HW. Other-

334   *Horizontal Wells*

wise, the tubing strings can become firmly lodged in the HW pipe if sand fills the wellbore. The tubing strings can be removed more easily if initially located close to the vertical ends of the horizontal section. It was for this second reason that the tubing strings in the above study were moved, rather than initiating steam injection along the entire length of the well. Areally, the reservoir characteristics were the same for each layer so the likelihood of nonuniform steam breakthrough at the producer would have been diminished, for the idealized reservoir description.

6. Work by Camilleri and Mobarak (1987) assessed the variation of recovery relative to vertical permeability for horizontal displacement. Figure 7-45 depicts the well configurations simulated and the element of symmetry employed. The HW injector and producer were spaced 165 ft apart, each at the base of the reservoir unless otherwise noted. Table 7-7 contains the reservoir characteristics associated with the simulations. Of particular note in Table 7-7 is the lower oil phase viscosity compared to the previous examples described. As a consequence of the lower oil viscosity, a mobile water zone is not required to establish injectivity.

Figure 7-70 illustrates the impact of absolute vertical permeability on recovery performance. At 0.5 pore volume of steam injected, performance is comparable for k_v/k_h equal to 1 and 0.1, respectively. However, for the k_v/k_h = 0.1 case at 1.0 PV of steam injected, performance begins to lag the k_v/k_h = 1.0 case. The difference in recovery may be attributed to a slower building of the steam zone upward in the k_v/k_h = 0.1 case. As a consequence, steam travels more at the base of the reservoir and channels to the HW producer to a greater extent for the k_v/k_h = 0.1 case.

Placing the HW injector at the top of the reservoir improves performance in the case where k_v/k_h = 1.0, because the steam zone is able to build more quickly at the top of the reservoir then displace oil downward effectively to the producer HW location (Figure 7-71). For the k_v/k_h = 0.1 case, placing the HW injector at the top of the reservoir delays performance further vs the placement of the HW injector at the base, because the steam zone is not capable of growing downward to the horizontal producer at the same rate. Essentially, because of the lower k_v the displacement of cold oil in advance of the steam front is slower with the injector (push point) at a different horizon to the producer well. The permeability between injector and producer is lower when the injector is at the top of the reservoir for k_v/k_h = 0.1. For k_v/k_h = 0.1 with the injector at the base, the injector (push point) does not have to push cold oil through lower k_v reservoir, but pushes instead through the higher horizontal permeability rock. Thus the rapid dis-

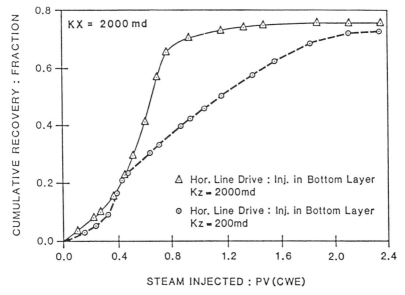

Figure 7-70. Cumulative recovery for different well patterns (adapted from Camilleri and Mobarak, 1987).

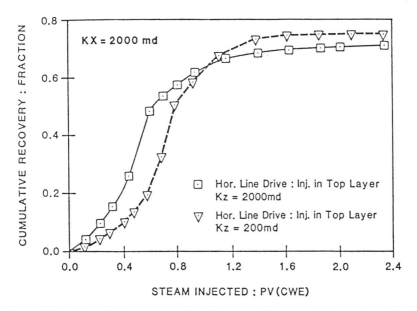

Figure 7-71. Cumulative recovery for different well patterns in isotropic and anisotropic reservoirs (adapted from Camilleri and Mobarak, 1987).

placement for the $k_v/k_h = 0.1$ with injector at base establishes the steam zone faster.

For displacement HW processes, an assessment of k_v/k_h variations is important in determining where to locate the injector HWs.

The work of Camilleri and Mobarak (1987) also compared performance of HW displacement with SAGD and VW nine spot patterns. Figure 7-45 depicts the SAGD and VW elements of symmetry simulated. Figure 7-46 compares performance for all three processes in an isotropic reservoir. Figure 7-71a compares performance in a reservoir where k_v/k_h ratio = 0.1.

For the isotropic reservoir, the SAGD process is the superior process up to the point of 0.4 PV of steam injected. Where $k_v/k_h = 0.1$, a vertical well displacement process is indicated to be superior. In addition, with $k_v/k_h = 0.1$ the SAGD process performs most poorly. SAGD processes may be considered superior in reservoirs where no reservoir mobility exists. In reservoirs where acceptable mobility exists, alternate recovery processes must be considered.

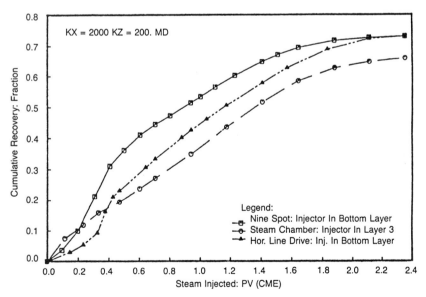

Figure 7-71a. Cumulative recovery for different well patterns in an anisotropic reservoir (after Camilleri and Mobarak, 1987).

Thermal Recovery and Primary Production for Heavy Oils   337

Combination HW and VW Processes for Mobile Oils

For reservoirs that are capable of undertaking displacement processes at original reservoir conditions, a combination of VWs and HWs may improve conformance. The ability to inject steam uniformly along an entire HW with lengths greater than 1,000 ft has yet to be established in field trials. If steam is not uniformly distributed along a HW injector, conformance and hence ultimate recovery of OOIP could be impinged. The previous section describes the impact of nonuniform steam distribution on recovery efficiency. VW injectors alleviate the concern of conformance along a HW. Huang and Hight (1986) describe 3-pattern configurations where VWs are employed as injectors and HWs as producers. Figures 7-72 through 7-74 describe the patterns. Each pattern was simulated using the shaded area depicted in each figure as the element of symmetry. Each simulation comprised a bounded pattern with no flow across the boundaries of the symmetrical element. The first configuration (Figure 7-72) consists of 3 VW injectors and 2 HW producers per repeatable pattern. The pattern size is 18.5 acres. Completed interval totals 1,500 ft from the VWs and HWs over the 18.5-acre area. The second configuration consists of 2 VW injectors and 2 HW producers per repeatable pattern. The pattern size is 18.5 acres and

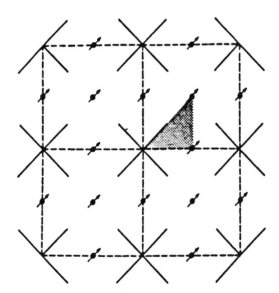

SHADED AREA REPRESENTS 1/8 OF AN 18.5 ACRE PATTERN

Figure 7-72. Combination of vertical and horizontal well patterns: configuration C-1 (after Huang and Hight, 1986).

338 Horizontal Wells

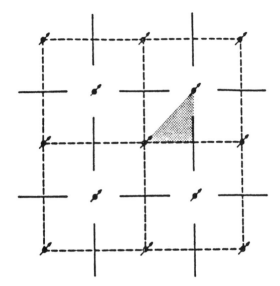

SHADED AREA REPRESENTS 1/8 OF AN 18.5 ACRE PATTERN

Figure 7-73. Combination of vertical and horizontal well patterns: configuration C-2 (after Huang and Hight, 1986).

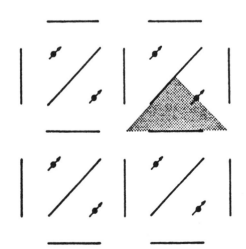

SHADED AREA REPRESENTS 1/4 OF A 26 ACRE PATTERN

Figure 7-74. Combination of vertical and horizontal well patterns: configuration C-3 (after Huang and Hight, 1986).

Thermal Recovery and Primary Production for Heavy Oils 339

completed interval totals 1,050 ft over the area. The third configuration consists of 3 HW producers and 2 VW injectors on 26 acres per repeatable pattern spacing. Completed interval totals 1,950 ft over the 26-acre area. Reservoir characteristics used in the simulations are depicted in Table 7-12. The reservoir thickness for all cases is 125 ft. Five 25-ft-thick simulation blocks were used in the vertical direction. The HWs were placed in the lowest vertical grid block. In the simulator, maximum steam injection pressure was set to 1,200 psi which corresponded to slightly below formation fracture pressure. The production HWs were shut in when the instantaneous steam-oil ratio reached 15 bbl/bbl.

Results of the 3 cases are summarized in Table 7-13. The results in the table suggest that Case C-1, relating to Figure 7-72, provides the best performance due to the shortest time for recovery of the oil in place (7 yr vs 14 yr for the other 2 cases). The apparent reason for the faster recovery is due to having 3 injectors vs 2 injectors for the other cases. The additional injector allows for higher injection rates per pattern which provides for higher displacement rates for the oil phase. (Injection rates per pattern were 3,900

Table 7-12
Reservoir and Fluid Properties—Horizontal Well Simulation

Porosity, fraction	0.39
Initial fluid saturations, fraction: Oil	0.589
Water	0.411
Gas	0
Initial reservoir temperature, °F	100
Initial reservoir pressure, psi	50
Permeability, md: Horizontal	3,000
Vertical	900
Reservoir thermal conductivity BTU/day-ft-°F	31.2
Reservoir heat capacity, BTU/ft^3-°F	37.0
Cap and base rock thermal conductivity, BTU/day-ft-°F	24.0
Cap and base rock heat capacity, BTU/ft^3-°F	46.0
Oil viscosity, cp @ °F	
1,230 @ 100	
10 @ 300	
3.99 @ 400	
Quality of injected steam, fraction (at sand face)	0.65
Residual oil saturation, fraction	
to water: 0.25	
to steam: 0.15	

After Huang and Hight, 1986.

Table 7-13
Combination VW and HW Displacement Performance Results

Case	Pattern Spacing Size (Acres)	Number of VW Injectors	Recovery of OOIP (%)	Time required to Recover Oil (yr)	PV of Steam Injected to Recover Oil (PV)	Oil to Steam Ratio (bbl/bbl)
C-1	18.5	3	72.2	7	1.4	0.304
C-2	18.5	2	73.1	14	1.8	0.242
C-3	26.0	2	72.9	14	1.4	0.300
Run 1 (9 spot)	18.5	1	64.7	15	1.8	0.200
Run 2 (13 spot VW infill)	18.5	1	63.2	11	1.3	0.286
Run 5 (13 spot HWs drilled after 6 years)	32	1	70.0	15	1.1	0.372

Adapted from Huang and Hight, 1986.

bbl/d CWE steam for Case C-1, 3,400 bbl/d CWE for Case C-2, and 2,800 bbl/d CWE steam for Case C-3.) Thus the HWs in each case are capable of producing large volumes of fluid because of their large contact area with the reservoir. It is necessary to establish the optimum number of VW injectors that can match the drawdown characteristics of the HWs to the injectivity characteristics of all of the VW injectors. This optimum number must be constrained to a maximum bottomhole injection pressure so that the formation is not fractured. The previous example shows that 3 VW injectors proved to be more effective at displacing oil to the HW than 2 VW injectors, *provided that conformance between injectors and producers can be maintained as ideally represented in the simulation model for the injector and production rates predicted from the studies.*

The shorter period of time for recovery in Case C-1 provides the highest OSR of the three cases, because the heat lost to the over and underburden is not as great as in the remaining two cases with twice as long a recovery period. Refer to Table 7-13 for the OSR for each case.

Combination HW and VW Processes for Mobile Oils in Established Steam Flooded Fields

Huang and Hight (1986) describe the benefits of drilling HWs in VW patterns previously flooded with steam. Figure 7-75 depicts 3 patterns investigated. The 18.5-acre nine spot pattern was infilled with either VWs or HWs (runs 2, 3, and 4). The reservoir characteristics for the infill drilling cases were the same as in the previous section. Figure 7-76 depicts the benefit of infill drilling the nine spot pattern for the various cases, in terms of recovery of OOIP. Note that the HW cases provide the highest recovery of OOIP. The difference between runs 3 and 4 for the HW cases is that in run 4 the infill VWs drilled midway between the injector and producer HWs are converted to steam injection rather than being used as VW producers.

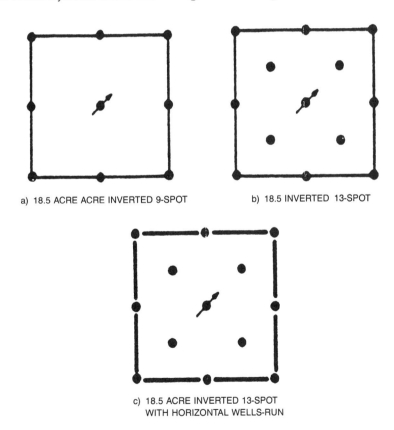

a) 18.5 ACRE ACRE INVERTED 9-SPOT

b) 18.5 INVERTED 13-SPOT

c) 18.5 ACRE INVERTED 13-SPOT WITH HORIZONTAL WELLS-RUN

Figure 7-75. Well configurations used for mature steamflood study (after Huang and Hight, 1986).

342 Horizontal Wells

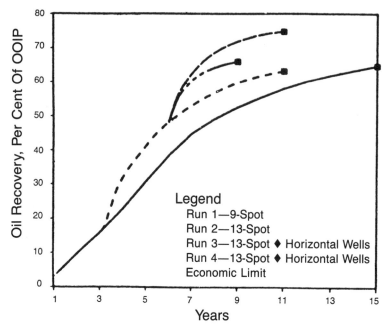

Figure 7-76. Comparison of horizontal well oil reservoirs (after Huang and Hight, 1986).

The HWs provide for a higher recovery as a result of being located near the base of the reservoir, which minimizes the tendency for steam override to occur and short-circuit the displacement process. By virtue of the large contact area with the reservoir for the HWs relative to infill VWs, the likelihood of bypassing oil in "blind spots" is reduced, provided that short-circuiting of steam to the HW does not occur due to reservoir heterogeneities. Figure 7-77 depicts the incremental recovery to be derived from drilling infills over that of the base case nine spot pattern. For bounded patterns on 18.5-acre spacing, drilling an additional 4 infill VWs and 4 HWs can increase annual recovery in the range of 650,000 bbl per year before gradually declining, relative to the nine spot well configuration. The additional incremental recovery may be sufficient to warrant the drilling of the additional wells.

For undeveloped reservoirs it may be preferable to drill combined HW and VW patterns at the start rather than to drill VWs first then infill later with HWs and/or VWs. Table 7-13 depicts that by infill drilling later in the pattern life, recovery of OOIP is reduced and the time to recover the oil is increased compared to drilling initially with a combination of HWs and VWs (run number 5 in the table).

Thermal Recovery and Primary Production for Heavy Oils 343

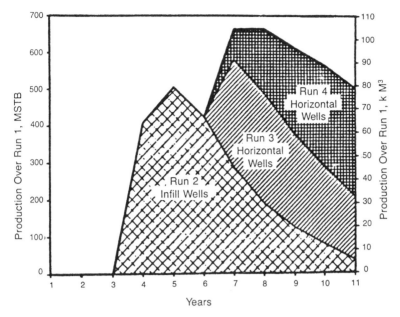

Figure 7-77. Additional cumulative oil production over that predicted for an 18.5-acre, nine spot pattern (after Huang and Hight, 1986).

The work of Dietrich (1987) also compared infill HW performance to VW steam flood performance for a reservoir with an average oil saturation of 0.5. The results indicated that the HW Heavy Oil Process (HOP) could recover 70% of the OOIP compared with 50% of the OOIP for a conventional 5 spot steamflood (Figure 7-78). The characteristics of the reservoir are provided in Table 7-14.

Kern River Field Performance for Combination HW and VW Processes

From 1982 through 1983 an 8 HW and 8 VW combination process was evaluated on a 25-acre pilot site at Kern River (Dietrich, 1987). To access the reservoir with the HWs a 7-ft-diameter shaft was sunk to below the base of the reservoir, approximately 350 ft from surface. Eight 8⅝-inch-diameter HWs were radially drilled out from the shaft, as depicted in Figure 7-79. The HWs ranged in length from 430 ft to 700 ft. The HWs were produced conventionally from surface. Eight VW injectors were drilled from the surface as depicted in Figure 7-79.

The oil saturation averaged 0.5 of pore space within the pilot area. Twenty-one VWs were drilled over the pilot area during 1947 and 1948 and

text continued on page 346

344 Horizontal Wells

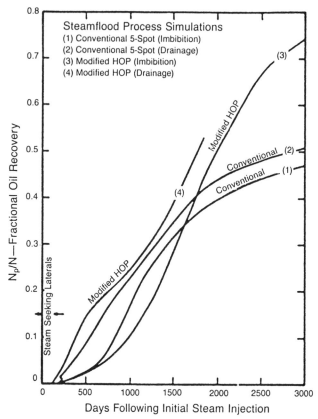

Figure 7-78. Comparison of simulated oil recovery—conventional five spot vs horizontal well steamfloods (after Dietrich, 1987).

Table 7-14
Measured Oil Viscosity vs Temperature on Updip Kern River Q-Zone Oil Sample

Temperature (°F)	Oil Viscosity (Centipoise)
80	5,046.8
140	302.0
180	78.7
200	44.3
250	19.9
300	8.9

Thermal Recovery and Primary Production for Heavy Oils 345

Table 7-14
Continued

Rock & Fluid Properties for Simulator Input	
Pore volume compressibility	0.00025 vol/vol/psi
Irreducible water saturation	0.30 fraction pore volume
Critical gas saturation	0.05 fraction pore volume
Residual oil saturation to steam	0.01 to 0.08 frac. pore vol.
Initial reservoir temperature	110 °F
Rock heat capacity	35.0 BTU/cu ft °F
Rock thermal conductivity	28.0 BTU/day ft °F
Relative permeability	Temperature independent
Oil density at 60°F	59.5 lb/cu ft
Oil density at 100°F	58.9 lb/cu ft
Oil gravity at 60°F	15.9°API
Oil gravity at 110°F	18.4°API
Oil compressibility	0.000008 vol/vol/psi
Oil thermal expansion coeff.	0.00034 vol/vol/deg
Oil specific heat	0.44 BTU/lb °F
Reservoir depth	350 ft
Reservoir thickness	77 ft
Porosity	0.26
Permeability	300 to 700 md
Reservoir pressure	Atmospheric

After Dietrich, 1987.

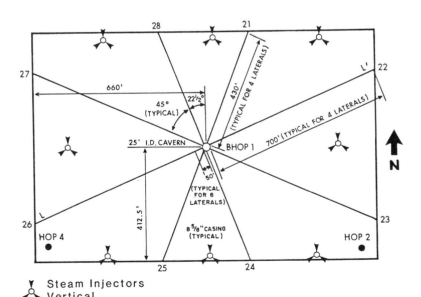

Figure 7-79. Position and numbering of horizontal wells (after Dietrich, 1987).

346 Horizontal Wells

text continued from page 343

had been produced intermittently on primary until 1968, when steam soak operations were initiated. Because of the prior production from the area and the general reservoir quality, oil saturation was low.

The 8 HWs were thermally stimulated with slugs of steam ranging from 4,000 bbl to 37,000 bbl of CWE steam prior to commencing injection of steam into the 8 VW injectors. Over a period of 8 months 606,010 bbl of CWE steam were injected into the VWs. Approximately 47,000 bbl oil were recovered from the operation, resulting in a project cumulative OSR of 0.06 bbl/bbl at the point of terminating the project. The project was stopped because of the low OSR.

Figures 7-80 and 7-81 provide the overall pilot response on a cumulative basis and rate basis, respectively. Note, in the figures, that before steam injection was stopped, OSR was increasing with time, albeit the OSR was exceedingly low (0.06 bbl/bbl). At the point when steam injection was stopped, oil production rate decreased substantially, from 178 bbl/d to 123 bbl/d, indicating that steam injection was contributing to oil productivity. The low initial oil saturation over the pilot area tended to provide generally low oil productivity. In reservoirs with low oil saturation careful consideration must be given to the viability of any recovery process whether it be with HWs, VWs, or a combination of HWs and VWs.

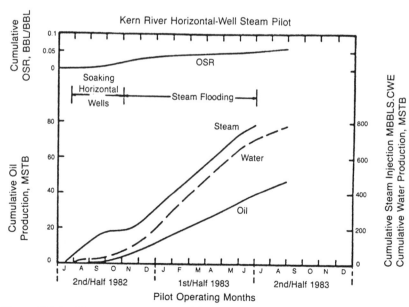

Figure 7-80. Observed pilot response—cumulative basis (after Dietrich, 1987).

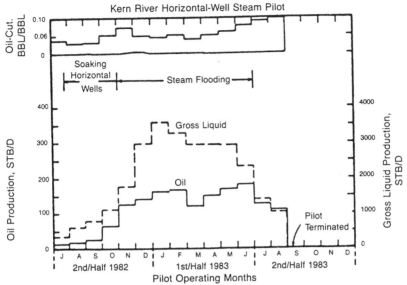

Figure 7-81. Observed pilot response—rate basis (after Dietrich, 1987).

The positive note of the Kern River project was that oil cut and OSR were improving with time. Simulations of the pilot area had indicated that oil productivity would be low for the first two years and then increase rapidly with time as the steam zone became established above the HWs and gravity drainage would dominate (Figure 7-82). The pilot was not operated long enough to establish whether the predictions were correct. As discussed in the introduction of this chapter, patience is required to wait for oil productivity to improve over time. Often, 1 to 2 years are required for acceptable productivity to be established. Economic evaluations of thermal processes must frequently take into account the initial low oil productivity.

Employing Cyclic Steam Stimulation with HWs

Cyclic steam stimulation consists of the injection and production phases occurring at separate times from one single wellbore. The injection of steam allows the pressuring up of the reservoir which during the production phase results in fluid displacement to the wellbore due to reservoir depressurization. Typically, for VWs, many cycles of steam and production may be employed until uneconomic production ends the well life. Figure 7-83 depicts a typical injection and production sequence for cyclic steam.

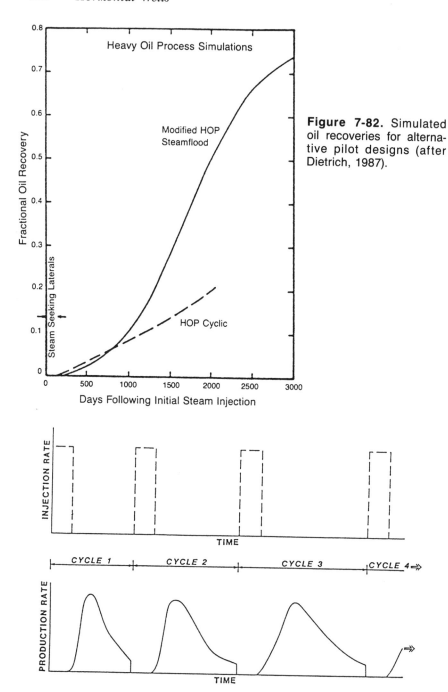

Figure 7-82. Simulated oil recoveries for alternative pilot designs (after Dietrich, 1987).

Figure 7-83. Cyclic steam stimulation, injection, and production sequence.

Thermal Recovery and Primary Production for Heavy Oils 349

For HWs to effectively use the cyclic steam approach, ways to ensure uniform placement of steam along the HW need to be developed. Without uniform placement the benefits of HWs for the process diminish.

Camilleri and Mobarak (1987) assessed the benefits of cyclic steam with HWs and compared performance to that of a VW. The volumes of reservoir that the HW and VW could access were set equal. It was assumed that steam could be uniformly placed along the length of the HW. Three cycles of steam stimulation were employed for each case. The reservoir characteristics associated with the study are provided in Table 7-7.

Figure 7-84 depicts the comparative performance between HW and VW in an isotropic reservoir. In separate simulations the HW was placed in each 20-ft-thick layer and it was noted that the best location for the HW was near the base of the reservoir, in layer 3 or 4. After three cycles, the HW had recovered approximately twice as much oil as the VW, when the HW was located near the base of the reservoir.

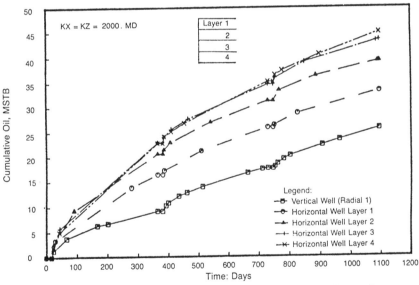

Figure 7-84. Cumulative oil production for cyclic steam in an isotropic reservoir (after Camilleri and Mobarak, 1987).

Figure 7-85 depicts the comparative performance for an anisotropic reservoir where horizontal to vertical absolute permeability differs by a factor of ten. Compared to the VW case the HW can at best recover 1.5 times the volume of oil from a VW over three steam cycles. As with the SAGD process, vertical permeability has an influence on HW performance for cyclic steam stimulation.

350   Horizontal Wells

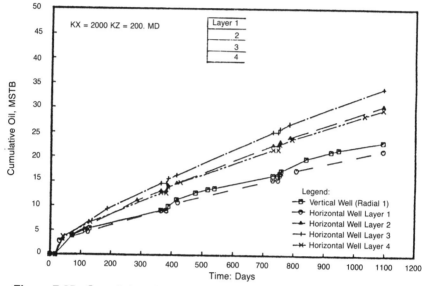

Figure 7-85. Cumulative oil production for cyclic steam in an anisotropic reservoir (after Camilleri and Mobarak, 1987).

Cyclic steam stimulation may be an acceptable process to employ to initiate SAGD or steam displacement. However, on its own for HWs in a multi-well configuration, deleterious linkup by cyclic HWs could affect performance. As discussed throughout this chapter, it is preferable to balance steam injection to the withdrawal of fluids from the reservoir such that sweep conformance can potentially be as high as possible. Cyclic steam, as a separate process on its own is not considered a strong HW candidate for thermal processes. (Mini steam injection and production cycles might be viable for an extended period of time because, by employing mini-cycles, the injection of steam begins to approach the balancing of fluids produced.)

PLACEMENT OF STEAM ALONG HW INJECTORS

For a thermal HW process that employs a HW as an injector a means is required to ensure that steam is capable of being injected along the entire length of the HW. Without the capability of injecting along the entire well, productivity could suffer because the benefit of employing a HW injector may be diminished to a large extent.

To ensure that steam is placed along the entire length of the HW (as opposed to all steam condensing at a point along the length of the HW) it is proposed that tubing be run to the end of the HW. Attached to the end of the tubing is a thermocouple string and a means to measure pressure (perhaps a bubbletube). Injection of steam occurs through the tubing string and fluid is injected into the reservoir through the annular space. A small volume of steam is continually produced on surface through the annulus to ensure that steam has the capability of being injected along the entire HW length. If a small volume of steam were not produced on surface it may be that only a portion of the HW is taking the volume of injected steam. Producing a small volume of steam on surface ensures that steam exists throughout the annular space and should be capable of flowing into the formation.

The thermocouple and bubbletube, at the end of the HW, provide an indication as to whether quality steam exists at the end of the HW or only saturated water exists. By noting the pressure and temperature from the end of the HW and referring to steam tables it can be determined whether the water is undersaturated or steam quality exists. In this manner, at the furthest point along the HW steam is assured. In the annular space from the end of the HW to the beginning point of the HW, steam is assured because heat is transferred from the tubing to the annular space. Figure 7-86 pictorially depicts the concept of steam placement.

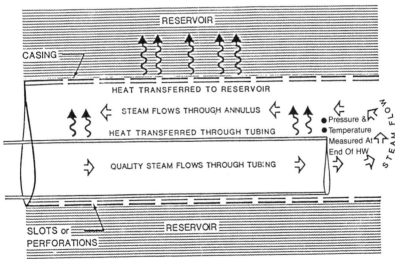

Provided that quality steam exists at the end of the HW, as measured by the pressure and temperature, because of heat transfer between the tubing and annulus, quality steam should exist throughout the entire annular space length even though heat is being transferred to the reservoir.

Figure 7-86. Schematic of heat transfer between tubing, annulus, and formation for steam circulation through HW.

352 Horizontal Wells

For steam circulation purposes as opposed to steam injection, Edmunds et al. (1988) indicated at the AOSTRA UTF site that low steam flow rates are better than high flow rates. This is because heat transfer between the tubing and annulus is driven by the temperature difference across the tubing, which in a saturated steam environment is a function of the pressure drop. As the steam flow rates increase, the pressure gradients go up, hence increasing the temperature differential. The result is that the heat transfer rate increases more quickly than the steam flow rate which indicates that lower steam circulation rates provide less heat transfer per unit mass of flow than high rates. The net result is that heat is not transferred as quickly to the annulus and as a consequence gaseous steam is able to reach the far end of the well and exit as steam into the annular space.

PRIMARY RECOVERY OF HEAVY OILS

Primary recovery of heavy oils from HWs is similar to primary recovery from HWs associated with conventional oil. A major difference between heavy oil and conventional oil is that the productivity increase over VWs for heavy oil is greater than that of conventional oil. Figure 7-87 depicts the ratio of HW productivity (PI_{HW}) to VW productivity (PI_{VW}) for a series of

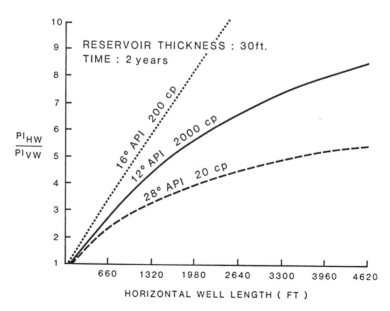

Figure 7-87. Unsteady state productivity ratio comparison between HW and VW.

Thermal Recovery and Primary Production for Heavy Oils 353

HW lengths, and all are related to oil viscosity. All reservoir characteristics are the same for each oil viscosity, except for the viscosity. For a 2000-ft-long HW, the instantaneous PI_{HW}/PI_{VW} ratio for a 2,000 cp oil is 9, for a 200 cp oil it is 6, and for a 20 cp oil it is 4, after 2 years on production for an infinite reservoir. The productivity ratio is more than double for 2,000 cp oil compared to 20 cp oil. On a productivity ratio basis there is greater incentive to drill HWs in heavy oil reservoirs compared to conventional oil reservoirs. However, for 20 cp oil, VW production could be in the thousands of barrels of oil per day with a fourfold increase in the thousands of barrels of oil per day for a HW drilled into a conventional oil reservoir. For a 2,000 cp oil, VW production could be in the tens of barrels of oil per day, whereas a HW might achieve productivity in the low hundreds of barrels of oil per day. The absolute increase in productivity is greater for conventional oil HWs relative to heavy oil HWs.

Field primary productivity has been reported (Coffin, 1989; Freeborn et al., 1989) and indications are that the theory is valid for projecting HW performance compared to VW performance.

NOTATION

a	radius of HW	GF	geometric shape factor used in simulations
A	surface area open to flow		
B_o	formation volume factor for oil (reservoir barrel/stock tank barrel)	h	reservoir thickness
		HAS drive	heated annulus steam drive
BTU	British Thermal Unit	HW	horizontal well
c	heat capacity of oil sand	HWs	horizontal wells
C_o	specific heat of oil	K	permeability
crf	cumulative recovery fraction	k_h	horizontal permeability
		k_o	oil permeability or oil relative permeability
c_s	heat capacity of sand		
C_w	specific heat of water	k_v	vertical permeability
CWE	cold water equivalent	k_w	water permeability or water relative permeability
D	well diameter		
f	pipe friction factor		
f_w	frictional flow of water	L	length
g	acceleration due to gravity	m	dimensionless value determined from temperature integral of Equation 7-11
g_c	gravity conversion constant in Imperial units		
G	factor used in determining water density	M	volumetric heat capacity of oil sand

354   Horizontal Wells

n	timestep counter of equal divisions	S_o	oil saturation
OOIP	original oil in place	SOR	steam oil ratio
OSR	oil to steam ratio	S_w	water saturation
P	point on steam oil interface	t_1, t_2	time at t_1 or t_2
		T	temperature
ΔP	pressure drop across pipe or reservoir	T_r	original reservoir temperature
P_e	reservoir pressure at external boundary	T_s	steam temperature
		U	velocity of steam-oil interface normal to the interface
PI	productivity index		
P_o	oil reservoir pressure	UTF	underground test facility
PV	pore volume		
P_w	reservoir pressure at wellbore	v	temperature at radius r minus temperature of reservoir
P_{wa}	water reservoir pressure		
q	flow rate (bbl/d)	V	temperature at HW
q_o	oil flow rate	VW	vertical well
q_r	flow rate at original reservoir conditions	VWs	vertical wells
		w	width of steam chamber
q_t	total flow rate	X	distance between two points in reservoir where flow rate and pressure is measured
q_w	water flow rate		
R_e	Reynolds number		
r_e	radius of reservoir at external boundary		
		x	distance in x direction
r_w	radius of wellbore	y	distance in y direction
SAGD	steam assisted gravity drainage	z	distance in z direction

Greek Symbols

α	thermal diffusivity	μ	viscosity
Δ	changes in parameters at two different points in time	μ_{os}	viscosity difference between oil and steam
		μ_o	oil viscosity
θ	angle to the horizontal	μ_w	water viscosity
$\varkappa$	thermal diffusivity	∂	partial derivative sign
λ	thermal conductivity of oil sand at temperature T	ν	kinematic viscosity
		ν_r	kinematic viscosity of oil at original reservoir conditions
λ_s	thermal conductivity of oil sand at standard condition		
		ξ	distance from the steam-oil interface

ρ_g gas or steam density
ρ_o oil density
$\rho_{o,1}$ dimensionless oil density at standard conditions
$\rho_o, T, 1$ dimensionless oil density related to datum water density at temperature T

ρ_{steam} density of steam
$\rho_{w,T}$ density of water at temperature T
τ dimensionless time starting at beginning of time period
ϕ porosity fraction

REFERENCES

1. Meyer, R. F. and Fulton, P. A., "Toward an Estimate of World Heavy Crude Oil and Tar Sands Resources," Proc. 2nd Intl. Conference on Heavy Crude and Tar Sands. United Nations Inst. for Training and Research, Caracas (1982).
2. Carrigy, M. A., "Thermal Recovery from Tar Sands," *Journal of Petroleum Technology*, Vol. 35, No. 13., paper #2149-2157, Dec. 1983.
3. Butler, R. M., "The Potential of Horizontal Wells," Advances in Petroleum Recovery & Upgrading Technology Conference, Calgary, Alberta, Canada (June 14-15, 1984).
4. Tam, E. S., Edmunds, N. R., and Redford, D. A., "Methods of Enhancing Fluid Communication in Oil Sands," SPE/DOE Fourth Symposium on Enhanced Oil Recovery, Tulsa, OK (April 15-18, 1984) SPE/DOE 12659.
5. Butler, R. M., McNab, G. S., and Lo, H. Y., "Theoretical Studies on the Gravity Drainage of Heavy Oil During In Situ Steam Heating," presented at the 29th Canadian Chemical Engineering Conference, Sarnia, Ontario (Oct. 1, 1979). *The Canadian Journal of Chemical Engineering*, Vol. 59, Aug. 1981, pp. 455-460.
6. Edmunds, N. R., Haston, J. A., and Best, D. A., "Analysis and Implementation of the Steam Assisted Gravity Drainage Process at the AOSTRA UTF," paper No. 125., presented at the Fourth Unitar/UNDP Conference on Heavy Crude and Tar Sands, Edmonton, Alberta (Aug. 7-12, 1988).
7. Best, D. A., Cordell, G. M., and Haston, J. A., "Underground Test Facility: Shaft and Tunnel Laboratory for Horizontal Well Technology," presented at the 60th Annual Technical Conference and Exhibition of the Society of Petroleum Engineers (SPE), Las Vegas, NV (September 22-25, 1985) SPE 14333.

8. Mainland, Glenn G., and Lo, H. Y., "Technological Basis for Commercial in Situ Recovery of Cold Lake Bitumen," presented at the 11th World Petroleum Congress—Session: Recovery of Extra Heavy Oil and Bitumen by Non-mining Methods, London, U.K., pp. 235–242 (1983).
9. MacDonald, R. R., "Drilling the Cold Lake Horizontal Well Pilot No. 2," presented at the 60th Annual Technical Conference and Exhibition of the Society of Petroleum Engineers (SPE), Las Vegas, NV (September 22–25, 1985) SPE 14428.
10. Jesperson, P., "Horizontal Well Application at the Tangleflags North Steamflood Pilot Project," Horizontal Well Seminar, Calgary Convention Centre (April 20, 1989) sponsored by the Petroleum Society of the CIM.
11. Joshi, S. D., "A Laboratory Study of Thermal Oil Recovery Using Horizontal Wells," presented at the SPE/DOE Fifth Symposium on Enhanced Oil Recovery of the Society of Petroleum Engineers and the Department of Energy held in Tulsa, OK (April 20–23, 1986) SPE/DOE 14916.
12. Butler, R. M., "The Potential for Horizontal Wells for Petroleum Production," presented at the 39th Annual Technical Meeting of the Petroleum Society of CIM held in Calgary, Alberta (June 12–16, 1988) paper no. 88-39-22.
13. Butler, R. M. and Stephens, D. J., "The Gravity Drainage of Steam-Heated Heavy Oil to Parallel Horizontal Wells," 31st Annual Technical Meeting of the Petroleum Society of CIM, Calgary, Alberta (May 25–28, 1980) paper no. 80-31-31.
14. Edmunds, N. R., Wong, A., McCormack, M. E., and Suggett, J. C., "Design of Horizontal Well Completions, AOSTRA Underground Test Facility," presented at the Fourth Annual Heavy Oil and Oil Sands Symposium at the The University of Calgary (February 18, 1987).
15. Joshi, S. D. and Threlkeld, C. B., "Laboratory Studies of Thermally-Aided Gravity Drainage Mechanism Using Horizontal Wells," presented at the Fifth Annual Advances in Petroleum Recovery and Upgrading Technology Conference, Calgary, Alberta (June 14–15, 1984).
16. Griffin, P. J. and Trofimenkoff, P. N., "Laboratory Studies of Steam-Assisted Gravity Drainage Process," presented at the Fifth Annual Advances in Petroleum Recovery and Upgrading Technology Conference, Calgary, Alberta (June 14–15, 1984).
17. Toma, P., Redford, D., and Livesey, D., "The Laboratory Simulation of Bitumen Recovery by Steam Simulation of Horizontal Wells," presented at the WRI-DOW Tar-Sand Symposium, Session 3, Vail, Colorado (June 26–29, 1984).
18. Carslaw, H. S. and Jaeger, J. C., *Conduction of Heat in Solids*, Oxford 2nd Edition (1959).

19. Khosla, A. and Cordell, G., "Potential of Horizontal Well Processes," Advances in Petroleum Recovery and Upgrading Technology Conference, Calgary, Alberta (June 14–15, 1984).
20. Camilleri D. and Mobarak, S. A., "Application of a Numerical Simulator to Investigate the Use of Horizontal Wells in Steamfloods," II Simposio Internacional Sobre Recuperacion Mejorada de Crudo, Maracaibo, Venezuela, Del 24 al 27 de Febrero de 1987.
21. Aziz, K., Ramesh, A. B., and Woo, P. T., "Fourth SPE Comparative Solution Project; Comparison of Steam Injection Simulators," presented at the 8th Reservoir Simulation Symposium, Dallas (February 10–13, 1985) SPE 13510.
22. Huang, W. S. and Hight, M. A., "Evaluation of Steamflood Processes Using Horizontal Wells," presented at the SPE 1986 International Meeting on Petroleum Engineering held in Beijing, China (March 17–20, 1986) SPE 14130.
23. Hsueh, L., "Numerical Simulation of the HAS Drive Process," paper prepared for the 56th California Regional Meeting of the Society of Petroleum Engineers, Oakland, CA (April 2–4, 1986) SPE 15088.
24. Anderson, D. J., "The Heated Annulus Steam Drive Process for Immobile Tar Sands," presented at the Fourth Unitar/UNDP Conference on Heavy Crude and Tar Sands, Edmonton, Alberta, Canada (Aug. 7–12, 1988) paper no. 104.
25. Duerksen, J. H., "Simulation of HASDrive UTF Pilot," presented at the Fourth Unitar/UNDP Conference on Heavy Crude and Tar Sands, Edmonton, Alberta (Aug. 7–12, 1988) paper no. 125.
26. Jain, S. and Khosla, A., "Predicting Steam Recovery of Athabasca Oil Through Horizontal Wells," presented at the 36th Annual Technical Meeting of the Petroleum Society of CIM held jointly with the Canadian Society of Petroleum Geologists, Edmonton, Alberta (June 2–5, 1985) paper no. 85-36-28.
27. Dietrich, J. K., "The Kern River Horizontal Well Steam Pilot," presented at the SPE California Regional Meeting, Ventura, CA (April 8–10, 1987) SPE 16346.
28. Butler, R. M., "New Interpretation of the Meaning of the Exponent "m" in the Gravity Drainage Theory for Continuously Steamed Wells," *AOSTRA Journal of Research*, 2, 67–74 (1985).
29. Suggett, J. C., Cordell, G. M., and Winestock, A. G., "Underground Test Facility Horizontal Well Process Study," (September 1985) File No. 7903.20.20.
30. Wu, D. H., "A Critical Review of Steamflood Mechanisms," (1977) SPE 6550.
31. Perry, R. H. and Chilton, C. H., *Chemical Engineers' Handbook, Fifth Edition* (1973).

32. Somerton, W. H., Keese, J. A., and Chu, S. L., "Thermal Behaviour of Unconsolidated Oil Sands," *Society Petroleum Engineering Journal*, Vol. 14 (1974) pp. 513–521.
33. Edmunds, N. R., "A Model of the Steam Drag Effect in Oil Sands," *The Journal of Canadian Petroleum Technology*, Sept.–Oct., 1984.
34. Butler, R. M., Stephens, D. J., and Weiss, M., "The Vertical Growth of Steam Chambers in the In-Situ Thermal Recovery of Heavy Oils," Proc. 30th. Canadian Chemical Engineering Conference, Edmonton, Alberta, 1152–1160 (Oct. 19–22, 1980).
35. Coffin, P., "Horizontal Well Application in the Pelican Lake and Winter Reservoirs," presented at the Horizontal Well Seminar held in association with the Heavy Oil Special Interest Group of the Petroleum Society of CIM, Calgary, Alberta (April 20, 1989).
36. Freeborn, W. R., Russell, B., and MacDonald, A. J., "South Jenner Horizontal Wells: A Water Coning Case Study," presented at the Horizontal Well Seminar held in association with the Heavy Oil Special Interest Group of the Petroleum Society of CIM, Calgary, Alberta (April 20, 1989).
37. Moody, L. W., "Friction Factors for Pipe Flow," *Trans ASME 66*, (1944) pp. 671–684.

8
Naturally Fractured Reservoirs

GEOLOGIC ASPECTS

Naturally fractured reservoirs represent excellent targets for horizontal wells. This fact has been recognized by the oil and gas industry, as more than 70% of horizontal wells drilled to date have been experienced in naturally fractured reservoirs. The idea is to drill these wells in such a way that they go perpendicular to the orientation of the natural fractures.

Petroleum engineering literature has exploded with information and discussions regarding the advantages and disadvantages of horizontal wells. Researchers have placed special emphasis on comparing horizontal wells and vertical wells that are hydraulically fractured. There are arguments on both sides. But on one point most publications agree. Horizontal wells provide better production and recoveries in reservoirs that contain vertical natural fractures. The key to success is properly determining the natural fractures' azimuth.

Various sources of information can be used for determining the preferential natural fracture orientations, including borehole televiewers, microscanners, oriented cores, interference tests, tracers, and in some instances, aerial photography and land satellite information.

Estimating Optimum Drilling Direction

In this method the assumption is made that fractures occur in sets with common orientation, uniform closest spacings S^*, and a representative fracture width e, which control the flow capacity of each fracture set (Nolen and Howard, 1987).

Figure 8-1 shows a schematic of one fracture set in plan view. Angle θ is the azimuth of direction of closest fracture spacing, K is the azimuth of direction of random fracture spacing, S^* is the closest fracture spacing, and S is an apparent fracture spacing given by:

$$S = S^* (1 + \tan^2 \beta) \qquad (8\text{-}1)$$
$$= S^* [1 + \tan^2 (\theta - K)]^{1/2}$$

In this work the convention is taken for measuring the azimuth that north is zero degrees with angles increasing clockwise to 360°.

The fracture interception rate P is given by the reciprocal of the apparent fracture spacing in the K direction, or

$$P = 1/S = n/L \qquad (8\text{-}2)$$

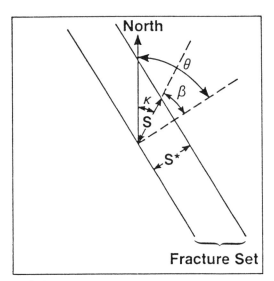

Definitions:
$S^* \equiv$ Closest fracture spacing
$S \equiv$ Apparent fracture spacing
$\equiv S^* [1 + \tan^2 \beta]^{1/2} = S^* [1 + \tan^2 (\theta - \kappa)]^{1/2}$
$\theta \equiv$ Azimuth of direction of closest fracture spacing
$\kappa \equiv$ Azimuth of direction of random fracture spacing

Figure 8-1. Definition diagram in plan view for one fracture set. Source: Nolen and Howard (1987).

where n is the number of fractures intercepted per unit length of horizontal wellbore L.

Figure 8-2 shows a schematic for the case of two fracture sets in plan view. Distances S_1^* and S_2^* are the closest fracture spacings for sets 1 and 2, angles θ_1 and θ_2 are the azimuths of the S_1^* and S_2^* directions, and the fracture interception rate P_k is given by:

$$P_k = \frac{1}{S_1^*} + \frac{1}{S_2^*} = \frac{n_1 + n_2}{L} \tag{8-3}$$

where n_1 and n_2 are the number of fractures of the first and second set intercepted per unit length of horizontal wellbore.

The geometry presented in Figure 8-2 leads to calculation of the drilling direction K in which the rate of natural fractures interception is the greatest from the equation:

$$K = \arctan \frac{\frac{1}{S_1^*}\sin\theta_1 + \frac{1}{S_2^*}\sin\theta_2}{\frac{1}{S_1^*}\cos\theta_1 + \frac{1}{S_2^*}\cos\theta_2} \tag{8-4}$$

where all nomenclature has been defined previously.

Notice, in Figure 8-2, that the maximum P_k cuts the acute angle between the closest spacing directions θ_1 and θ_2.

Example 1

This problem is extracted directly from the work by Nolen and Howard (1987). Assume that there are two fracture sets, each normal to bedding and forming traces on a horizontal surface. The azimuth of fracture set 1 is 300° with a closest spacing S_1^* of 2 ft. The azimuth of fracture set 2 is 0° with a closest spacing S_2^* of 1 ft. Determine the optimum drilling direction to intercept the major number of natural fractures per unit length of wellbore.

Solution. Ideally, the well should go perpendicular to the natural fractures. Thus, angle θ_1 is equal to 30° and angle θ_2 is equal to 90°. Inserting these numbers into Equation 8-4 leads to an angle K equal to 70.9°, i.e., the optimum drilling direction to intercept the most fractures per unit length of wellbore is N70.9°E.

362 Horizontal Wells

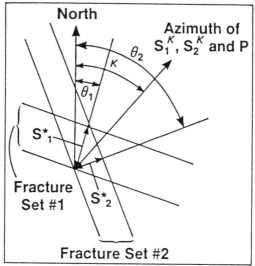

Definitions:
$S^*_1 \equiv$ Closest fracture spacing for Set 1
$S^*_2 \equiv$ Closest fracture spacing for Set 2
P $\equiv$ Fracture interception rate

$$= \frac{1}{S_1^K} + \frac{1}{S_2^K}$$

$\theta_1 \equiv$ Azimuth of S^*_1 direction
$\theta_2 \equiv$ Azimuth of S^*_2 direction

Figure 8-2. Definition diagram in plan view for two fracture sets. Source: Nolen and Howard (1987).

Effect of Fracture Aperture

Equation 8-4, used in Example 1-1, assumes that the fracture apertures e are identical in sets 1 and 2. If these apertures are different they can be incorporated into Equation 8-4, and the optimum drilling orientation is calculated from (Nolen and Howard, 1987):

$$K_f = \text{arc tan} \frac{\dfrac{e_1^3}{S_1^*} \sin\theta_1 + \dfrac{e_2^3}{S_2^*} \sin\theta_2}{\dfrac{e_1^3}{S_1^*} \cos\theta_1 + \dfrac{e_2^3}{S_2^*} \cos\theta_2} \tag{8-5}$$

where e_1 and e_2 are the fracture apertures of zones 1 and 2, respectively. Because fracture permeability is equal to a constant times e^3/S^*, Equation 8-5 can also be written as:

$$K_f = \arctan\left[\frac{K_1 \sin\theta_1 + K_2 \sin\theta_2}{K_1 \cos\theta_1 + K_2 \cos\theta_2}\right] \tag{8-6}$$

This is an important consideration, as usually there is strong anisotropy in naturally fractured reservoirs. If, in the previous example, we consider that fracture sets 1 and 2 have apertures of 0.002 in. and 0.001 in., respectively, the optimum fracture orientation is calculated to be N40.9°E from Equation 8-5; a shift of 30° from the optimum orientation determined in Example 1-1.

If three or more fracture sets are present, then the above equations can be extended to (Nolen and Howard, 1987):

$$P_K = \sum_{i=1}^{N} \frac{e_i^3}{S_i^*[1 + \tan^2(\theta_i - K_f)]^{1/2}} \tag{8-7}$$

$$K_f = \arctan\left[\frac{\sum_{i=1}^{N} \frac{e_i^3}{S_i^*}\sin\theta_i}{\sum_{i=1}^{N} \frac{e_i^3}{S_i^*}\cos\theta_i}\right] \tag{8-8}$$

where N = number of fracture sets
i = index to designate each set

Site Selection

Drilling of horizontal wells is technically and economically feasible as costs have decreased to levels lower than originally expected. (Bosio and Reiss, 1988). The key now is determining where these wells should be drilled to ensure economic success. Some of the possibilities include (Bosio and Reiss, 1988; Economides et al., 1989; de Montigny and Combe, 1988):

1. **Naturally Fractured Reservoirs:** This is accomplished by drilling wells perpendicular to the orientation of the fractures, as discussed in the previous section. More than 70% of the horizontal wells drilled to date have been experienced in naturally fractured reservoirs.
2. **Higher and More Rapid Recoveries by Drilling Infill Horizontal Wells:** This is achieved by intercepting natural fractures that other-

364 Horizontal Wells

wise would be inaccessible. Higher recoveries would also be obtained by reaching pockets of hydrocarbons in lenticular reservoirs and in regular formations.
3. **Better Results in Secondary Recovery Projects:** This is achieved by improving areal sweep efficiencies. The idea is to drill the horizontal well parallel to the natural fractures in such a way that the direction of flood goes perpendicular to the orientation of the fractures.
4. **Reduction of Gas and Water Coning:** This is the result of lower pressure drawdowns at the same production rate due to a larger contact area.
5. **Reduction of Sand Production Problems:** This is also due to a lower pressure drawdown.
6. **Environmental Protection:** This can be achieved by drilling extended wells under cities, rivers, etc. from limited areas. A good geologic knowledge of the area is, of course, required to drill profitable horizontal wells.
7. **Low Permeability Reservoirs:** Horizontal wells might be very useful in this type of reservoir, as they can easily reach > 2,000 feet horizontally. Resistance to flow is almost negligible in horizontal wells, making them approach the idealized condition of infinite conductivity. Often, low permeability reservoirs contain localized networks of vertical natural fractures that can be intercepted by the horizontal well.
8. **Decrease in Formation Damage:** This is the result of a substantial decrease in pressure drops in the horizontal well.
9. **Gas Reservoirs:** The presence of turbulence is quite common in gas reservoirs due to high fluid velocities around the wellbore. In a long horizontal hole these velocities are smaller, leading to better productiveness.
10. **Dipping Beds:** Figure 8-3 shows the effect of dip on horizontal wells. The vertical distance between the gas-oil and water-oil contacts is 30 m. On the average, vertical wells spaced 115 m apart would intercept three sandstone beds in the oil column. It is likely, however, that only the middle bed would be perforated, as the lower one would lead very quickly to water coning and production from the upper one would lead to premature gas coning. For the schematic of Figure 8-3, a single horizontal well 575 m long would be more efficient than five vertical wells.

Enhancement of Reservoir Appraisal

Vertical wells provide localized information that must be extended by constructing cross sections with data from various wells. A horizontal well

Naturally Fractured Reservoirs 365

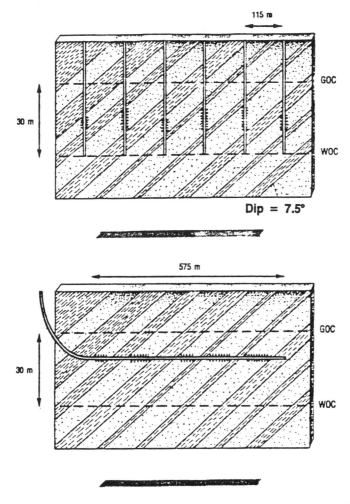

Figure 8-3. Effect of dip on horizontal wells. Source: de Montigny and Combe (1988).

enhances reservoir appraisal because it enables lateral surveys of the reservoir.

The collection of significant information from horizontal wells in various naturally fractured reservoirs has been reported by de Montigny and Combe (1988). For example, wells Lacq 90 and 91 in southwest France intercepted unexpectedly large naturally fractured corridors which had never been crossed by any of the many vertical wells drilled in that chalk facies.

In the Castera Lou reservoir in southwest France, a well was stopped after 1,130 feet of horizontal section due to complete circulation loss while crossing a fault (de Montigny et al., 1988). The exact location of this fault was one of the objectives of the well. One surprising finding was the presence of two fracturing directions not observed previously in vertical wells. Furthermore, the well came out of the reservoir into the caprock at two different locations. This was caused by two minigrabens not observable in seismic studies.

An excellent example of enhanced reservoir description is provided by well Rospo Mare 6d in the Italian Adriatic sea (Montigny et al., 1988). The logging program of this well included the spherical focused induction resistivity, borehole compensated sonic, cement bond, variable density, dual laterolog, and gamma ray logs. Production logs included the full bore spinner as well as sensors for pressure and temperature measurements. The comparison of all these logs with cores indicated that the fluid entry occurred at intervals characterized by natural fractures and Karstic voids.

In other wells the logging programs included the sonic, natural gamma ray, formation microscanner, dual laterolog, microspherical focused, compensated neutron, and lithodensity logs, auxiliary measurement tool (AMT) for detection of theft zones by measuring temperature, the general purpose inclination tool (GPIT), which was used for pad orientation of tools in the lower part of the hole, and the measurement-while-drilling (MWD) gamma ray tool.

Figure 8-4 shows an interesting correlation of well logs and geology. The upper portion of the figure shows the percent flow from each one of the horizontal production intervals. Next is the geological interpretation. The well

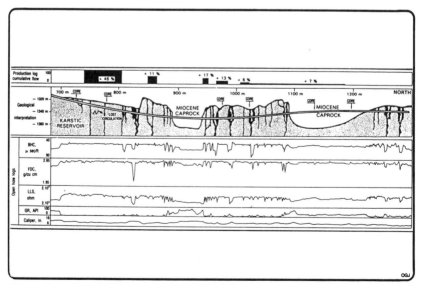

Figure 8-4. Correlation of geology and well logs, Rospo Mare 6d. Source: de Montigny and Combe (1988).

Naturally Fractured Reservoirs 367

intercepted many of the vertical fractures and allowed a reasonable estimate of the distance between large natural fractures (approximately 100 ft). Total loss of circulation occurred very soon after the well crossed the first fracture. The well came out of the reservoir twice: first, when it passed through a sinkhole, and second, when it went through a depression. It is interesting to note that nearly one third of the horizontal length, or approximately 620 ft, ended up being outside the reservoir.

These results indicated that a conventional seismic survey using a 1,500 ft grid is not good enough for this kind of reservoir, as it was not able to pick up the depressions shown on Figure 8-4. Consequently, before further field development, a three-dimensional, high-resolution seismic survey was carried out using an 80 ft grid. This survey gave a good description of the irregularities at the top of the reservoir.

DRILLING AND COMPLETION

On the average, one or two horizontal wells per year were drilled between 1980 and 1984. By 1988, the figure had reached at least 100 horizontal wells (Giannesini, 1989) in fractured and non-fractured reservoirs. At the end of 1989 there were approximately 600 wells.

Horizontal wells can be drilled using a short, medium, or long radius of curvature. All of these methods have been used in naturally fractured reservoirs. The radius of curvature is related to the rate of angle build used to go from a vertical to horizontal borehole (Mahony, 1988). Chapter 2 discusses these drilling methods in detail.

Completions in naturally fractured reservoirs have varied widely. In some cases the horizontal portions have to be completed open-hole (short radius of curvature). In some instances they are completed with perforated liners, and sometimes with cemented casing. Stimulations are not uncommon. A thorough discussion of completions methods is presented in Chapter 3.

WELL TEST ANALYSIS

The model under consideration is presented in Figure 8-5. The horizontal well is located at the center of a semi-infinite, anisotropic, naturally fractured reservoir. The average distance between natural fractures is equal to h_m. The reservoir thickness and the reservoir width are constant. There is no fluid flow across the lateral boundaries, the overburden, or the underburden. Gravity effects are neglected.

The fluid is slightly compressible. It flows isothermally towards a horizontal well of infinite conductivity. Fluid properties are independent of pressure. Fluid movement towards the wellbore occurs only in the fractures.

368 Horizontal Wells

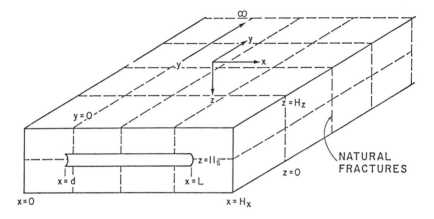

Figure 8-5. Model of a horizontal well in a naturally fractured reservoir. Source: Aguilera and Ng (1991).

This can be handled mathematically by withdrawing the matrix blocks from the composite system as suggested by de Swaan (1976).

The matrix effects on pressure changes in the fractures are taken into account by use of functions uniformly distributed as sources (or sinks) in the fractured medium.

Pressure distribution in Darcy units for this model is given by the diffusivity equation in three dimensions (Goode and Thambynayagam, 1987) with a source term (de Swaan, 1976):

$$\frac{k_x}{k_y}\frac{\delta^2 p}{\delta x^2} + \frac{\delta^2 p}{\delta y^2} + \frac{k_z}{k_y}\frac{\delta^2 p}{\delta z^2} = \frac{\phi \mu c}{k_y}\frac{\delta p}{\delta t} - \frac{\mu Q}{k_y} \quad (8\text{-}9)$$

The reservoir pressure is initially constant. As an initial approximation, the horizontal well is replaced by a thin strip source of width (b – a) and length (L – d). Wellbore storage is zero. The reservoir width is much greater than the reservoir thickness.

Pressure drawdown and buildup solutions for these initial and boundary conditions in the case of single-porosity reservoirs have been published by Goode and Thambynayagam (1987). These solutions were obtained by successive applications of Laplace and finite Fourier integral cosine transforms.

Aguilera and Ng (1991) have extended these solution to include functions that account for the matrix contribution resulting from pressure changes in the natural fractures.

Drawdown

The above initial and boundary conditions lead to the following drawdown solution:

$$p_i - p_{wf} = \frac{282.4 \, q\mu B}{H_x H_z k_y} \left[\sqrt{\frac{\pi A_t}{N_t}} + H_x h_x \sum_{n=1}^{\infty} \frac{1}{n} \text{erf}\left(\frac{n}{h_x}\sqrt{\frac{A_t}{N_t}}\right) \epsilon_n \cos(nX) \right.$$

$$+ \frac{H_x H_z h_z}{(L-d)} \sum_{m=1}^{\infty} \frac{1}{m} \text{erf}\left(\frac{m}{h_z}\sqrt{\frac{A_t}{N_t}}\right) \epsilon_m \cos(mZ)$$

$$\left. + \frac{H_z H_x (k_y/k_z)^{1/2} S_m}{2(L-d)} \right] \quad (8\text{-}10)$$

where $\epsilon_n = \{\sin[n\pi L/H_x] - \sin[n\pi d/H_x]\} \div \{n[L-d]\}$ (8-11)

and

$\epsilon_m = \{\sin[m\pi (H_s + 2r_w')/H_z] - \sin[m\pi (H_s - 2r_w')/H_z]\}$

$\qquad \div \{m(4r_w')\}$ (8-12)

$N_t = \omega + (1-\omega) \times f(t, \tau)$ (8-13)

$f(t, \tau) = 1 - \exp(-t/\tau)$ (8-14)

for pseudo steady state interporosity flow, and

$f(t, \tau) = (t/\tau)^{1/2} \tanh(\tau/t)^{1/2}$ (8-15)

for transient interporosity flow.

Functions of time and τ for tectonic, regional and contractional natural fractures have been published by Aguilera (1987).

$A_t = 0.000264 \, k_y \, t/(\phi \mu c)$ (8-16)

$1/h_x = (\pi/H_x)(k_x/k_y)^{1/2}$ (8-17)

$1/h_z = (\pi/H_z)(k_z/k_y)^{1/2}$ (8-18)

$X = (\pi/H_x)(0.131 L + 0.869 d)$ (8-19)

$$Z = (\pi/H_z)(Hs + 1.47 r_w') \quad (8\text{-}20)$$

$$r_w' = r_w (k_z/k_y)^{1/4} \quad (8\text{-}21)$$

The solution represented by Equation 8-10 assumes an effective wellbore radius equal to (Prats, 1961):

$$r_w = (b - a)/4 \; (k_y/k_z)^{1/4} \quad (8\text{-}22)$$

The effective average pressure is taken at a dimensionless distance of 0.869 from any extremity of the well (Goode and Thambynayagam, 1987).

τ is a real time in hours that corresponds to approximately the beginning of the radial or linear flow period in the composite system. Its value depends on the size of the matrix blocks and the hydraulic diffusivity of the matrix. Mathematically, τ can be represented by:

$$\tau = \text{constant} \; (h_m^2/\eta_m) \quad (8\text{-}23)$$

where

$$\eta_m = k_m/(\phi\mu c) \quad (8\text{-}24)$$

and the constant has a value equal to 2,370 according to Streltsova (1983) and 532.2 based on work by Najurieta (1980).

Validation

The data presented in Table 8-1 have been inserted into Equations 8-10 through 8-24, together with a value of ω equal to 1.0, to generate the pressure drawdown shown on Figure 8-6. Results compare well with simulated and analytical pressure drawdowns published by Goode and Thambynayagam (1987).

Flow Periods

The information in Table 8-1, together with $\omega = 0.1$, $\tau = 0.1$ hours, and the assumption of pseudosteady state interporosity flow has been used to produce the drawdown semilog crossplot presented in Figure 8-7. Various flow periods can be identified:

Naturally Fractured Reservoirs 371

Table 8-1
Reservoir and Fluid Characteristics: Drawdown Example

ω	= 0.10
k_x	= 100 md
k_y	= 100 md
k_z	= 100 md
ϕ	= 0.10
c	= 3 × 10^{-5} psi^{-1}
H_z	= 220 ft
H_x	= 2100 ft
H_y	= infinite
r_w	= 0.354 ft
(L-d)	= 500 ft
y	= 1050 ft
H_s	= 110 ft
q	= 3000 STB/D
μ	= 1.5 cp
B	= 1.0 RB/STB
s_m	= 0

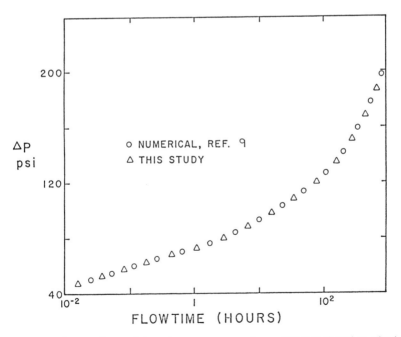

Figure 8-6. Comparison of drawdown pressures from different theories, single-porosity reservoir. Source: Aguilera and Ng (1991).

372 *Horizontal Wells*

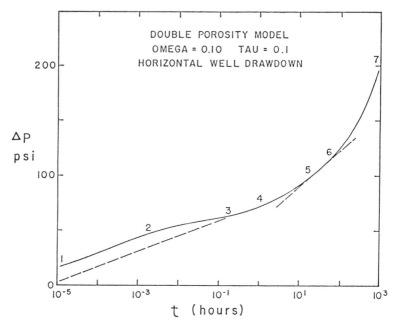

Figure 8-7. Drawdown flow periods in a horizontal well, dual-porosity model. Source: Aguilera and Ng (1991).

1. A *first radial flow period* at early times from the natural fractures into the horizontal well. This occurs in a vertical plane and is recognized by a straight line in a semilogarithmic crossplot. This is represented by the segment between points 1 and 2 on Figure 8-7.
2. A *transition period* due to flow from the matrix into the fractures. The path of this transition depends on the size and shape of the matrix blocks, the hydraulic diffusivity of the matrix, and the pseudosteady state or transient interporosity flow regime. The transition period is represented by the curved segment between points 2 and 3 on Figure 8-7.
3. A *second radial flow period*, still in a vertical plane, which starts when pressures in matrix and fractures reach an equilibrium. This is recognized by a second straight line in semilogarithmic coordinates parallel to the previous one. This segment is located between points 3 and 4 on Figure 8-7.
4. A *first linear flow period* still in a vertical plane, when the pressure transient reaches the upper and lower boundaries of the reservoir. This period is not present in Figure 8-7. The same linear flow period pressure data would yield a straight line on a Cartesian crossplot of pressure differential vs square root of time. This plot is presented in Figure 8-8. Note that the straight line is nonexistent in this example.

Naturally Fractured Reservoirs 373

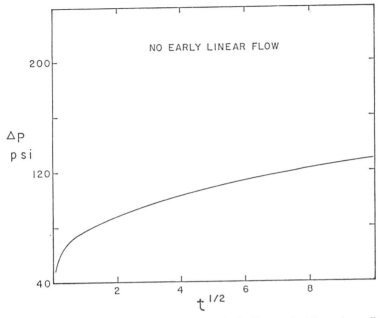

Figure 8-8. Square root of time vs Δp crossplot indicates that there is no linear flow at early times in this example. Source: Aguilera and Ng (1991).

5. A *pseudo radial flow period* towards the wellbore in a horizontal (rather than vertical) plane. This is recognized by a straight line in a semilogarithmic crossplot and corresponds to the segment between points 5 and 6 on Figure 8-7.
6. A *late linear flow period* which occurs when the pressure transient reaches the outer parallel boundaries. This is represented by the curved segment between points 6 and 7 on Figure 8-7. The same data yield a straight line on a Cartesian plot of Δp vs square root of time as shown on Figure 8-9.

Goode and Thambynayagam (1987) have presented approximate times at which each one of the flow periods begins and ends. These times can be compared with τ to estimate in what flow period the double-porosity system starts to behave as a conventional reservoir.

Variations in the Transition Period

The transition period discussed previously is due to flow from the matrix into the fractures. In Figure 8-7, this occurs between 2 radial flow periods. The transition, however, can occur in any part depending on the size of the matrix blocks and the hydraulic diffusivity of the matrix, i.e., depending on the value of τ.

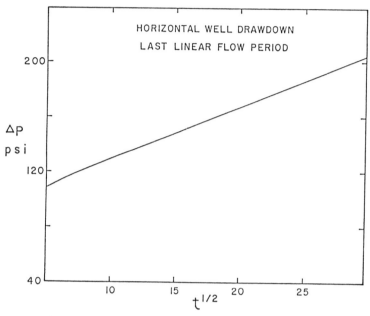

Figure 8-9. Straight line indicates late linear flow in a horizontal plane. Source: Aguilera and Ng (1991).

This situation is illustrated in Figure 8-10, which shows drawdown data generated on the basis of $\omega = 0.1$ and $\tau = 0.1$ and 100 hr.

The bottom curve shows the drawdown for the case of the conventional reservoir. The other curves are for dual-porosity systems. Notice that, as already discussed, the transition period can begin and end at different flow periods.

Anisotropic Fracture Permeability

Average permeability of all fractures intercepting the wellbore in the z-y plane is given by:

$$(k_z k_y)^{1/2} = 162.6\ q\mu B/\{m_{1r}\ (L - d)\} \qquad (8\text{-}25)$$

where m_{1r} is the slope of the semilog straight line in the first or third radial flow period (the two lines are parallel).

For example, an average permeability $(k_z k_y)^{1/2}$ equal to 100 md is calculated from Figure 8-7 by using a slope of 14.63 psi/cycle, a rate of 3,000 STB/D, a formation volume factor equal to 1.0, a viscosity of 1.5 cp, and a horizontal well length $(L - d)$ of 500 ft.

Naturally Fractured Reservoirs 375

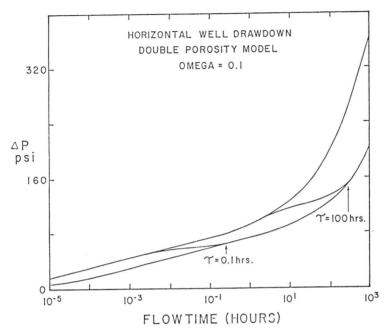

Figure 8-10. Drawdown pressures from dual-porosity model. Source: Aguilera and Ng (1991).

Permeability, k_y, is calculated from the first linear flow period using the equation:

$$k_y = \left(\frac{8.128 \, q \, B}{H_z \, (L - d) \, m_{1L}} \sqrt{\frac{\mu}{\phi \, c}} \right)^2 \quad (8\text{-}26)$$

where m_{1L} is the slope of the straight line on a Cartesian crossplot of pressure differential vs square root of time. This first linear flow period is nonexistent in this example.

The average permeability $(k_x \, k_y)^{1/2}$ is calculated from:

$$(k_x \, k_y)^{1/2} = 162.6 \, q\mu B/(m_{3r} \, H_z) \quad (8\text{-}27)$$

where m_{3r} is the slope of the semilog crossplot in the last radial flow period, i.e., the one that occurs in a horizontal plane. For this example, an average permeability $(k_z \, k_y)^{1/2}$ of 100 md is calculated from Figure 8-7, using a slope of 33.2 psi/cycle and the same rock and fluid properties discussed previously.

The above results indicate that the reservoir is isotropic because k_x, k_y, and k_z are all equal to 100 md. In an anisotropic reservoir, different values of permeabilities would have been obtained.

376 *Horizontal Wells*

The same value of ky can also be calculated from the last linear flow period with the use of Equation 8-26 having (L − d) m_{1L} replaced by $H_x m_{2L}$ where m_{2L} is the slope of the crossplot of pressure differential vs square root of time on Cartesian coordinates. This represents the last linear flow period (sixth flow period, Figure 8-9).

Storage Capacity Coefficient

The fraction of the total storage within the fracture is calculated from:

$$\omega = t_1/t_2 \qquad (8\text{-}28)$$

where t_1 is any time in the first straight line and t_2 is a time in the second straight line; both times taken at the same pressure. From Figure 8-7, for example, ω is calculated to be 10%. Notice, in Figure 8-10, that the ratio t_1/t_2 between the upper and lower bounding curves is 0.1 at all values of $p_i - p_{wf}$.

Skin Effect

Van Everdingen and Hurst mechanical skin (S_m) can be calculated from various flow periods:

First Flow Period. This corresponds to segment 1–2 in Figure 8-7 and represents strictly radial flow towards the horizontal wellbore via the natural fractures. Consequently, only fracture porosity ϕ_2 and fracture compressibility c_f must be introduced in the equations. Skin from this period is calculated by extrapolating the 1–2 straight line to 1 hr, reading $p_i - p_{1hr}$, and using the equation:

$$S_m = 1.151 \left[\frac{p_i - p_{1h}}{m_{1r}} - \log \left(\frac{(k_z k_y)^{1/2}}{\phi_2 \mu c_f r_w^2} \right) + 3.23 \right] \qquad (8\text{-}29)$$

Second Flow Period. Calculation of skin from the transition period between the two parallel straight lines (segment 2–3 in Figure 8-7) is not recommended, as the path of the transition period changes depending on the size of the matrix blocks, hydraulic diffusivity of the matrix, and type of interporosity flow. In some instances, the transition period produces a straight line with a slope equal to half the early and late parallel straight lines. Unfortunately, this occurs in very rare occasions.

Third Flow Period. This corresponds to radial flow in a vertical plane from the composite system towards the horizontal wellbore. Skin is calculated from Equation 8.29 having $\phi_2 c_f$ replaced by total porosity and total compressibility of the composite system. The value of $p_i - p_{1hr}$ is read from the extrapolation of the second parallel straight line (segment 3–4 in Figure 8-7) to 1 hr.

For example, a zero skin is calculated from the second parallel straight line using $p_i - p_{1hr}$ equal to 73.4 psi. In this case the total porosity is 0.1 and the total compressibility 3×10^{-5} psi^{-1}.

Fourth Flow Period. This corresponds to linear flow when the pressure transient reaches the upper and lower boundaries of the reservoir. Skin is calculated from the equation:

$$S_m = \frac{0.058}{H_z} \sqrt{\frac{k_z}{\pi \mu c}} \left[\frac{p_i - p_{ohr}}{m_{1L}} \right] - S_z \tag{8-30}$$

where s_z is a pseudo skin due to partial penetration in the z direction. The pressure differential $p_i - p_{ohr}$ is read at zero time in the extrapolation of a straight line given by a crossplot of Δp vs square root of time.

If this linear flow period happens to occur when the flow is still dominated by the natural fractures, the values of porosity and compressibility in Equation 8-30 have to be replaced by $\phi_2 c_f$. If this linear flow period occurs when the pressures in matrix and factures have reached an equilibrium, total porosity and total compressibility of the composite system are used in the analysis.

The pseudo skin, S_z, due to partial penetration in the vertical direction is calculated from (Goode and Thambynayagam, 1987):

$$S_z = (0.07958 \ H_z/r_w') \cdot [\psi(\eta_1) + \psi(\eta_2) - \psi(\eta_3) - \psi(\eta_4)] \tag{8-31}$$

where $\psi(\eta)$ is the Spence function (Clausen, 1832) evaluated numerically and tabulated by Ashour and Sabri (1956). The values of η are given by:

$$\eta_1 = 0.52 \ (\pi/H_z) \ r_w' \tag{8-32}$$

$$\eta_2 = (\pi/H_z) \ (2H_s + 3.48 r_w') \tag{8-33}$$

$$\eta_3 = -3.48 \ (\pi/H_z \ r_w') \tag{8-34}$$

$$\eta_4 = (\pi/H_z) \ (2H_s - 0.52 r_w') \tag{8-35}$$

A graphic representation of the Spence function between -2π and $+2\pi$ is displayed in Figure 8-11. The maximum and minimum values of the function are $+1.01494$ and -1.01494. The function becomes equal to zero at -2π, $-\pi$, $+\pi$ and $+2\pi$.

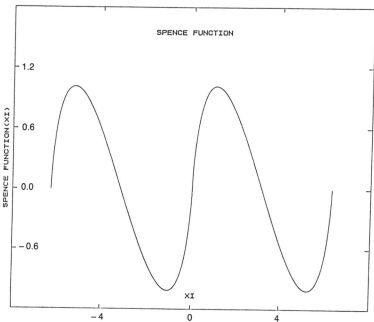

Figure 8-11. Spence function used for calculating skin in horizontal well due to partial vertical penetration.

Fifth Flow Period. This corresponds to pseudo radial flow in a horizontal plane towards the wellbore. Skin is calculated from:

$$S_m = 1.151\,[(L-d)/H_z]\,[k_z/k_x]^{1/2} \times$$

$$\left[\frac{p_i - p_{1hr}}{m_{3r}} - \log\left(\frac{k_x}{\phi\mu c(L-d)^2}\right) + 2.02\right] - S_z \quad (8\text{-}36)$$

where $p_i - p_{1hr}$ is obtained by extrapolating this radial flow period straight line in semilog paper to 1 hr; and m_{3r} is the slope in psi/cycle. If this period occurs when the flow is still dominated by the natural fractures, the values of porosity and compressibility in Equation 8-36 have to be replaced by ϕ_2 and c_f.

Based on our basic data, values of η are calculated from Equations 8-32 through 8-35 as follows:

$\eta_1 = 0.00263$ rad $= 0.1507$ deg.
$\eta_2 = 3.14780$ rad $= 180.356$ deg.
$\eta_3 = -0.0176$ rad $= 1.0084$ deg.
$\eta_4 = 3.13900$ rad $= 179.85$ deg.

Sixth Flow Period. This corresponds to the late linear flow period which occurs when the pressure transient reaches the outer parallel boundaries. Skin is calculated from the equation:

$$S_m = \frac{0.058 (L - d)}{H_x H_z} \sqrt{\frac{k_y}{\phi \mu c}} \left[\frac{p_i - p_{0hr}}{m_{2L}}\right] - S_z - S_x \qquad (8\text{-}37)$$

If this period occurs when the flow is still dominated by the natural fractures, the values of porosity and compressibility have to be replaced by ϕ_2 and c_f.

The pseudo skin S_x due to partial penetration in the x direction is calculated from (Goode and Thambynayagam, 1987):

$$S_x = 0.3183 \, H_x/[(k_z/k_x)^{1/2} H_z] \cdot [\psi(\xi_1) + \psi(\xi_2) - \psi(\xi_3) - \psi(\xi_4)] \qquad (8\text{-}38)$$

where

$$\xi_1 = 0.13 \, (\pi/H_x)(L - d) \qquad (8\text{-}39)$$

$$\xi_2 = (\pi/H_x)(1.87L + 0.13d) \qquad (8\text{-}40)$$

$$\xi_3 = -0.87 \, (\pi/H_x)(L - d) \qquad (8\text{-}41)$$

$$\xi_4 = (\pi/H_x)(0.87L + 1.13d) \qquad (8\text{-}42)$$

The slope m_{2L} is obtained from the straight line in the Cartesian crossplot of pressure differential vs square root of time (Figure 8-9), and is equal to 3.73 psi/hr$^{1/2}$.

From Equation 8-38, the pseudo skin S_x is calculated to be 2.28. The pseudo skin S_z was calculated previously to be 4.6. Thus, an actual skin S_m of zero is calculated from Equation 8-37 by utilizing a $p_i - p_{0hr}$ equal to 87.3 psi.

Size of Matrix Blocks

There is a high degree of uncertainty in determining the size of matrix blocks on the basis of well testing data alone. In practice, we should resort to additional supporting information from geological models, logs, cores, and outcrops (Aguilera, 1980). Different values will be obtained depending on the model used and the degree of approximation. Consequently, the values of h_m obtained from well testing data should be taken only as orders of magnitude. One possibility is to calculate h_m from Equations 8-23 and 8-24.

From Figure 8-7, for example, the value of τ is approximately 0.1 hour. Based on a matrix permeability of 0.01 md, a porosity of 0.1, a matrix compressibility of 3×10^{-5}, and a viscosity of 1.5 cp, the matrix hydraulic diffusivity is calculated from Equation 8-24 to be:

$$n_m = 0.01/(0.10 \times 1.5 \times 3 \times 10^{-5}) = 2{,}222.2 \text{ md psi/cp}$$

The average distance between natural fractures, h_m, is determined from Equation 8-23 to be:

$$h_m = (2{,}222.2 \times 0.1/532.3)^{1/2} = 0.65 \text{ ft}$$

with the use of Najurieta's constant (Najurieta, 1980) and 0.31 ft with Streltsova's constant (Streltsova, 1983). These numbers must be considered only as orders of magnitude.

Fracture Porosity

Fracture porosity ϕ_2, attached to bulk properties, is calculated from:

$$\phi_2 = \omega \phi_m c_m / [c_f (1 - \omega)] \tag{8-43}$$

where the value of ω is obtained from Equation 8-28 and the other parameters must be determined, preferably from laboratory work.

The fracture aperture (Aguilera, 1987) can be approximated as the product of ϕ_2, and the reservoir thickness by assuming that the fracture porosity attached to single properties is equal to 100%.

Calculation Strategy

A computer program has been written to handle the previous equations. Our strategy has been to use a regression package, utilizing the reservoir pressure data on the appropriate parameters—for example, the anisotropic permeability tensor.

Once a solution is obtained, it will be used to calculate the beginning and the end of the various flow periods (Goode and Thambynayagam, 1987).

Naturally Fractured Reservoirs 381

With these data we go on to conventional Cartesian and semilog crossplots to determine permeabilities in x, y, and z directions, skin, and pseudo skins due to partial penetration in the x and z direction. These parameters are calculated with the use of equations presented earlier in this chapter.

Finally, a pressure curve based on these calculated values will be plotted, together with the real pressure data to validate the regression results.

Regression

The regression package is based on the constrained simplex method of Box (1965). The simplex method's advantage is that one can start with almost any initial guessed values. It has the versatility to adapt itself to the local landscape of the regression surface. It will elongate down inclined planes, it will change direction on encountering a valley at an angle, and it will contract in the neighborhood of an extremum. The method is quite effective. It is computationally compact and it will converge quickly to a good approximate solution.

A second regression package has been used based on the Gauss Newton method (Fletcher, 1971). It converges very rapidly but suffers the drawback of needing good initial estimates of the regression variables, as pointed out by Abbaszadeh and Kamal (1988).

The Box method has been found to match the field data reasonably well (primarily pressure data) with theoretical reservoir models using a constrained, nonlinear least square regression technique by means of the simplex method. We have used it in drawdowns and buildups of horizontal wells in infinite or semi-infinite reservoirs.

The least square regression minimizes an objective function defined as $\phi(x)$ = sum to N terms of the squares of $(\phi(x) - \phi^*)$ where $\phi(x)$ is pressure from the model and ϕ^* is the observed pressure. The vector x contains the reservoir parameters to be estimated. Typically, they are anisotropic permeabilities, mechanical skin, and pseudo skins due to partial penetration in the x and z directions.

The parameters are constrained between maximum and minimum values supplied by the user. Imposing bounds on parameters with physically meaningful intervals offers the user some control of the final outcome.

Buildup

The solution for a buildup test is developed by taking the difference between the drawdown solution at a time equal to $(t + \Delta t)$ and the drawdown solution at a time Δt. This leads to the following buildup equation:

382 Horizontal Wells

$$p_i - p_{ws} = \frac{282.4 \, q\mu B}{H_x H_z k_y} \left[\sqrt{\frac{\pi A_{t+\Delta t}}{N_{t+\Delta t}}} - \sqrt{\frac{\pi A_{\Delta t}}{N_{\Delta t}}} + H_x h_x \sum_{n=1}^{\infty} \frac{1}{n} \left[\text{erf} \left(\frac{n}{h_x} \sqrt{\frac{A_{t+\Delta t}}{N_{t+\Delta t}}} \right) - \text{erf} \left(\frac{n}{h_x} \sqrt{\frac{A_{\Delta t}}{N_{\Delta t}}} \right) \right] \epsilon_n \cos(nX) + \frac{H_x H_z h_z}{(L-d)} \sum_{m=1}^{\infty} \frac{1}{m} \left[\text{erf} \left(\frac{m}{h_z} \sqrt{\frac{A_{t+\Delta t}}{N_{t+\Delta t}}} \right) - \text{erf} \left(\frac{m}{h_z} \sqrt{\frac{A_{\Delta t}}{N_{\Delta t}}} \right) \right] \epsilon_m \cos(mZ) \quad (8\text{-}44)$$

where

$$N_{t+\Delta t} = \omega + (1 - \omega) \cdot f(t + \Delta t, \tau) \qquad (8\text{-}45)$$

$$N_{\Delta t} = \omega + (1 - \omega) \cdot f(\Delta t, \tau) \qquad (8\text{-}46)$$

$$f(t + \Delta t, \tau) = 1 - \exp[-(t + \Delta t)/\tau] \qquad (8\text{-}47)$$

$$f(\Delta t, \tau) = 1 - \exp(-\Delta t/\tau) \qquad (8\text{-}48)$$

$$A_{t+\Delta t} = 0.000264 \, k_y (t + \Delta t)/(\phi\mu c) \qquad (8\text{-}49)$$

$$A_{\Delta t} = 0.000264 \, k_y \, \Delta t/(\phi\mu c) \qquad (8\text{-}50)$$

Equations 8-47 and 8-48 are for the case of pseudosteady state interporosity flow. The buildup solution can also be worked out for the case of transient interporosity flow using functions published by Aguilera (1987) for tectonic, regional, and contractional fractures.

τ, in the above equations, is approximately the shut-in time (hr) at which the effects of the natural fractures are not felt any longer, i.e., τ corresponds to that time in which an equilibrium is reached between pressure in the matrix and pressure in the natural fractures.

The above buildup solutions are valid only for analyzing cases where τ is up to a maximum of 10% of the flow period.

Validation

The data presented in Table 8-2 has been inserted into Equations 8-44 through 8-50, together with ω equal to 1.0, to generate the Horner plot presented in Figure 8-12. Our results compare well with those published by Goode and Thambynayagam (1987). Note that the data in Table 8-2 is for the case of an anisotropic reservoir.

Naturally Fractured Reservoirs

Table 8-2
Reservoir and Fluid Characteristics: Buildup Example

ω	= 0.10
k_x	= 50 md
k_y	= 100 md
k_z	= 25 md
ϕ	= 0.10
c	= 3×10^{-5} psi^{-1}
H_z	= 60 ft
H_x	= infinite
H_y	= infinite
r_w	= 0.354 ft
(L-d)	= 1000 ft
H_s	= 30 ft
q	= 3000 STB/D
μ	= 1.5 cp
B	= 1.5 RB/STB
S_m	= 5

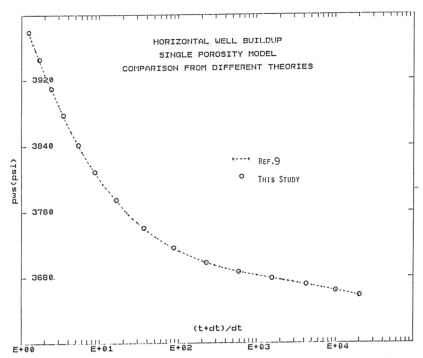

Figure 8-12. Comparison of buildup pressures from different theories. Horizontal well in dual-porosity reservoir. Source: Aguilera and Ng (1991).

Flow Periods

The same six flow periods discussed in the section on drawdowns can be present in a buildup. This is illustrated in Figure 8-13, which was generated using the data of Table 8-2, $\omega = 0.1$, and various values of τ. The flow time previous to shut-in was 200 hours. Note that the values of τ in Figure 8-13 represent up to a maximum of 10% of the flow time, i.e., a maximum of 20 hours.

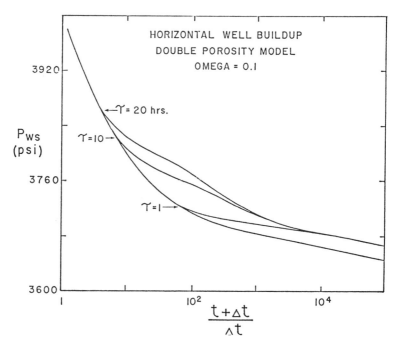

Figure 8-13. Buildup pressures from horizontal well in dual-porosity reservoir. Source: Aguilera and Ng (1991).

Anisotropic Permeabilities

The average permeability $(k_z k_y)^{1/2}$ can be calculated from the semilog straight lines in the first or third flow period with the use of Equation 8-25. Permeability k_y is determined from Equation 8-26 based on a crossplot of $p_{ws} - p_{wf}$ vs square root of shut-in time, or $p_{ws} - p_{wf}$ vs $(t + \Delta t)^{1/2} - (\Delta t)^{1/2}$. The product $(k_x k_y)^{1/2}$ is calculated from Equation 8-27 based on the last radial flow period. From the previous results one determines the values of k_x, k_y, and k_z.

Storage Capacity Coefficient

The fraction of the total storage within the fractures is calculated from:

$$\omega = \frac{[(t + \Delta t)/\Delta t]_1}{[(t + \Delta t)/\Delta t]_2} \tag{8-51}$$

where the Horner time$_1$ is read in the first semilog straight line and the Horner time$_2$ is read in the last straight line, both times taken at the same pressure. This assumes the presence of the first and third flow periods.

A similar approach can be used for other flow periods. In fact, notice, in Figure 8-13, that the time ratio between the two bounding curves is 0.1 throughout all values of Δp.

Lambda(λ)

Warren and Root (1963) introduced the concept of λ: a dimensionless matrix to fracture permeability ratio. λ is given by the equation:

$$\lambda = \frac{\alpha\,k_m}{k_2}\,r_w^2 \tag{8-52}$$

where k_m = matrix permeability, md
k_2 = fracture permeability, md
r_w = wellbore radius, ft
α = a shape factor given by $\alpha = 4n(n + 2)/h_m^2$

In the case of a stratum model, $n = 1$, and consequently, $\alpha = 12/h_m^2$. There is a relationship between λ and parameter τ discussed in previous sections. This relationship, in the case of a stratum model, is given by:

$$\lambda = \frac{6{,}387.6\,\phi_m\,\mu c_m\,r_w^2}{k_2\,\tau} \tag{8-54}$$

In the case of a model made out of matchsticks, the constant is changed to 17,033.6. For cubes with spaces in between, the constant is equal to 31,938.

Skin Effect

Equations 8-29 through 8-42 can be used for calculating skin with $p_i - p_{1hr}$ replaced by $p_{1hr} - p_{wf}$ and $p_i - p_{0hr}$ replaced by $p_{0hr} - p_{wf}$.

Size of the Matrix Blocks

Equations 8-23 and 8-24 can be used for calculating h_m based on knowledge of τ.

Fracture Porosity

It is calculated from Equation 8-43 based on knowledge of ω, fracture compressibility, matrix compressibility, and matrix porosity.

Initial Pressure

The initial pressure, or the value of p^*, in older wells is calculated by conventional extrapolations from a Horner plot using the fifth flow period (last radial) or from a crossplot of $(t + \Delta t)^{1/2} - (\Delta t)^{1/2}$ using data from the sixth flow period (last linear).

Wellbore Storage Effects

A solution in Laplace space that handles wellbore storage and skin in horizontal wells has been presented by Carvalho and Rosa (1988):

$$\bar{p}_{wD}(C_D, S, s) = \frac{\bar{p}_D(s) + \dfrac{h_D}{2s} S}{1 + \dfrac{h_D}{2} SC_D s + C_D s^2\, \bar{p}_D(s)} \qquad (8\text{-}55)$$

where $\bar{p}_D(s)$ = the Laplace transform of the dimensionless wellbore pressure drop for constant rate, without wellbore storage and without skin (Rosa and Carvalho, 1987)
S = skin
s = Laplace variable
h_D = given by:

$$h_D = \frac{h}{L} \sqrt{\frac{k_h}{k_v}} \qquad (8\text{-}56)$$

The formation thickness is represented by h, L is the horizontal well half-length, and k_h and k_v are horizontal and vertical permeabilities, respectively.

FRACTURED SHALES

The United States Geological Survey (USGS) has estimated that there are between 577 and 1,100 Tscf of gas-in-place in the Devonian shales of the Appalachian Basin; 85–160 Tscf in areas of historical gas production.

Numerical simulation work conducted by Mercer et al. (1988) has indicated that gas recoveries from these shales can be improved by drilling horizontal wells. The study used a three-dimensional, dual-porosity simulator. The model was fine-tuned using well test data, production rates, and cores taken from the actual horizontal well.

Table 8-3 shows basic parameters used during the simulation work. Notice the very low porosity and matrix permeability. Table 8-4 shows parameters determined by history matching. The natural fracture permeability ranges between 0.02 and 0.16 md. This compares with a matrix permeability equal to 5×10^{-6} md.

The work of Mercer et al. (1988) led to the conclusion that four unstimulated 2,000-ft horizontal wells could produce 1.5 times as much gas as 16 vertical shot wells. Furthermore, two hydraulically fractured horizontal

Table 8-3
Single Well Analysis Parameters Held Constant (Wayne County, West Virginia, Original Simulation)

Parameter	Value
Gas desorption rate	0.005 scf/ft³/psia
Drainage radius	1,000 ft
Natural fracture spacing	5 ft
Matrix porosity	0.01 (1%)
Matrix permeability	5×10^{-6} md
Fracture porosity	0.0009 (0.09%)
Permeability anisotropy ratio	1:1
Average initial rock pressure	375 psia

Source: Mercer et al., 1988.

Table 8-4
Parameters Determined by History Matching (Wayne County, West Virginia, Original Simulation)

Parameter	Range
Current rock pressure	150–300 psia
Natural fracture permeability, k_f	.02–.16 md
Net producing thickness, h	40–310 ft
Flow capacity, $k_f \cdot h$	1–42 md-ft

Source: Mercer et al., 1988.

wells nearly matched the production of the four unstimulated horizontal wells.

With respect to recoverable reserves, Mercer et al. (1988) found that in a virgin reservoir, four unstimulated horizontal wells accounted for 95% of the total production from 23 vertical wells.

Reservoir Performance

Theory and practice demonstrate that horizontal wells can enhance considerably the production rates and recoverable reserves from naturally fractured reservoirs.

Da Prat et al. (1981) and Sageev et al. (1985) presented the following solution to the dual porosity diffusivity equation under the assumptions of a vertical well, constant pressure, and single phase flow:

$$\bar{q}_D = \{\sqrt{sf(s)}[I_1\sqrt{sf(s)}\ r_{eD})K_1(\sqrt{sf(s)}) - K_1(\sqrt{sf(s)}r_{eD}))I_1(\sqrt{sf(s)})]\} /$$
$$s\{K_1(\sqrt{sf(s)}\ r_{eD})\ I_0(\sqrt{sf(s)}) + I_1(\sqrt{sf(s)}\ r_{eD})\ K_0(\sqrt{sf(s)}) - \sqrt{sf(s)}\ S$$
$$[K_1(\sqrt{sf(s)}\ r_{eD})\ I_1(\sqrt{sf(s)}) - K_1(\sqrt{sf(s)})\ I_1)\sqrt{sf(s)}\ r_{eD})]\} \qquad (8\text{-}57)$$

where S is skin and f(s) is a function given by:

$$f(s) = \frac{\omega(1-\omega)s + \lambda}{(1-\omega)s + \lambda} \qquad (8\text{-}58)$$

for the case of pseudosteady state interporosity flow. ω is the fraction of the total storage which is within the fractures and λ is defined by Equation 8-52.

For early times, the above solution becomes:

$$q_D = \frac{1}{\left(\ln r_{eD} - \frac{3}{4}\right)} \exp\left|\frac{-2}{r_{eD}^2\left(\ln r_{eD} - \frac{3}{4}\right)\omega}\ t_{Df}\right| \qquad (8\text{-}59)$$

For late times, the solution is:

$$q_D = \left(\frac{r_{eD}^2 - 1}{2}\right) \lambda \exp\left(\frac{-\lambda}{1-\omega}\ t_{Df}\right) \qquad (8\text{-}60)$$

Naturally Fractured Reservoirs

where $q_D = \dfrac{141.3 \, q\mu B}{kh\Delta p}$ \hfill (8-61)

$$t_{Df} = \frac{0.000264 \, k_2 \, t}{\phi_t \, \mu \, c_t \, r_{we}}$$ \hfill (8-62)

$r_{eD} = r_e/r_{we}$ \hfill (8-63)

The dimensionless radius r_{eD} is given by the ratio between drainage radius r_e and effective wellbore radius r_{we} of the horizontal well. The effective wellbore radius concept has been developed by Joshi (1988) and is given by:

$$r_{we} = \frac{r_{eh} \, (L/2)}{a \, \{1 + [1 - (L/2a)^2]^{1/2}\} \, (h/2r_w)^{\beta h/L}}$$ \hfill (8-64)

where r_{eh} = horizontal drainage radius, ft
L = length of horizontal well, ft
h = reservoir height, ft
$\beta = (k_h/k_v)^{1/2}$
k_h = horizontal permeability, md
k_v = vertical permeability, md

Half the major axis of the drainage ellipse a is given by:

$$a = (L/2) \, [0.5 + \sqrt{0.25 + (2r_{eh}/L)^4}]^{0.5}$$ \hfill (8-65)

Production forecasting of a constant pressure horizontal well in a naturally fractured reservoir can be obtained by inserting r_{we} (calculated from Equation 8-64) into Equations 8-59 and 8-60. Knowledge of r_{we} allows calculation of the skin factor S from the relationship:

$$r_{we} = r_w e^{-S}$$ \hfill (8-66)

The above approximation has been found useful for production forecast in dual porosity reservoirs. For a more rigorous approach, a numerical simulator would have to be used.

NOTATION

A group defined by Equations 8-16, 8-49, and 8-50, ft²
a vertical distance from the top reservoir boundary to the top edge of the horizontal strip source, ft
B FVF, RB/STB

b   vertical distance from the top reservoir boundary to the bottom of the horizontal strip source, ft
c   compressibility, psi^{-1}
d   lateral distance from the left reservoir boundary to the near end of the well, ft
e   fracture width
f   function
H   reservoir thickness, ft
h_m matrix thickness (height), ft
h_x group defined by Equation 8-17, ft
h_z group defined by Equation 8-18, ft
k   permeability, md
L   lateral distance from the left reservoir boundary to the far end of the well, ft
$m_{1,2,3r}$ semilog radial flow straight line slope, psi/cycle
$m_{1,2L}$ Cartesian linear flow straight line slope, $psi/(hour)^{1/2}$
m   subindex for summation
N   group defined by Equations 8-13, 8-45, and 8-46, dimensionless
n   subindex for summation (also number of fractures)
p   pressure, psi
p_i initial pressure, psi
p_{wf} flowing wellbore pressure, psi
p_{ws} wellbore pressure at shut-in, psi
p^* semilog true slope pressure extrapolated to infinite shut-in time, psi
P   fracture interception rate
q   production rate, STB/D
Q   source term, $seconds^{-1}$
r   radial distance, ft

r_w wellbore radius, ft
r_w' radius defined by Equation 8-21, ft
S   fracture spacing
S_m mechanical (van Everdingen and Hurst) skin factor
S_x skin due to partial penetration in the x direction
S_z skin due to partial penetration in the z direction
t   time, hours
Δt shut-in time, hours spacing
$\psi(\xi)$ Spence function
μ viscosity
τ approximate time when pressure equilibrium between matrix and fractures is reached, hours
ϕ porosity, fraction
x   lateral coordinate, ft
X   dimensionless lateral coordinate defined by Equation 8-19
y   longitudinal coordinate, ft
z   vertical coordinate, ft
Z   dimensionless vertical coordinate defined by Equation 20
δ partial derivative
δ^2 second derivative
K   azimuth of random fracture
ϕ_f fracture porosity attached to single point properties, fraction
ϕ_2 fracture porosity attached to bulk properties, fraction
ω storativity coefficient, fraction
Θ azimuth of closest fracture spacing

Subscripts

c	composite	w	water
f	fracture	x, y, z	coordinate indicator
i	initial	0hr	extrapolated to zero hours
ma	matrix	1hr	extrapolated to 1 hour
m	mechanical	1L	first linear flow period
o	oil	2L	second linear flow period
p	pseudo	1r	first radial flow period
t	total	2r	second radial flow period
t	time	3r	third radial flow period
Δt	shut-in time		

REFERENCES

1. Nolen-Hoeksema, R. C. and Howard, J. H., "Estimating Drilling Direction for Optimum Production in a Fractured Reservoir," *AAPG Bulletin*, V.71, No. 8 (August 1987) 958–966.
2. Bosio, J. and Reiss, L. H., "Site Selection Remains Key to Success in Horizontal Well Operations," *Oil and Gas Journal* (March 21, 1988) 71–76.
3. Economides, M. J. et al., "Performance and Stimulation of Horizontal Wells," *World Oil* (June 1989) 41–45.
4. de Montigny, O. and Combe, J., "Hole Benefits, Reservoir Types Key to Profit," *Oil and Gas Journal* (April 11, 1988) 50–56.
5. de Montigny, O. et al., "Horizontal Well Drilling Data Enhance Reservoir Appraisal," *Oil and Gas Journal* (July 4, 1988) 40–48.
6. Giannesini, J. F., "Horizontal Drilling Becoming Commonplace: Here is How It's Done," *World Oil* (March 1989) 35–40.
7. Mahony, B. J., "Horizontal Drilling Use on the Rise," *World Oil* (October 1988) 45–57.
8. de Swaan, A., "Analytic Solutions for Determining Naturally Fractured Reservoir Properties by Well Testing," *SPEJ* (June 1976) 117–122; Trans. *AIME*, 261.
9. Goode, P. A. and Thambynayagam, R. K. M., "Pressure Drawdown and Buildup Analysis of Horizontal Wells in Anisotropic Media," *SPEFE* (December 1987) 683–697; Trans. *AIME*, 283.
10. Aguilera, R. and Ng, M. C., "Transient Pressure Analysis of Horizontal Wells in Anisotropic Naturally Fractured Reservoirs," SPEFE (March 1991).
11. Prats, M., "Effect of Vertical Fractures on Reservoir Behavior-Incompressible Fluid Case," *SPE Journal* (June 1961).

12. Streltsova, T. D., "Well Pressure Behavior of a Naturally Fractured Reservoir," *SPEJ* (October 1983) 769–780.
13. Najurieta, H. L., "A Theory for Transient Pressure Analysis in Naturally Fractured Reservoirs," *JPT* (July 1980) 1241–50.
14. Clausen, T., Uber Die Zerlegung Reeller Gebrochener Funktionen, J. f. d. reine u. angew, Math, Crelle (1832) vo. 8, p. 298–300.
15. Ashour, A. and Sabri, A., "Tabulation of the Spence Function Mathematical Tables and Other Aids to Computation," National Academy of Sciences (April 1956) *vol. 10*, no. 54.
16. Aguilera, R., *Naturally Fractured Reservoirs*, PennWell Books, Tulsa, OK (1980).
17. Aguilera, R., "Multiple-Rate Analysis for Pressure Buildup Tests in Reservoirs with Tectonic, Regional and Contractional Natural Fractures," *SPEFE* (September 1987).
18. Box, M. J., "A New Method of Constrained Optimization and a Comparison with Other Methods," *Computer Journal* (1965) vol. 8, p.42.
19. Fletcher, R., "A Modified Marquardt Subroutine for Nonlinear Least Squares," *Atomic Energy Research Establishment Report 6799* (May 1971).
20. Abbaszadeh, M. and Kamal, M. M., "Automatic Type Curve Matching for Well Test Analysis," *SPEFE* (September 1988) 567–577.
21. Mercer, J. C. et al., "Infill Drilling Using Horizontal Wells: A Field Development Strategy for Tight Fractured Formations," SPE paper 17727 presented at the SPE Gas Technology Symposium, Dallas, TX (June 13–15, 1988).
22. Warren, J. E. and Root, P. J., "The Behavior of Naturally Fractured Reservoirs," *SPE Journal* (September 1963) 245–255.
23. Da Prat, G. et al., "Decline Curve Analysis Using Type Curves for Two-Porosity Systems," *SPE Journal* (June 1981) 354–362.
24. Rosa, A. J. and Carvalho, R. de S., "A Mathematical Model for Pressure Evaluation in an Infinite Conductivity Horizontal Well," unsolicited SPE paper 15967 (1987).
25. Carvalho, R. de S. and Rosa, A. J., "Transient Pressure Behavior for Horizontal Wells in Naturally Fractured Reservoirs," SPE paper 18302 presented at the 63rd Annual Technical Conference and Exhibition, Houston, TX October 2–5, 1988.
26. Sageev, A. et al., "Decline Curve Analysis for Double-Porosity Systems," SPE paper 13630 presented at the California Regional Meeting, Bakersfield, CA (March 27–29, 1985).
27. Joshi, S. D., "Horizontal Well Production Forecasting Methods and a Comparison of Horizontal Wells and Stimulated Vertical Wells," paper presented at the NPD Seminar on Thin Oil Zone Development, Stavanger, Norway (April 21–22, 1988).

Index

Abrasive formation, causes bit wear, 41
Absolute vertical permeability (mobile oil reservoir), 309, 310
Acid stimulation, 77, 79
Acid wash, 30
Additive, steam, 261
Alberta Oil Sands Technology and Research Authority (AOSTRA), 246, 324
Angle, tangent, 34
Aperture, fracture, 362
Applications of horizontal wells, 2
Austin chalk, 9, 148, 149
Azimuth, 1, 2, 7
   determination of, 11, 13
   of natural fracture, 359–361

Bed boundaries, 138, 140, 143
Bending stress, maximum
   to calculate, 62, 63
   of casing, 61, 62
Bending stress, on downhole assembly, 38
Bima field, Indonesia, 147, 148
Bit, penetration rate, 16

Bit life, 19, 20
Bit wear, 41
Blowdown, 290
Borehole rugosity, 136
Boundary, no-flow, 275–279
Braided stream system, 9
Breakout, 13
Bridge plug
   permanent, 85
   retrievable, 85
Bridging agent, 103
Buckling
   critical buckling load, 92
   of tubing, 58, 92
Buildup pressure, 167, 169–174
Buildup rate (BUR), 20ff
Buildup, naturally fractured reservoir, 381–386

Caddo Pine Island field, Louisiana, 6
Carbonate reef, 8
Carbonate system, 7
Cardium sandstone, Alberta, 6
Case histories, well logging, 143–154
Cased hole log, 101

Casing, 26–30
  collapsed, 29
  maximum bending stress, 61, 62
Casing alternatives in completion, 79–83
Casing design, 61–64
Castera Lou reservoir, France, 365, 366
Cemented lateral, 98
Cemented liner, 77, 80–82
Centralization of tools, 133
Centralizer, 78
Channel point bar system, 1, 9
Clearance, radial, 59
Coiled tubing, advantages, 97
  critical buckling load, 97
Coiled-tubing-conveyed logging system, 100, 132–135
  advantages, 100, 133, 134
Cold Lake, Alberta, 247, 248
Cold water equivalent (CWE) volume of steam, 226–229
Collapsed casing, prevention, 29
Completion, cased and cemented, 82, 83
Compressive service drillpipe, 58
Computer model of drag, 46, 47, 50
Computer model of torque, 46, 47, 50
Computer simulation of reservoir, 216, 217
Continuous marker bed, 31
Conventional longitudinal fracture, 122
Conventional oil, limited resources, 214
Core, to determine azimuth, 11, 13
  to identify natural fractures, 11
Critical buckling load, calculation, 92
Critical production rate, 4

Crystal growth on fracture surface, 11, 15
Cutting bed, 48
  development, 49
Cutting transport, 64ff
  reduced, 72
Cyclic steam stimulation, 347–350

Damage removal, 103, 104
Damage types, 103
Darcy's Law, 262–265
Deep Basin, Alberta, 30
Density, steam, 269
Design of casing, 61–64
Design of drillstring, 50ff
Design of fractures, 111–123
Devonian shale, 387
Dimensionless matrix to fracture permeability ratio (λ), 385
Dipmeter, 138–141
Dipping beds, 364
Directionally focused tool measurement, 140, 141
Discontinuous marker bed, 31
Discontinuous reservoir, 32
Displacement and SAGD, 311ff
Displacement, thermal process, 324ff
Drag, 43ff
  computer model of, 46, 47, 50, 52
  from cutting bed, 46
Drawdown, naturally fractured reservoir, 369–381
Drawdown pressure, 167–169
Drill cutting log, 13ff
Drilling fluid, 64ff
  evaluating rheology of, 65, 66
  incompatible, 72
  induces formation damage, 72, 73
Drilling fluid velocity profile, 65

Drilling technique, steerable, 19
Drillpipe analysis, 58–61
Drillpipe tension design, 52
Drillpipe
  compressive service, 58
  heavy-walled, 58
Drillstring, inverted, 51
Drillstring design, 50ff
Drillstring fatigue, 54
Drillstring tension, 45
Dual induction, 142, 143
Dual laterolog, 143
Dual porosity reservoir, 388, 389
Duration of flow regime, 161–167

Early time radial flow, 162, 166, 169, 171
Electric submersible pump, 87
End of curve (EOC), 21
External casing packer, 77, 83

Fann rotating viscometer, 65, 66
Fatigue, drillpipe, 50
  drillstring, 54
Fatigue damage, 56
  prevention of, 56
Fault, 32, 33
Filter cake, 48
Finite difference method, 202
Flow regime, 158–161
  calculation of skin factor in, 178
  duration, 161–167
  naturally fractured reservoir, 371–373, 384
Flow time equations, 164–167
Flow, laminar, 242
Flow, turbulent, 243
Flowmeter, 102
  petal basket, 102
Fluid, drilling, 64ff

Fluid, fracturing, 119
Fluid, gelled, 119
Fluid flow problems
  gas coning, 3, 4
  low permeability reservoir, 5, 6
  water coning, 3, 4
Fluid loss, 72, 73
Formation
  heterogeneous, 76
  homogeneous, 76, 77
  naturally fractured, 77, 78
Formation damage, 364
  induced by drilling fluid, 72, 73
Formation dip
  calculation of, 33, 34
  correct, 33
  incorrect, 33, 34
Fracture
  hydraulic, 77
  induced, 11, 79
  longitudinal, 108, 109, 121–123
    conventional, 122
    quasilongitudinal, 122
  natural, 138, 139
  quasilongitudinal, 122
  transverse, 109, 115–121
  undamaged, 77
  vertical, 258, 259
Fracture aperture, 362
Fracture design, 111–123
Fracture permeability, 374–376
Fracture porosity, calculation of, 380
Fracture system, 9
Fractured shale, 387–389
Fracturing fluid, 119
Fracturing
  hydraulic, 5
  multiple (sequential), 115–121
  stimulation option, 108, 109
Friction force, sliding, 43–49

Gas coning, 3, 4, 364
Gas lift pump, 86
Gas reservoir, 364
Gelled fluid systems, 119
Geological model of reservoir, 2
Geologist, as team member, 1, 17
Geolograph, 54
Geology, naturally fractured reservoir, 359–367
Grain size, 13, 15
Guar gum, 89
Gum
    guar, 89
    Xanthum, 89
Gun, articulated casing, 92
Gun, perforating, 92
Gun clearance (standoff), 94
Gun conveyance methods, 96
    coiled tubing, 97
    tubing, 96, 97
Gun orientation, 93, 94
Gun size, 94

HAS drive. *See* Heated annulus steam drive.
HAS drive steam oil ratio (SOR), 317
Heat capacity of oil sand, 221–224
Heated annulus steam (HAS) drive, 313ff
    advantages, 314–317
    field applications, 324
    laboratory experiments, 321–323
Heavy oil, 213
    primary recovery increase, 352, 353
    resources, 214
Heavy oil sands, 76
Heavy-walled drillpipe, 58
Heterogeneous formation, 76

Heterogeneous reservoir, 6ff
Heterogeneous reservoir types
    braided stream, 9
    carbonate system, 7
    channel point bar, 9
    fractured, 9
Hole cleaning, 88, 89
    drilling rate, 72
    formation change during, 72
    pipe rotation in, 71
    problems, 48
Hole inclination at target zone, 35, 36
Homogeneous formation, 76, 77
Horizontal drilling, medium radius, 18ff
Horizontal well
    applications,
    cyclic steam stimulation, 347–350
    injector, 246ff, 280ff
    pressure drop, 239–246
    producer, 246ff, 280ff
    productivity, 185ff
    well, stimulation, 102–107
Horizontal well model, 158
Hydraulic fracture, 5, 77
Hydraulic jet pump, 87

Induced fracture, 11, 79
Initial pressure, 386
Injection rate, steam, 308
    calculation, 221
Injector well, horizontal, 246ff, 280ff, 310
Interface velocity, 266–270
Intermediate time linear flow, 163, 166, 167, 169, 171
Invasion profile, 137
Invert oil mud, 29, 30
Inverted drillstring, 51

Index 397

Junk, remove with magnet, 88

Kern River field, 343–347
Kickoff point (KOP), 21

Lag time, determining, 15
Lagoon, 8
Laminar flow, 64, 71, 72, 242
Laplace transform, 202–204
Late time radial flow, 162, 167, 169, 172
Least linear squares method, 195–200
Limited entry stimulation, 115
Linear flow, intermediate time, 163, 166, 167
Liner
   cemented, 77
   perforated, 77
   preperforated, 77
   slotted, 80
   uncemented, 77
Location of wells, 109–111
Log
   drill cutting, 13ff
   mud gas, 15
   total gas, 13ff
Log interpretation, 136–143
Logging, naturally fractured reservoir, 365–367
Logging system
   coiled-tubing conveyed, 132–134
   pipe conveyed, 127–132
Logging tools, 101, 139–141
   cased hole, 101
   openhole, 101
Long radius horizontal well, 18
Longitudinal fracture, 108, 109, 121–123
Low permeability reservoir, 5, 6, 364

stimulation, 105–107
Low speed motor, 19

Magnet, used to remove downhole junk, 88
Marker bed, 31
   continuous, 31
   dip angle, 31
   discontinuous, 31
Marker sand, 15
Material balance, 195, 265, 266
Matrix acidizing, stimulation option, 108
Matrix blocks, size, 379, 380, 386
Maximum bending stress of casing, 61, 62
   to calculate, 62, 63
Maximum permeability, 11
McMurray formation, 246, 247
Measurement while drilling (MWD), 135, 136
   tools, 19
Mechanical plugging, 73
Medium radius horizontal drilling, 18ff
Microfracture, 110
Mobile oil, combined horizontal and vertical process, 337ff
Mobile oil reservoir, SAGD, 304–311
Mobile water zone, 327
Model
   geological, of reservoir, 2
   horizontal well, 158
Moineau pump, 36
Motor
   low speed, 19
   steerable, 37–42
Mud
   invert oil, 29, 30
   polymer, 103

Mud cake, 137
Mud gas log, 15
Multiphase flow, 101, 102
Multiple (sequential) fracturing, 115–119

Naphtha (steam additive), 261
Natural fracture, 138, 139
   azimuth, 359–361
Naturally fractured reservoir, 77, 78
   geology, 359–367
   logging, 365–367
   optimum drilling direction, 359–361
   skin effect, 376–379
   well test analysis, 367ff
Net present value (NPV), 115
No-flow boundary, 275–279
Numerical simulation
   reservoir, 216, 217
   SAGD, 282ff

Oil, conventional
   limited resources, 214
Oil, heavy, 213
   resources, 214
Oil column, 245
Oil relative permeability, 283
Oil sand
   heat capacity, 221–224
Oil to steam ratio (OSR), 238, 239
Openhole completion, 79–82
   with segmented liner, 80
   with slotted (preperforated) liner, 80
   true, 80
Openhole log, 101
Orientation of gun, 93, 94
Orientation of well, 110
Oriented core, 11
Oriented core tools, 11

Packer, 85, 86
   external casing, 77
Particle shape, 64
Particle size, 64, 68
Perforated liner, 77
Perforation density, 96, 97
Perforating
   to initiate multiple fractures, 118, 119
   stimulation option, 107
Perforating gun, 92
Perforating technique, 98, 99
   cemented laterals, 98
   uncemented laterals, 98, 99
Permanent bridge plug, 85
Permeability, fracture, 374–376
Permeability, maximum, 11
Permeability anisotrophy, 105
Petal basket flowmeter, 102
Pipe, stuck, 46–48
Pipe conveyed logging system, 127–132
   advantages, 129, 131
   operation, 131, 132
Plug
   hollow aluminum, 80, 99
   permanent bridge, 85
   retrievable bridge, 85
Plugging, mechanical, 73
Polymer mud, 103
Porosity, fracture
   calculation of, 380
Pressure
   buildup, 167, 169–174
   drawdown, 167–169
Pressure drop
   horizontal flow, 230–233, 239, 246
   vertical flow, 234–236
Pressure gradient, 3, 4
Pressure transient analysis, 156ff
Preperforated liner, 77, 80–82

Primary recovery, heavy oil vs
  conventional oil, 352, 353
Producer well, horizontal, 246ff,
  280ff
Production logging tools, 100, 101
Productivity in horizontal wells,
  185ff
Productivity index, 293, 294
  calculation of, 200, 201
Proppant, 123
  selection, 119–121
Prudhoe Bay, Alaska, 4
Pseudo openhole completion, 79
Pseudo radial flow, 163, 164, 167,
  169, 172
Pseudo skin, 178
Pumping equipment, 86, 87
  electric submersible, 87
  gas lift, 86
  hydraulic jet, 87
  sucker rod, 87

Quality steam, 228–230, 351
Quasilongitudinal fracture, 122

Radial clearance, 59
Radial flow
  early time, 162, 166, 169
  late time, 162, 167, 169
Radially averaged tool
  measurement, 139, 140
Rate of bit penetration (ROP), 16
Recorded total gas, 16
Recovery, thermal, 215ff
Reeled-tubing-conveyed logging
  system. *See* coiled-tubing-
  conveyed logging system.
Regression, 381
Reservoir
  discontinuous, 32
  dual porosity, 388, 389
  gas, 364

  geological model of, 2
  heterogeneous, 6ff
  low permeability, 5, 6, 364
    stimulation, 105–107
  naturally fractured, 359ff
    geology, 359–367
    logging, 365–367
    skin effect, 376–379
    well test analysis, 367ff
  sandstone, 105
Reservoir permeability
  anisotrophy, 105
Reservoir simulation, numerical,
  216, 217
Reservoir type, 79
Resistivity, 142, 143
Resources of heavy oil (world), 214
Retrievable bridge plug, 85
Rospo Mare field, Italy, 147, 366
Rugosity, of borehole, 136

SAGD. *See* Steam assisted gravity
  drainage.
Salt zone, 29
Saltation, 64, 68–71
Sand transport, 115–116
Sandstone reservoir, 105
Saskatchewan, Sceptre
  Tangleflags, 248–250
Sequential (multiple) fracturing,
  115–119
Shale, fractured, 387–389
Shale drape, 1, 9
Shear stress, evaluated with
  viscometer, 66
Short radius horizontal well, 18
Skin, 201, 202
Skin effect, 177, 178
  calculation of, 178
  naturally fractured reservoir,
    376–379, 385
Sliding friction force, 43–49

Slotted liner, 80
Spacing of wells, tight, 5
Spalling, 13
Spraberry field, Texas, 148
Star guide, 85
Steam
   CWE, 226–229
   quality, 228–230
   volume, 224–227
Steam additive, 261
Steam assisted gravity drainage
   (SAGD), 210–213, 236ff
   advantages, 238, 239
   and displacement, 311ff
   field applications, 246–250
   laboratory experiments, 250–262
   limitations, 239, 241
   mobile oil reservoir, 304–311
   numerical simulation, 282ff
   vs vertical well cyclic steam,
      299–304
Steam chamber development, 250ff
Steam channel, 244, 245
Steam column, 244, 245
Steam density, 269
Steam oil ratio, HAS drive, 317
Steam placement, 350–352
Steam injection rate, 308
   calculation of, 221
Steam zones, multiple, 275
Steerable drilling technique, 19
Steerable motor, 37–42
Stimulation
   acid, 77, 79
   cyclic steam, 347–350
   horizontal wells, 79, 102–107
   limited entry, 115
   low permeability reservoirs,
      105–107
   options, 107–111
      fracturing, 108, 109
      matrix acidizing, 108
      perforating, 107

Stinger, 99
Storage, wellbore, 386
Storage capacity, calculation of,
   376, 385
Stress
   tubing, 91
   yield, 91
Stuck pipe, 46–48
Sucker rod pump, 87
Surfactant, 323
Swivel sub, 94

TANDRAIN, 279
Tangent, 21ff
Tangent angle, 34
Target zone, hole inclination at,
   35, 36
Teamwork, 1, 17, 20
Tension, drillstring, 45
Thermal recovery, 215ff
Tools, 85, 86
   cased hole log, 101
   centralization, 133
   cup-type, 85
   directionally focused, 140, 141
   logging, 139–141
   MWD, 19
   openhole log, 101
   operation, 84, 85
   radially averaged, 139, 140
   size, 84
Torque, 43ff
   computer model of, 46, 47, 50,
      52
Total gas log, 13ff
Transition period, 372–374
Transverse fractures, 109, 115–121
Treating pressures, 118, 119, 122
True vertical depth (TVD), 21ff,
   138
Tubing stress, 91, 92
Tubular selection, 50ff
Turbulent flow, 64, 71, 72, 243

Index 401

Two-phase flow, 15
  in horizontal pipe, 68–71

Uncased wellbore, failure in, 80
Uncemented lateral, 98, 99
Uncemented liner, 77, 80–82
Undamaged fractures, 77

Velocity, interface, 266–270
Velocity profile, drilling fluid, 65
Vertical fracture, 258, 259
Vertical location of well, 110
Vertical permeability, absolute (mobile oil reservoir), 309, 310
Vertical well cyclic steam vs SAGD, 299–304
Viscometer, Fann rotating, 65, 66

Water coning, 3, 4, 364
Well configuration
  HAS drive, 315–321
  for SAGD in mobile oil reservoir, 305, 306, 310
  for SAGD recovery efficiency, 254, 255, 258

simulation, 337–343
tight spacing, 5
Well design, cost effective, 30
Well location, 109–111
Well logging, naturally fractured reservoir, 365–367
Well orientation, 110
Well stimulation, 79
Well test analysis, naturally fractured reservoir, 367ff
Wellbore failure, in uncased wellbore, 80
Wellbore storage, 201–204, 386
Wellbore storage effect, 176, 177
Western Canadian Basin, 13
Williston Basin, North Dakota, 29, 30, 80, 98
Wireline logging tools, conveyance, 99–102
World heavy oil resources, 214

Xanthum gum, 89

Yarega, USSR, 250
Yield strength, 62, 63
Yield stress, 91, 92